CISM COURSES AND LECTURES

The series presents lecture notes, monographs, edited works and proceedings in the field of Mechanics, Engineering, Computer Science and Applied Mathematics.
Purpose of the series is to make known in the international scientific and technical community results obtained in some of the activities organized by CISM, the International Centre for Mechanical Sciences.

INTERNATIONAL CENTRE FOR MECHANICAL SCIENCES

COURSES AND LECTURES - No. 434

PRINCIPLES OF NONPARAMETRIC LEARNING

EDITED BY

LÁSZLÓ GYÖRFI

BUDAPEST UNIVERSITY OF TECHNOLOGY
AND ECONOMICS

Springer-Verlag Wien GmbH

This volume contains 13 illustrations

Originally published by Springer-Verlag Wien New York in 2002

SPIN 10880232

In order to make this volume available as economically and as rapidly as possible the authors' typescripts have been reproduced in their original forms. This method unfortunately has its typographical limitations but it is hoped that they in no way distract the reader.

ISBN 978-3-211-83688-0 ISBN 978-3-7091-2568-7 (eBook)
DOI 10.1007/978-3-7091-2568-7

PREFACE

This volume contains the course material of the summer school PRINCIPLES OF NONPARAMETRIC LEARNING, held at the International Centre for Mechanical Sciences (CISM), Udine, Italy, July 9-13, 2001.

Modern nonparametric methods have become the most important tools in various fields of applications of pattern recognition, density and regression function estimation, data compression, on-line learning, and prediction. The common feature in these problems is that some unknown underlying system generates data and the best action is to be learnt from these data.

The purpose of the course was to teach the basic principles of nonparametric inference with emphasis on the cited areas.

The volume consists of six independent, but closely related areas of nonparametric learning:

- *The first chapter summarizes the basics of statistical learning theory and pattern recognition, including a self-contained study of empirical risk minimization, Vapnik-Chervonenkis theory, complexity regularization, and error estimation.*
- *The second chapter reviews nonparametric regression function estimation, such as local averaging estimates (partitioning, kernel, and nearest neighbor estimates), empirical risk minimization applied to spline estimates, penalized least squares estimates.*
- *In Chapter 3 the theory of on-line prediction of individual sequences is summarized. The performance of predictors under various loss functions are discussed in detail and the results are applied is different learning scenarios, including universal prediction, universal coding, and universal portfolio selection.*
- *Tools of learning theory may be applied to derive novel techniques for lossy data compression. Chapter 4 describes learning in vector quantization, and clustering, both for fixed-rate and variable-rate data compression.*
- *Chapter 5 is dedicated to the study of the fundamentals of nonparametric distribution and density estimation. The* L_1*-theory of density estimation is summarized, putting the emphasis on the kernel estimate and histogram estimates and paying special attention to the automatic optimal selection of density estimates.*
- *The last chapter is concerned with constrained non-parametric learning based on Evolution Algorithms, Genetic Programming in particular. Examples of real-world constraints*

and applications are taken from Macro-mechanical Modeling and concern the identification of materials behavioral laws (rheological models, dimensionally consistent models) from experimental data.

László Györfi

Contents

1

PATTERN CLASSIFICATION AND LEARNING THEORY

G. Lugosi
Pompeu Fabra University, Barcelona, Spain

1.1 A binary classification problem

Pattern recognition (or *classification* or *discrimination*) is about guessing or predicting the unknown class of an observation. An *observation* is a collection of numerical measurements, represented by a d-dimensional vector x. The unknown nature of the observation is called a *class*. It is denoted by y and takes values in the set $\{0,1\}$. (For simplicity, we restrict our attention to binary classification.) In pattern recognition, one creates a function $g(x) : \mathcal{R}^d \to \{0,1\}$ which represents one's guess of y given x. The mapping g is called a *classifier*. A classifier errs on x if $g(x) \neq y$.

To model the learning problem, we introduce a probabilistic setting, and let (X,Y) be an $\mathcal{R}^d \times \{0,1\}$-valued random pair.

The random pair (X,Y) may be described in a variety of ways: for example, it is defined by the pair (μ,η), where μ is the probability measure for X and η is the regression of Y on X. More precisely, for a Borel-measurable set $A \subseteq \mathcal{R}^d$,

$$\mu(A) = \mathbb{P}\{X \in A\},$$

and for any $x \in \mathcal{R}^d$,

$$\eta(x) = \mathbb{P}\{Y = 1|X = x\} = \mathbb{E}\{Y|X = x\}.$$

Thus, $\eta(x)$ is the conditional probability that Y is 1 given $X = x$. The distribution of (X,Y) is determined by (μ,η). The function η is called the *a posteriori probability*.

Any function $g : \mathcal{R}^d \to \{0,1\}$ defines a *classifier*. An *error* occurs if $g(X) \neq Y$, and the *probability of error* for a classifier g is

$$L(g) = \mathbb{P}\{g(X) \neq Y\} .$$

The Bayes classifier given by

$$g^*(x) = \begin{cases} 1 & \text{if } \eta(x) > 1/2 \\ 0 & \text{otherwise.} \end{cases}$$

minimizes the probability of error:

Theorem 1.1. *For any classifier* $g : \mathcal{R}^d \to \{0,1\}$,

$$\mathbb{P}\{g^*(X) \neq Y\} \leq \mathbb{P}\{g(X) \neq Y\}.$$

PROOF. Given $X = x$, the conditional probability of error of any decision g may be expressed as

$$\begin{aligned}
&\mathbb{P}\{g(X) \neq Y | X = x\} \\
&= 1 - \mathbb{P}\{Y = g(X) | X = x\} \\
&= 1 - (\mathbb{P}\{Y = 1, g(X) = 1 | X = x\} + \mathbb{P}\{Y = 0, g(X) = 0 | X = x\}) \\
&= 1 - \left(\mathbb{I}_{\{g(x)=1\}}\mathbb{P}\{Y = 1 | X = x\} + \mathbb{I}_{\{g(x)=0\}}\mathbb{P}\{Y = 0 | X = x\}\right) \\
&= 1 - \left(\mathbb{I}_{\{g(x)=1\}}\eta(x) + \mathbb{I}_{\{g(x)=0\}}(1 - \eta(x))\right),
\end{aligned}$$

where $\mathbb{I}_A$ denotes the indicator of the set A. Thus, for every $x \in \mathcal{R}^d$,

$$\begin{aligned}
&\mathbb{P}\{g(X) \neq Y | X = x\} - \mathbb{P}\{g^*(X) \neq Y | X = x\} \\
&= \eta(x)\left(\mathbb{I}_{\{g^*(x)=1\}} - \mathbb{I}_{\{g(x)=1\}}\right) + (1 - \eta(x))\left(\mathbb{I}_{\{g^*(x)=0\}} - \mathbb{I}_{\{g(x)=0\}}\right) \\
&= (2\eta(x) - 1)\left(\mathbb{I}_{\{g^*(x)=1\}} - \mathbb{I}_{\{g(x)=1\}}\right) \\
&\geq 0
\end{aligned}$$

by the definition of g^*. The statement now follows by integrating both sides with respect to $\mu(dx)$. □

L^* is called the Bayes probability of error, Bayes error, or Bayes risk. The proof above reveals that

$$L(g) = 1 - \mathbb{E}\left\{\mathbb{I}_{\{g(X)=1\}}\eta(X) + \mathbb{I}_{\{g(X)=0\}}(1 - \eta(X))\right\},$$

and in particular,

$$L^* = 1 - \mathbb{E}\left\{\mathbb{I}_{\{\eta(X)>1/2\}}\eta(X) + \mathbb{I}_{\{\eta(X)\leq 1/2\}}(1 - \eta(X))\right\} = \mathbb{E}\min(\eta(X), 1 - \eta(X)).$$

Note that g^* depends upon the distribution of (X,Y). If this distribution is known, g^* may be computed. Most often, the distribution of (X,Y) is unknown, so that g^* is unknown too.

In our model, we have access to a data base of pairs (X_i,Y_i), $1 \leq i \leq n$, observed in the past. We assume that $(X_1,Y_1),\ldots,(X_n,Y_n)$, the *data*, is a sequence of independent identically distributed (*i.i.d.*) random pairs with the same distribution as that of (X,Y).

A classifier is constructed on the basis of $X_1,Y_1,\ldots,X_n,Y_n$ and is denoted by g_n: Y is guessed by $g_n(X;X_1,Y_1,\ldots,X_n,Y_n)$. The process of constructing g_n is called *learning*, *supervised learning*, or *learning with a teacher*. The performance of g_n is measured by the conditional *probability of error*

$$L_n = L(g_n) = \mathbb{P}\{g_n(X;X_1,Y_1,\ldots,X_n,Y_n) \neq Y | X_1,Y_1,\ldots,X_n,Y_n)\}\ .$$

This is a random variable because it depends upon the data. So, L_n averages over the distribution of (X,Y), but the data is held fixed. Even though averaging over the data as well is unnatural, since in a given application, one has to live with the data at hand, the number $\mathbb{E}L_n = \mathbb{P}\{g_n(X) \neq Y\}$ which indicates the quality on an average data sequence, provides useful information, especially if the random variable L_n is concentrated around its mean with high probability.

1.2 Empirical risk minimization

Assume that a class C of classifiers $g : \mathcal{R}^d \to \{0,1\}$ is given and our task is to find one with a small probability of error. In the lack of the knowledge of the underlying distribution, one has to resort to using the data to estimate the probabilities of error for the classifiers in C. It is tempting to pick a classifier from C that minimizes an estimate of the probability of error over the class. The most natural choice to estimate the probability of error $L(g) = \mathbb{P}\{g(X) \neq Y\}$ is the error count

$$\widehat{L}_n(g) = \frac{1}{n}\sum_{j=1}^{n} \mathbb{I}_{\{g(X_j)\neq Y_j\}}.$$

$\widehat{L}_l(g)$ is called the *empirical error* of the classifier g.

A good method should pick a classifier with a probability of error that is close to the minimal probability of error in the class. Intuitively, if we can estimate the error probability for the classifiers in C *uniformly* well, then the classification function that minimizes the estimated probability of error is likely to have a probability of error that is close to the best in the class.

Denote by g_n^* the classifier that minimizes the estimated probability of error over the class:

$$\widehat{L}_n(g_n^*) \leq \widehat{L}_n(g) \quad \text{for all } g \in C.$$

Then for the probability of error

$$L(g_n^*) = \mathbb{P}\{\, g_n^*(X) \neq Y | D_n\}$$

of the selected rule we have:

Lemma 1.1.

$$L(g_n^*) - \inf_{g\in C} L(g) \leq 2\sup_{g\in C} |\widehat{L}_n(g) - L(g)|,$$

$$|\widehat{L}_n(g_n^*) - L(g_n^*)| \leq \sup_{g\in C} |\widehat{L}_n(g) - L(g)|.$$

PROOF.

$$\begin{aligned} L(g_n^*) - \inf_{g\in C} L(g) &= L(g_n^*) - \widehat{L}_n(g_n^*) + \widehat{L}_n(g_n^*) - \inf_{g\in C} L(g) \\ &\leq L(g_n^*) - \widehat{L}_n(g_n^*) + \sup_{g\in C} |\widehat{L}_n(g) - L(g)| \\ &\leq 2\sup_{g\in C} |\widehat{L}_n(g) - L(g)|. \end{aligned}$$

The second inequality is trivially true. □

We see that upper bounds for $\sup_{g\in C} |\widehat{L}_n(g) - L(g)|$ provide us with upper bounds for two things simultaneously:

(1) An upper bound for the suboptimality of g_n^* within C, that is, a bound for $L(g_n^*) - \inf_{g\in C} L(g)$.

(2) An upper bound for the error $|\widehat{L}_n(g_n^*) - L(g_n^*)|$ committed when $\widehat{L}_n(g_n^*)$ is used to estimate the probability of error $L(g_n^*)$ of the selected rule.

It is particularly useful to know that even though $\widehat{L}_n(g_n^*)$ is usually optimistically biased, it is within given bounds of the unknown probability of error with g_n^*, and that no other test sample is needed to estimate this probability of error. Whenever our bounds indicate that we are close to the optimum in C, we must at the same time have a good estimate of the probability of error, and vice versa.

The random variable $n\widehat{L}_n(g)$ is binomially distributed with parameters n and $L(g)$. Thus, to obtain bounds for the success of empirical error minimization, we need to study uniform deviations of binomial random variables from their means. In the next two sections we summarize the basics of the underlying theory.

1.3 Concentration inequalities

1.3.1 Hoeffding's inequality

The simplest inequality to bound the difference between a random variable and its expected value is *Markov's inequality*: for any nonnegative random variable X, and $t > 0$,

$$\mathbb{P}\{X \geq t\} \leq \frac{\mathbb{E}X}{t}.$$

From this, we deduce *Chebyshev's inequality*: if X is an arbitrary random variable and $t > 0$, then

$$\mathbb{P}\{|X - \mathbb{E}X| \geq t\} = \mathbb{P}\{|X - \mathbb{E}X|^2 \geq t^2\} \leq \frac{\mathbb{E}\{|X - \mathbb{E}X|^2\}}{t^2} = \frac{\mathbf{Var}\{X\}}{t^2}.$$

As an example, we derive inequalities for $\mathbb{P}\{S_n - \mathbb{E}S_n \geq t\}$ with $S_n = \sum_{i=1}^n X_i$, where $X_1, \ldots, X_n$ are independent real-valued random variables. Chebyshev's inequality and independence immediately gives us

$$\mathbb{P}\{|S_n - \mathbb{E}S_n| \geq t\} \leq \frac{\mathbf{Var}\{S_n\}}{t^2} = \frac{\sum_{i=1}^n \mathbf{Var}\{X_i\}}{t^2}.$$

The meaning of this is perhaps better seen if we assume that the X_i's are i.i.d. Bernoulli(p) random variables (i.e., $\mathbb{P}\{X_i = 1\} = 1 - \mathbb{P}\{X_i = 0\} = p$), and normalize:

$$\mathbb{P}\left\{\left|\frac{1}{n}\sum_{i=1}^n X_i - p\right| \geq \varepsilon\right\} \leq \frac{p(1-p)}{n\varepsilon^2}.$$

To illustrate the weakness of this bound, let $\Phi(y) = \int_{-\infty}^{y} e^{-t^2/2}/\sqrt{2\pi}\,dt$ be the normal distribution function. The central limit theorem states that

$$\mathbb{P}\left\{\sqrt{\frac{n}{p(1-p)}}\left(\frac{1}{n}\sum_{i=1}^n X_i - p\right) \geq y\right\} \to 1 - \Phi(y) \leq \frac{1}{\sqrt{2\pi}}\frac{e^{-y^2/2}}{y},$$

from which we would expect something like

$$\mathbb{P}\left\{\frac{1}{n}\sum_{i=1}^n X_i - p \geq \varepsilon\right\} \approx e^{-n\varepsilon^2/(2p(1-p))}.$$

Clearly, Chebyshev's inequality is off mark. An improvement may be obtained by *Chernoff's bounding method*. By Markov's inequality, if s is an arbitrary positive number, then for any random variable X, and any $t > 0$,

$$\mathbb{P}\{X \geq t\} = \mathbb{P}\{e^{sX} \geq e^{st}\} \leq \frac{\mathbb{E}e^{sX}}{e^{st}}.$$

In Chernoff's method, we find an $s > 0$ that minimizes the upper bound or makes the upper bound small. In the case of a sum of independent random variables,

$$\begin{aligned}\mathbb{P}\{S_n - \mathbb{E}S_n \geq t\} &\leq e^{-st}\mathbb{E}\left\{\exp\left(s\sum_{i=1}^n (X_i - \mathbb{E}X_i)\right)\right\} \\ &= e^{-st}\prod_{i=1}^n \mathbb{E}\left\{e^{s(X_i - \mathbb{E}X_i)}\right\} \quad \text{(by independence).}\end{aligned}$$

Now the problem of finding tight bounds comes down to finding a good upper bound for the moment generating function of the random variables $X_i - \mathbb{E}X_i$. There are many ways of doing this. For bounded random variables perhaps the most elegant version is due to Hoeffding (1963):

Lemma 1.2. *Let X be a random variable with $\mathbb{E}X = 0$, $a \leq X \leq b$. Then for $s > 0$,*

$$\mathbb{E}\left\{e^{sX}\right\} \leq e^{s^2(b-a)^2/8}.$$

PROOF. Note that by convexity of the exponential function

$$e^{sx} \leq \frac{x-a}{b-a}e^{sb} + \frac{b-x}{b-a}e^{sa} \quad \text{for } a \leq x \leq b.$$

Exploiting $\mathbb{E}X = 0$, and introducing the notation $p = -a/(b-a)$ we get

$$\begin{aligned}
\mathbb{E}e^{sX} &\leq \frac{b}{b-a}e^{sa} - \frac{a}{b-a}e^{sb} \\
&= \left(1 - p + pe^{s(b-a)}\right)e^{-ps(b-a)} \\
&\stackrel{\text{def}}{=} e^{\phi(u)},
\end{aligned}$$

where $u = s(b-a)$, and $\phi(u) = -pu + \log(1 - p + pe^u)$. But by straightforward calculation it is easy to see that the derivative of ϕ is

$$\phi'(u) = -p + \frac{p}{p + (1-p)e^{-u}},$$

therefore $\phi(0) = \phi'(0) = 0$. Moreover,

$$\phi''(u) = \frac{p(1-p)e^{-u}}{(p + (1-p)e^{-u})^2} \leq \frac{1}{4}.$$

Thus, by Taylor series expansion with remainder, for some $\theta \in [0, u]$,

$$\phi(u) = \phi(0) + u\phi'(0) + \frac{u^2}{2}\phi''(\theta) \leq \frac{u^2}{8} = \frac{s^2(b-a)^2}{8}.$$

Now we may directly plug this lemma into the bound obtained by Chernoff's method:

$$\begin{aligned}
&\mathbb{P}\{S_n - \mathbb{E}S_n \geq \varepsilon\} \\
&\leq e^{-s\varepsilon} \prod_{i=1}^{n} \mathbb{E}\left\{e^{s(X_i - \mathbb{E}X_i)}\right\} \\
&\leq e^{-s\varepsilon} \prod_{i=1}^{n} e^{s^2(b_i - a_i)^2/8} \quad \text{(by Lemma 1.2)} \\
&= e^{-s\varepsilon} e^{s^2 \sum_{i=1}^{n} (b_i - a_i)^2/8} \\
&= e^{-2\varepsilon^2 / \sum_{i=1}^{n} (b_i - a_i)^2} \quad \left(\text{by choosing } s = 4\varepsilon / \textstyle\sum_{i=1}^{n} (b_i - a_i)^2\right).
\end{aligned}$$

The result we have just derived is generally known as *Hoeffding's inequality*. For binomial random variables it was proved by Chernoff (1952) and Okamoto (1952). Summarizing, we have:

Theorem 1.2. (HOEFFDING'S INEQUALITY). *Let $X_1,\ldots,X_n$ be independent bounded random variables such that X_i falls in the interval $[a_i,b_i]$ with probability one. Denote their sum by $S_n=\sum_{i=1}^n X_i$. Then for any $\varepsilon>0$ we have*

$$\mathbb{P}\{S_n-\mathbb{E}S_n\geq\varepsilon\}\leq e^{-2\varepsilon^2/\sum_{i=1}^n(b_i-a_i)^2}$$

and

$$\mathbb{P}\{S_n-\mathbb{E}S_n\leq-\varepsilon\}\leq e^{-2\varepsilon^2/\sum_{i=1}^n(b_i-a_i)^2}.$$

If we specialize this to the binomial distribution, that is, when the X_i's are i.i.d. Bernoulli(p), we get

$$\mathbb{P}\{S_n/n-p\geq\varepsilon\}\leq e^{-2n\varepsilon^2},$$

which is just the kind of inequality we hoped for.

We may combine this inequality with that of Lemma 1.1 to bound the performance of empirical risk minimization in the special case when the class C contains finitely many classifiers:

Theorem 1.3. *Assume that the cardinality of C is bounded by N. Then we have for all $\varepsilon>0$,*

$$\mathbb{P}\left\{\sup_{g\in C}|\widehat{L}_n(g)-L(g)|>\varepsilon\right\}\leq 2Ne^{-2n\varepsilon^2}.$$

An important feature of the result above is that it is completely distribution free. The actual distribution of the data does not play a role at all in the upper bound.

To have an idea about the size of the error, one may be interested in the expected maximal deviation

$$\mathbb{E}\sup_{g\in C}|\widehat{L}_n(g)-L(g)|.$$

The inequality above may be used to derive such an upper bound by observing that for any nonnegative random variable X,

$$\mathbb{E}X=\int_0^\infty\mathbb{P}\{X\geq t\}dt.$$

Sharper bounds result by combining Lemma 1.2 with the following simple result:

Lemma 1.3. *Let $\sigma>0$, $n\geq2$, and let $Y_1,\ldots,Y_n$ be real-valued random variables such that for all $s>0$ and $1\leq i\leq n$, $\mathbb{E}\left\{e^{sY_i}\right\}\leq e^{s^2\sigma^2/2}$. Then*

$$\mathbb{E}\left\{\max_{i\leq n}Y_i\right\}\leq\sigma\sqrt{2\ln n}.$$

If, in addition, $\mathbb{E}\left\{e^{s(-Y_i)}\right\}\leq e^{s^2\sigma^2/2}$ for every $s>0$ and $1\leq i\leq n$, then for any $n\geq1$,

$$\mathbb{E}\left\{\max_{i\leq n}|Y_i|\right\}\leq\sigma\sqrt{2\ln(2n)}.$$

PROOF. By Jensen's inequality, for all $s > 0$,

$$e^{s\mathbb{E}\{\max_{i\le n} Y_i\}} \le \mathbb{E}\left\{e^{s\max_{i\le n} Y_i}\right\} = \mathbb{E}\left\{\max_{i\le n} e^{sY_i}\right\} \le \sum_{i=1}^{n} \mathbb{E}\left\{e^{sY_i}\right\} \le n e^{s^2\sigma^2/2} .$$

Thus,

$$\mathbb{E}\left\{\max_{i\le n} Y_i\right\} \le \frac{\ln n}{s} + \frac{s\sigma^2}{2} ,$$

and taking $s = \sqrt{2\ln n/\sigma^2}$ yields the first inequality. Finally, note that

$$\max_{i\le n} |Y_i| = \max(Y_1, -Y_1, \ldots, Y_n, -Y_n)$$

and apply the first inequality to prove the second. □

Now we obtain

$$\mathbb{E}\sup_{g\in C} |\widehat{L}_n(g) - L(g)| \le \sqrt{\frac{\ln(2N)}{2n}}.$$

1.3.2 Other inequalities for sums

Here we summarize some other useful inequalities for the deviations of sums of independent random variables from their means.

Theorem 1.4. BENNETT'S INEQUALITY. *Let $X_1, \ldots, X_n$ be independent real-valued random variables with zero mean, and assume that $|X_i| \le c$ with probability one. Let $\sigma^2 = \frac{1}{n}\sum_{i=1}^n \mathbf{Var}\{X_i\}$. Then that for any $t > 0$,*

$$\mathbb{P}\{S_n > t\} \le \exp\left(-\frac{n\sigma^2}{c^2} h\left(\frac{ct}{n\sigma^2}\right)\right),$$

where the function h is defined by $h(u) = (1+u)\log(1+u) - u$ for $u \ge 0$.

SKETCH OF PROOF. We use Chernoff's method as in the proof of Hoeffding's inequality. Write

$$\mathbb{E}\left\{e^{sX_i}\right\} = 1 + s\mathbb{E}\{X_i\} + \sum_{r=2}^{\infty} \frac{s^r \mathbb{E}\{X_i^r\}}{r!} = 1 + s^2 \mathbf{Var}\{X_i\} F_i \le e^{s^2 \mathbf{Var}\{X_i\} F_i}$$

with $F_i = \sum_{r=2}^{\infty} s^{r-2} \mathbb{E}\{X_i^r\} / (r!\,\mathbf{Var}\{X_i\})$. We may use the boundedness of the X_i's to show that $\mathbb{E}\{X_i^r\} \le c^{r-2}\mathbf{Var}\{X_i\}$, which implies $F_i \le (e^{sc} - 1 - sc)/(sc)^2$. Choose the s which minimizes the obtained upper bound for the tail probability. □

Theorem 1.5. BERNSTEIN'S INEQUALITY. *Under the conditions of the previous exercise, for any $t > 0$,*

$$\mathbb{P}\{S_n > t\} \le \exp\left(-\frac{t^2}{2n\sigma^2 + 2ct/3}\right).$$

PROOF. The result follows from Bennett's inequality and the inequality $h(u) \geq u^2/(2+2u/3)$, $u \geq 0$. □

Theorem 1.6. *Let $X_1,\ldots,X_n$ be independent random variables, taking their values from $[0,1]$. If $m = \mathbb{E}S_n$, then for any $m \leq t \leq n$,*

$$\mathbb{P}\{S_n \geq t\} \leq \left(\frac{m}{t}\right)^t \left(\frac{n-m}{n-t}\right)^{n-t}.$$

Also,

$$\mathbb{P}\{S_n \geq t\} \leq \left(\frac{m}{t}\right)^t e^{t-m},$$

and for all $\varepsilon > 0$,

$$\mathbb{P}\{S_n \geq m(1+\varepsilon)\} \leq e^{-mh(\varepsilon)},$$

where h is the function defined in the previous theorem. Finally,

$$\mathbb{P}\{S_n \leq m(1-\varepsilon)\} \leq e^{-m\varepsilon^2/2}.$$

1.3.3 The bounded difference inequality

In this section we give some powerful extensions of concentration inequalities for sums to to general functions of independent random variables.

Let A be some set, and let $g : A^n \to \mathcal{R}$ be some measurable function of n variables. We derive inequalities for the difference between $g(X_1,\ldots,X_n)$ and its expected value when $X_1,\ldots,X_n$ are arbitrary independent random variables taking values in A. Sometimes we will write g instead of $g(X_1,\ldots,X_n)$ whenever it does not cause any confusion.

We recall the elementary fact that if X and Y are arbitrary bounded random variables, then $\mathbb{E}\{XY\} = \mathbb{E}\{\mathbb{E}\{XY|Y\}\} = \mathbb{E}\{Y\mathbb{E}\{X|Y\}\}$.

Te first result of this section is an improvement of an inequality of Efron and Stein (1981) proved by Steele (1986). We have learnt the short proof given here from Stéphane Boucheron.

Theorem 1.7. EFRON-STEIN INEQUALITY. *If $X_1',\ldots,X_n'$ form an independent copy of $X_1,\ldots,X_n$, then*

$$\mathbf{Var}(g(X_1,\ldots,X_n)) \leq \frac{1}{2}\sum_{i=1}^{n} \mathbb{E}\left\{(g(X_1,\ldots,X_n) - g(X_1,\ldots,X_i',\ldots,X_n))^2\right\}$$

PROOF. Introduce the notation $V = g - \mathbb{E}g$, and define

$$V_i = \mathbb{E}\{g|X_1,\ldots,X_i\} - \mathbb{E}\{g|X_1,\ldots,X_{i-1}\}, \qquad i = 1,\ldots,n.$$

Clearly, $V = \sum_{i=1}^n V_i$. Then

$$\begin{aligned}
\mathbf{Var}(g) &= \mathbb{E}\left\{\left(\sum_{i=1}^n V_i\right)^2\right\} \\
&= \mathbb{E}\sum_{i=1}^n V_i^2 + 2\mathbb{E}\sum_{i>j} V_i V_j \\
&= \mathbb{E}\sum_{i=1}^n V_i^2 ,
\end{aligned}$$

since, for any $i > j$,

$$\mathbb{E}V_i V_j = \mathbb{E}\mathbb{E}\left\{V_i V_j | X_1,\ldots,X_j\right\} = \mathbb{E}\left\{V_j \mathbb{E}\left\{V_i | X_1,\ldots,X_j\right\}\right\} = 0 .$$

To bound $\mathbb{E}V_i^2$, note that, by Jensen's inequality,

$$\begin{aligned}
V_i^2 &= (\mathbb{E}\{g|X_1,\ldots,X_i\} - \mathbb{E}\{g|X_1,\ldots,X_{i-1}\})^2 \\
&= \left(\mathbb{E}\left[\mathbb{E}\{g|X_1,\ldots,X_n\} - \mathbb{E}\{g|X_1,\ldots,X_{i-1},X_{i+1},\ldots,X_n\}\Big|X_1,\ldots,X_i\right]\right)^2 \\
&\le \mathbb{E}\left[(\mathbb{E}\{g|X_1,\ldots,X_n\} - \mathbb{E}\{g|X_1,\ldots,X_{i-1},X_{i+1},\ldots,X_n\})^2 \Big|X_1,\ldots,X_i\right] ,
\end{aligned}$$

and therefore

$$\begin{aligned}
\mathbb{E}V_i^2 &\le \mathbb{E}\left[(g - \mathbb{E}\{g|X_1,\ldots,X_{i-1},X_{i+1},\ldots,X_n\})^2\right] \\
&= \frac{1}{2}\mathbb{E}\left[(g(X_1,\ldots,X_n) - g(X_1,\ldots,X_i',\ldots,X_n\})^2\right] ,
\end{aligned}$$

where at the last step we used (conditionally) the elementary fact that if X and Y are independent and identically distributed random variables, then $\mathbf{Var}(X) = (1/2)\mathbb{E}\{(X-Y)^2\}$. □

Assume that a function $g : A^n \to \mathcal{R}$ satisfies the *bounded difference assumption*

$$\sup_{\substack{x_1,\ldots,x_n, \\ x_i' \in A}} |g(x_1,\ldots,x_n) - g(x_1,\ldots,x_{i-1},x_i',x_{i+1},\ldots,x_n)| \le c_i , \ 1 \le i \le n .$$

In other words, we assume that if we change the i-th variable of g while keeping all the others fixed, then the value of the function does not change by more than c_i. Then the Efron-Stein inequality implies that

$$\mathbf{Var}(g) \le \frac{1}{2}\sum_{i=1}^n c_i^2 .$$

For such functions is is possible to prove the following exponential tail inequality, a powerful extension of Hoeffding's inequality.

Theorem 1.8. THE BOUNDED DIFFERENCE INEQUALITY. *Under the bounded difference assumption above, for all* $t > 0$,

$$\mathbb{P}\{g(X_1,\ldots,X_n) - \mathbb{E}g(X_1,\ldots,X_n) \geq t\} \leq e^{-2t^2/\sum_{i=1}^n c_i^2},$$

and

$$\mathbb{P}\{\mathbb{E}g(X_1,\ldots,X_n) - g(X_1,\ldots,X_n) \geq t\} \leq e^{-2t^2/\sum_{i=1}^n c_i^2}.$$

McDiarmid (1989) proved this inequality using martingale techniques, which we reproduce here. The proof of Theorem 1.8 uses the following straightforward extension of Lemma 1.2:

Lemma 1.4. *Let* V *and* Z *be random variables such that* $\mathbb{E}\{V|Z\} = 0$ *with probability one, and for some function* h *and constant* $c \geq 0$

$$h(Z) \leq V \leq h(Z) + c.$$

Then for all $s > 0$

$$\mathbb{E}\{e^{sV}|Z\} \leq e^{s^2c^2/8}.$$

PROOF OF THEOREM 1.8. Just like in the proof of Theorem 1.7, introduce the notation $V = g - \mathbb{E}g$, and define

$$V_i = \mathbb{E}\{g|X_1,\ldots,X_i\} - \mathbb{E}\{g|X_1,\ldots,X_{i-1}\}, \qquad i = 1,\ldots,n.$$

Then $V = \sum_{i=1}^n V_i$. Also introduce the random variables

$$H_i(X_1,\ldots,X_i) = \mathbb{E}\{g(X_1,\ldots,X_n)|X_1,\ldots,X_i\}.$$

Then, denoting the distribution of X_i by F_i for $i = 1,\ldots,n$,

$$V_i = H_i(X_1,\ldots,X_i) - \int H_i(X_1,\ldots,X_{i-1},x)F_i(dx).$$

Define the random variables

$$W_i = \sup_u \left(H_i(X_1,\ldots,X_{i-1},u) - \int H_i(X_1,\ldots,X_{i-1},x)F_i(dx) \right),$$

and

$$Z_i = \inf_v \left(H_i(X_1,\ldots,X_{i-1},v) - \int H_i(X_1,\ldots,X_{i-1},x)F_i(dx) \right).$$

Clearly, $Z_i \leq V_i \leq W_i$ with probability one, and also

$$W_i - Z_i = \sup_u \sup_v (H_i(X_1,\ldots,X_{i-1},u) - H_i(X_1,\ldots,X_{i-1},v)) \leq c_i,$$

by the bounded difference assumption. Therefore, we may apply the lemma above to obtain, for all $i = 1,\ldots,n$,

$$\mathbb{E}\left\{e^{sV_i}|X_1,\ldots,X_{i-1}\right\} \le e^{s^2c_i^2/8}.$$

Finally, by Chernoff's bound, for any $s > 0$,

$$\begin{aligned}
&\mathbb{P}\{g - \mathbb{E}g \ge t\} \\
\le\ & \frac{\mathbb{E}\left\{e^{s\sum_{i=1}^n V_i}\right\}}{e^{st}} = \frac{\mathbb{E}\left\{e^{s\sum_{i=1}^{n-1} V_i}\mathbb{E}\left\{e^{sV_n}|X_1,\ldots,X_{n-1}\right\}\right\}}{e^{st}} \\
\le\ & e^{s^2c_n^2/8}\frac{\mathbb{E}\left\{e^{s\sum_{i=1}^{n-1} V_i}\right\}}{e^{st}} \\
\le\ & e^{-st}e^{s^2\sum_{i=1}^n c_i^2/8} \quad \text{(by repeating the same argument } n \text{ times).}
\end{aligned}$$

Choosing $s = 4t\Big/\sum_{i=1}^n c_i^2$ proves the first inequality. The proof of the second inequality is similar.

An important application of the bounded difference inequality shows that if C is any class of classifiers of form $g : \mathcal{R}^d \to \{0,1\}$, then

$$\mathbb{P}\left\{\left|\sup_{g\in C}|\widehat{L}_n(g) - L(g)| - \mathbb{E}\sup_{g\in C}|\widehat{L}_n(g) - L(g)|\right| > \varepsilon\right\} \le 2e^{-2n\varepsilon^2}.$$

Indeed, if we view $\sup_{g\in C}|\widehat{L}_n(g) - L(g)|$ as a function of the n independent random pairs (X_i,Y_i), $i = 1,\ldots,n$, then we immediately see that the bounded difference assumption is satisfied with $c_i = 1/n$, and Theorem 1.8 immediately implies the statement.

The interesting fact is that regardless of the size of its expected value, the random variable $\sup_{g\in C}|\widehat{L}_n(g) - L(g)|$ is sharply concentrated around its mean with very large probability. In the next section we study the expected value.

1.4 Vapnik-Chervonenkis theory

1.4.1 The Vapnik-Chervonenkis inequality

Recall from Section 1.3.1 that for any finite class C of classifiers, and for all $\varepsilon > 0$,

$$\mathbb{P}\left\{\sup_{g\in C}|\widehat{L}_n(g) - L(g)| > \varepsilon\right\} \le 2Ne^{-2n\varepsilon^2},$$

and

$$\mathbb{E}\sup_{g\in C}|\widehat{L}_n(g) - L(g)| \le \sqrt{\frac{\ln(2N)}{2n}}.$$

These simple bounds may be useless if the cardinality N of the class is very large, or infinite. The purpose of this section is to introduce a theory to handle such cases.

Let $X_1,\ldots,X_n$ be i.i.d. random variables taking values in $\mathcal{R}^d$ with common distribution

$$\mu(A) = \mathbb{P}\{X_1 \in A\} \quad (A \subset \mathcal{R}^d).$$

Define the empirical distribution

$$\mu_n(A) = \frac{1}{n}\sum_{i=1}^{n} \mathbb{I}_{[X_i \in A]} \quad (A \subset \mathcal{R}^d).$$

Consider a class $\mathcal{A}$ of subsets of $\mathcal{R}^d$. Our main concern here is the behavior of the random variable $\sup_{A\in\mathcal{A}} |\mu_n(A) - \mu(A)|$. We saw in the previous chapter that a simple consequence of the bounded difference inequality is that

$$\mathbb{P}\left\{\left|\sup_{A\in\mathcal{A}} |\mu_n(A) - \mu(A)| - \mathbb{E}\sup_{A\in\mathcal{A}} |\mu_n(A) - \mu(A)|\right| > t\right\} \le 2e^{-2nt^2}$$

for any n and $t > 0$. This shows that for any class $\mathcal{A}$, the maximal deviation is sharply concentrated around its mean. In the rest of this chapter we derive inequalities for the expected value, in terms of certain combinatorial quantities related to $\mathcal{A}$. The first such quantity is the VC *shatter coefficient*, defined by

$$\mathbb{S}_{\mathcal{A}}(n) = \max_{x_1,\ldots,x_n\in\mathcal{R}^d} |\{\{x_1,\ldots,x_n\}\cap A; A \in \mathcal{A}\}|.$$

Thus, $\mathbb{S}_{\mathcal{A}}(n)$ is the maximal number of different subsets of a set of n points which can be obtained by intersecting it with elements of $\mathcal{A}$. The main theorem is the following version of a classical result of Vapnik and Chervonenkis:

Theorem 1.9. VAPNIK-CHERVONENKIS INEQUALITY.

$$\mathbb{E}\left\{\sup_{A\in\mathcal{A}} |\mu_n(A) - \mu(A)|\right\} \le 2\sqrt{\frac{\log 2\mathbb{S}_{\mathcal{A}}(n)}{n}}.$$

PROOF. Introduce $X_1',\ldots,X_n'$, an independent copy of $X_1,\ldots,X_n$. Also, define n i.i.d. sign variables $\sigma_1,\ldots,\sigma_n$ such that $\mathbb{P}\{\sigma_1 = -1\} = \mathbb{P}\{\sigma_1 = 1\} = 1/2$, independent of $X_1,X_1',\ldots,X_n,X_n'$.

Then, denoting $\mu_n'(A) = (1/n)\sum_{i=1}^n \mathbb{I}_{[X_i' \in A]}$, we may write

$$\begin{aligned}
&\mathbb{E}\left\{\sup_{A\in\mathcal{A}} |\mu_n(A) - \mu(A)|\right\} \\
&= \mathbb{E}\left\{\sup_{A\in\mathcal{A}} \left|\mathbb{E}\{\mu_n(A) - \mu_n'(A)|X_1,\ldots,X_n\}\right|\right\} \\
&\le \mathbb{E}\left\{\sup_{A\in\mathcal{A}} \mathbb{E}\left\{\left|\mu_n(A) - \mu_n'(A)\right| \Big| X_1,\ldots,X_n\right\}\right\} \\
&\quad \text{(by Jensen's inequality)} \\
&\le \mathbb{E}\left\{\sup_{A\in\mathcal{A}} \left|\mu_n(A) - \mu_n'(A)\right|\right\} \\
&\quad \text{(since } \sup\mathbb{E}(\cdot) \le \mathbb{E}\sup(\cdot)\text{)} \\
&= \frac{1}{n}\mathbb{E}\left\{\sup_{A\in\mathcal{A}} \left|\sum_{i=1}^n \sigma_i \left(\mathbb{I}_{[X_i\in A]} - \mathbb{I}_{[X_i'\in A]}\right)\right|\right\} \\
&\quad \text{(because } X_1, X_1', \ldots, X_n, X_n' \text{ are i.i.d.)} \\
&= \frac{1}{n}\mathbb{E}\left\{\mathbb{E}\left\{\sup_{A\in\mathcal{A}} \left|\sum_{i=1}^n \sigma_i \left(\mathbb{I}_{[X_i\in A]} - \mathbb{I}_{[X_i'\in A]}\right)\right| \Bigg| X_1, X_1', \ldots, X_n, X_n'\right\}\right\}.
\end{aligned}$$

Now because of the independence of the σ_i's of the rest of the variables, we may fix the values of $X_1 = x_1, X_1' = x_1', \ldots, X_n = x_n, X_n' = x_n'$, and investigate

$$\mathbb{E}\left\{\sup_{A\in\mathcal{A}} \left|\sum_{i=1}^n \sigma_i \left(\mathbb{I}_{[x_i\in A]} - \mathbb{I}_{[x_i'\in A]}\right)\right|\right\}.$$

Denote by $\widehat{\mathcal{A}} \subset \mathcal{A}$ a collection of sets such that any two sets in $\widehat{\mathcal{A}}$ have different intersections with the set $\{x_1, x_1', \ldots, x_n, x_n'\}$, and every possible intersection is represented once. Thus, $|\widehat{\mathcal{A}}| \le \mathbb{S}_{\mathcal{A}}(2n)$, and

$$\mathbb{E}\left\{\sup_{A\in\mathcal{A}} \left|\sum_{i=1}^n \sigma_i \left(\mathbb{I}_{[x_i\in A]} - \mathbb{I}_{[x_i'\in A]}\right)\right|\right\} = \mathbb{E}\left\{\max_{A\in\widehat{\mathcal{A}}} \left|\sum_{i=1}^n \sigma_i \left(\mathbb{I}_{[x_i\in A]} - \mathbb{I}_{[x_i'\in A]}\right)\right|\right\}.$$

Observing that each $\sigma_i\left(\mathbb{I}_{[x_i\in A]} - \mathbb{I}_{[x_i'\in A]}\right)$ has zero mean and takes values in $[-1,1]$, we obtain from Lemma 1.2 that for any $s > 0$,

$$\mathbb{E}e^{s\sum_{i=1}^n \sigma_i\left(\mathbb{I}_{[x_i\in A]} - \mathbb{I}_{[x_i'\in A]}\right)} = \prod_{i=1}^n \mathbb{E}e^{s\sigma_i\left(\mathbb{I}_{[x_i\in A]} - \mathbb{I}_{[x_i'\in A]}\right)} \le e^{ns^2/2}.$$

Since the distribution of $\sigma_i\left(\mathbb{I}_{[x_i\in A]} - \mathbb{I}_{[x_i'\in A]}\right)$ is symmetric, Lemma 1.3 immediately implies that

$$\mathbb{E}\left\{\max_{A\in\widehat{\mathcal{A}}} \left|\sum_{i=1}^n \sigma_i \left(\mathbb{I}_{[x_i\in A]} - \mathbb{I}_{[x_i'\in A]}\right)\right|\right\} \le \sqrt{2n\log 2\mathbb{S}_{\mathcal{A}}(2n)}.$$

Conclude by observing that $\mathbb{S}_{\mathcal{A}}(2n) \leq \mathbb{S}_{\mathcal{A}}(n)^2$. □

Remark. The original form of the Vapnik-Chervonenkis inequality is

$$\mathbb{P}\left\{ \sup_{A\in\mathcal{A}} |\mu_n(A) - \mu(A)| > t \right\} \leq 4\mathbb{S}_{\mathcal{A}}(2n)e^{-nt^2/8}.$$

A combination of Theorem 1.9 with the concentration inequality for the supremum quickly yields an inequality of a similar form.

The main virtue of the Vapnik-Chervonenkis inequality is that it converts the problem of uniform deviations of empirical averages into a combinatorial problem. Investigating the behavior of $\mathbb{S}_{\mathcal{A}}(n)$ is the key to the understanding of the behavior of the maximal deviations. Classes for which $\mathbb{S}_{\mathcal{A}}(n)$ grows at a subexponential rate with n are managable in the sense that $\mathbb{E}\{\sup_{A\in\mathcal{A}} |\mu_n(A) - \mu(A)|\}$ converges to zero. More importantly, explicit upper bounds for $\mathbb{S}_{\mathcal{A}}(n)$ provide nonasymptotic distribution-free bounds for the expected maximal deviation (and also for the tail probabilities). Section 1.4.3 is devoted to some key combinatorial results related to shatter coefficients.

We close this section by a refinement of Theorem 1.9 due to Massart (2000). The bound below substantially improves the bound of Theorem 1.9 whenever $\sup_{A\in\mathcal{A}} \mu(A)(1-\mu(A))$ is very small.

Theorem 1.10. *Let* $\Sigma = \sup_{A\in\mathcal{A}} \sqrt{\mu(A)(1-\mu(A))}$. *Then*

$$\mathbb{E}\left\{ \sup_{A\in\mathcal{A}} |\mu_n(A) - \mu(A)| \right\} \leq \frac{16\log 2\mathbb{S}_{\mathcal{A}}(2n)}{n} + +\sqrt{\frac{32\Sigma^2 \log 2\mathbb{S}_{\mathcal{A}}(2n)}{n}}.$$

PROOF. From the proof of Theorem 1.9, we have

$$\begin{aligned}
&\mathbb{E}\left\{ \sup_{A\in\mathcal{A}} |\mu_n(A) - \mu(A)| \right\} \\
&\leq \frac{1}{n}\mathbb{E}\left\{ \mathbb{E}\left\{ \sup_{A\in\mathcal{A}} \left| \sum_{i=1}^{n} \sigma_i \left(\mathbb{I}_{[X_i\in A]} - \mathbb{I}_{[X_i'\in A]} \right) \right| \middle| X_1, X_1', \ldots, X_n, X_n' \right\} \right\}.
\end{aligned}$$

By Hoeffding's inequality, for each set A,

$$\mathbb{E}\left\{ e^{s\sum_{i=1}^{n} \sigma_i \left(\mathbb{I}_{[X_i\in A]} - \mathbb{I}_{[X_i'\in A]} \right)} \middle| X_1, X_1', \ldots, X_n, X_n' \right\} \leq e^{s^2 \sum_{i=1}^{n} \left(\mathbb{I}_{[X_i\in A]} - \mathbb{I}_{[X_i'\in A]} \right)^2 /2},$$

so by Lemma 1.3 we obtain

$$\mathbb{E}\left\{ \sup_{A\in\mathcal{A}} |\mu_n(A) - \mu(A)| \right\} \leq \frac{1}{n}\mathbb{E} \sup_{A\in\mathcal{A}} \sqrt{\sum_{i=1}^{n} \left(\mathbb{I}_{[X_i\in A]} - \mathbb{I}_{[X_i'\in A]} \right)^2} \sqrt{2\log 2\mathbb{S}_{\mathcal{A}}(2n)}.$$

To bound the right-hand side, note that

$$\begin{aligned}
& \mathbb{E}\sup_{A\in\mathcal{A}}\sqrt{\sum_{i=1}^{n}\left(\mathbb{I}_{[X_i\in A]}-\mathbb{I}_{[X_i'\in A]}\right)^2} \\
\leq & \sqrt{\mathbb{E}\sup_{A\in\mathcal{A}}\sum_{i=1}^{n}\left(\mathbb{I}_{[X_i\in A]}-\mathbb{I}_{[X_i'\in A]}\right)^2} \\
\leq & \sqrt{\mathbb{E}\sup_{A\in\mathcal{A}}\sum_{i=1}^{n}\left(\left(\mathbb{I}_{[X_i\in A]}-\mu(A)\right)+\left(\mu(A)-\mathbb{I}_{[X_i'\in A]}\right)\right)^2} \\
\leq & \sqrt{4\mathbb{E}\sup_{A\in\mathcal{A}}\sum_{i=1}^{n}\left(\mathbb{I}_{[X_i\in A]}-\mu(A)\right)^2} \\
= & 2\sqrt{\mathbb{E}\sup_{A\in\mathcal{A}}\sum_{i=1}^{n}\left[\left(\mathbb{I}_{[X_i\in A]}-\mu(A)\right)(1-\mu(A))+\mu(A)\left(\mu(A)-\mathbb{I}_{[X_i\in A]}\right)+\mu(A)(1-\mu(A))\right]} \\
\leq & 2\sqrt{n\Sigma^2}+2\sqrt{\mathbb{E}\sup_{A\in\mathcal{A}}\left|\sum_{i=1}^{n}\left(\mathbb{I}_{[X_i\in A]}-\mu(A)\right)\right|} \\
= & 2\sqrt{n\Sigma^2}+2\sqrt{n\mathbb{E}\sup_{A\in\mathcal{A}}|\mu_n(A)-\mu(A)|}\,.
\end{aligned}$$

Summarizing, if we denote $\mathbb{E}\sup_{A\in\mathcal{A}}|\mu_n(A)-\mu(A)|=M$, we have obtained

$$M\leq\sqrt{\frac{\log 2\mathbb{S}_{\mathcal{A}}(2n)}{2n}}\left(\Sigma+\sqrt{M}\right).$$

This is a quadratic inequality for $\sqrt{M}$, whose solution is just the statement of the theorem.

1.4.2 Inequalities for relative deviations

In this section we summarize some important improvements of the basic Vapnik-Chervonenkis inequality. The basic result is the following pair of inequalities, due to Vapnik and Chervonenkis (1974). The proof sketched here is due to Anthony and Shawe-Taylor (1993).

Theorem 1.11. *For every* $\varepsilon>0$,

$$\mathbb{P}\left\{\sup_{A\in\mathcal{A}}\frac{\mu(A)-\mu_n(A)}{\sqrt{\mu(A)}}>\varepsilon\right\}\leq 4\mathbb{S}_{\mathcal{A}}(2n)e^{-n\varepsilon^2/4}$$

and

$$\mathbb{P}\left\{\sup_{A\in\mathcal{A}}\frac{\mu_n(A)-\mu(A)}{\sqrt{\mu_n(A)}}>\varepsilon\right\}\leq 4\mathbb{S}_{\mathcal{A}}(2n)e^{-n\varepsilon^2/4}\,.$$

SKETCH OF PROOF. The main steps of the proof are as follows:

1. Symmetrization.

$$\mathbb{P}\left\{\sup_{A\in\mathcal{A}}\frac{\mu(A)-\mu_n(A)}{\sqrt{\mu(A)}}>\varepsilon\right\}\le 2\mathbb{P}\left\{\sup_{A\in\mathcal{A}}\frac{\mu_n'(A)-\mu_n(A)}{\sqrt{(1/2)(\mu_n'(A)+\mu_n(A))}}>\varepsilon\right\}.$$

2. Randomization, conditioning.

$$\begin{aligned}&\mathbb{P}\left\{\sup_{A\in\mathcal{A}}\frac{\mu_n'(A)-\mu_n(A)}{\sqrt{(1/2)(\mu_n'(A)+\mu_n(A))}}>\varepsilon\right\}\\ &= \mathbb{E}\left\{\mathbb{P}\left\{\sup_{A\in\mathcal{A}}\frac{(1/n)\sum_{i=1}^n\sigma_i(\mathbb{I}_{X_i'\in A}-\mathbb{I}_{X_i\in A})}{\sqrt{(1/2)(\mu_n'(A)+\mu_n(A))}}>\varepsilon\,\middle|\,X_1,X_1',\ldots,X_n,X_n'\right\}\right\}.\end{aligned}$$

3. Tail bound. Use the union bound and Hoeffding's inequality to bound the conditional probability inside.

Using the bounds above, we may derive other interesting inequalities. The first inequalities are due to Pollard (1995) and Haussler (1992).

COROLLARY 1.1. *For all $t\in(0,1)$ and $s>0$,*

$$\mathbb{P}\left\{\sup_{A\in\mathcal{A}}\frac{\mu(A)-\mu_n(A)}{\mu(A)+\mu_n(A)+s/2}>t\right\}\le 4\mathbb{S}_{\mathcal{A}}(2n)e^{-nst^2/4}$$

and

$$\mathbb{P}\left\{\sup_{A\in\mathcal{A}}\frac{\mu_n(A)-\mu(A)}{\mu(A)+\mu_n(A)+s/2}>t\right\}\le 4\mathbb{S}_{\mathcal{A}}(2n)e^{-nst^2/4}$$

SKETCH OF PROOF. Take $\alpha>0$. Considering the cases $\mu(A)<(\alpha+1)^2\varepsilon^2\alpha^{-2}$ and $\mu(A)\ge(\alpha+1)^2\varepsilon^2\alpha^{-2}$ separately, it is easy to show that $\mu(A)-\mu_n(A)\le\varepsilon\sqrt{\mu(A)}$ implies that $\mu(A)\le(1+\alpha)\mu_n(A)+\varepsilon^2(1+\alpha)/\alpha$. Then choosing $\alpha=2t/(1-t)$ and $\varepsilon^2=st^2/(1-t^2)$ we easily prove that the first inequality in Theorem 1.11 implies the first inequality. The second inequality follows similarly from the second inequality of Theorem 1.11.

Finally, we point out another corollary of Theorem 1.11 which has interesting applications in statistical learning theory:

COROLLARY 1.2.

$$\mathbb{P}\{\exists A\in\mathcal{A}:\mu(A)>\varepsilon \textit{ and } \mu_n(A)\le(1-t)\mu(A)\}\le 4\mathbb{S}_{\mathcal{A}}(2n)e^{-n\varepsilon t^2/4}\,.$$

In particular, setting $t=1$,

$$\mathbb{P}\{\exists A\in\mathcal{A}:\mu(A)>\varepsilon \textit{ and } \mu_n(A)=0\}\le 4\mathbb{S}_{\mathcal{A}}(2n)e^{-n\varepsilon/4}\,.$$

1.4.3 Shatter coefficients

Consider a class $\mathcal{A}$ of subsets of $\mathcal{R}^d$, and let $x_1,\ldots,x_n \in \mathcal{R}^d$ be arbitrary points. Recall from the previous section that properties of the finite set $\mathcal{A}(x_1^n) \subset \{0,1\}^n$ defined by

$$\mathcal{A}(x_1^n) = \{b = (b_1,\ldots,b_n) \in \{0,1\}^n : \\ b_i = \mathbb{I}_{[x_i \in A]},\ i = 1,\ldots,n \text{ for some } A \in \mathcal{A}\}$$

play an essential role in bounding uniform deviations of the empirical measure. In particular, the maximal cardinality of $\mathcal{A}(x_1^n)$

$$\mathbb{S}_{\mathcal{A}}(n) = \max_{x_1,\ldots,x_n \in \mathcal{R}^d} |\mathcal{A}(x_1^n)|$$

(i.e., the shatter coefficient) provides simple bounds via the Vapnik-Chervonenkis inequality. We begin with some elementary properties of the shatter coefficient.

Theorem 1.12. *Let $\mathcal{A}$ and $\mathcal{B}$ be classes of subsets of $\mathcal{R}^d$, and let $n,m \geq 1$ be integers. Then*

(1) $\mathbb{S}_{\mathcal{A}}(n+m) \leq \mathbb{S}_{\mathcal{A}}(n)\mathbb{S}_{\mathcal{A}}(m)$;

(2) If $C = \mathcal{A} \cup \mathcal{B}$, *then* $\mathbb{S}_C(n) \leq \mathbb{S}_{\mathcal{A}}(n) + \mathbb{S}_{\mathcal{B}}(n)$;

(3) If $C = \{C = A^c : A \in \mathcal{A}\}$, *then* $\mathbb{S}_C(n) = \mathbb{S}_{\mathcal{A}}(n)$;

(4) If $C = \{C = A \cap B : A \in \mathcal{A} \text{ and } B \in \mathcal{B}\}$, *then* $\mathbb{S}_C(n) \leq \mathbb{S}_{\mathcal{A}}(n)\mathbb{S}_{\mathcal{B}}(n)$;

(5) If $C = \{C = A \cup B : A \in \mathcal{A} \text{ and } B \in \mathcal{B}\}$, *then* $\mathbb{S}_C(n) \leq \mathbb{S}_{\mathcal{A}}(n)\mathbb{S}_{\mathcal{B}}(n)$;

(6) If $C = \{C = A \times B : A \in \mathcal{A} \text{ and } B \in \mathcal{B}\}$, *then* $\mathbb{S}_C(n) \leq \mathbb{S}_{\mathcal{A}}(n)\mathbb{S}_{\mathcal{B}}(n)$.

PROOF. Parts (1), (2), (3), and (6) are immediate from the definition. To show (4), fix $x_1,\ldots,x_n$, let $N = |\mathcal{A}(x_1^n)| \leq \mathbb{S}_{\mathcal{A}}(n)$, and denote by $A_1,A_2,\ldots,A_N$ the different sets of the form $\{x_1,\ldots,x_n\} \cap A$ for some $A \in \mathcal{A}$. For all $1 \leq i \leq N$, sets in $\mathcal{B}$ pick at most $\mathbb{S}_{\mathcal{B}}(|A_i|) \leq \mathbb{S}_{\mathcal{B}}(n)$ different subsets of A_i. Thus,

$$|\mathcal{A}(x_1^n)| \leq \sum_{i=1}^{N} \mathbb{S}_{\mathcal{B}}(|A_i|) \leq \mathbb{S}_{\mathcal{A}}(n)\mathbb{S}_{\mathcal{B}}(n).$$

(5) follows from (4) and (3). □

The VC *dimension* V of a class $\mathcal{A}$ of sets is defined as the largest integer n such that

$$\mathbb{S}_{\mathcal{A}}(n) = 2^n.$$

If $\mathbb{S}_{\mathcal{A}}(n) = 2^n$ for all n, then we say that $V = \infty$. Clearly, if $\mathbb{S}_{\mathcal{A}}(n) < 2^n$ for some n, then for all $m > n$, $\mathbb{S}_{\mathcal{A}}(m) < 2^m$, and therefore the VC dimension is always well-defined. If $|\mathcal{A}(x_1^n)| = 2^n$ for some points $x_1,\ldots,x_n$, then we say that $\mathcal{A}$ *shatters* the set $x_1^n = \{x_1,\ldots,x_n\}$. As the next basic result shows, the VC dimension provides a useful bound for the shatter coefficient of a class.

Theorem 1.13. SAUER'S LEMMA. *Let $\mathcal{A}$ be a class of sets with* VC *dimension $V < \infty$. Then for all n,*

$$\mathbb{S}_{\mathcal{A}}(n) \leq \sum_{i=0}^{V} \binom{n}{i}.$$

PROOF. Fix $x_1, \ldots, x_n$, such that $|\mathcal{A}(x_1^n)| = \mathbb{S}_{\mathcal{A}}(n)$. Denote $B_0 = \mathcal{A}(x_1^n) \in \{0,1\}^n$. We say that a set $B \subset \{0,1\}^n$ *shatters* a set $S = \{s_1, \ldots, s_m\} \subset \{1, 2, \ldots, n\}$ if the restriction of B to the components $s_1, \ldots, s_m$ is the full m-dimensional binary hypercube, that is,

$$\{(b_{s_1}, \ldots, b_{s_m}) : b = (b_1, \ldots, b_n) \in B\} = \{0,1\}^m.$$

It suffices to show that the cardinality of any set $B_0 \subset \{0,1\}^n$ that cannot shatter any set of size $m > V$, is at most $\sum_{i=0}^{V} \binom{n}{i}$. This is done by transforming B_0 into a set B_n with $|B_n| = |B_0|$ such that any set shattered by B_n is also shattered by B_0. Moreover, it will be easy to see that $|B_n| \leq \sum_{i=0}^{V} \binom{n}{i}$.

For every vector $b = (b_1, \ldots, b_n) \in B_0$, if $b_1 = 1$, then flip the first component of b to zero unless $(0, b_2, \ldots, b_n) \in B_0$. If $b_1 = 0$, then keep the vector unchanged. The set of vectors B_1 obtained this way obviously has the same cardinality as that of B_0. Moreover, if B_1 shatters a set $S = \{s_1, s_2, \ldots, s_m\} \subset \{1, \ldots, n\}$, then B_0 also shatters S. This is trivial if $1 \notin S$. If $1 \in S$, then we may assume without loss of generality that $s_1 = 1$. The fact that B_1 shatters S implies that for any $v \in \{0,1\}^{m-1}$ there exists a $b \in B_1$ such that $b_1 = 1$ and $(b_{s_2}, \ldots, b_{s_m}) = v$. By the construction of B_1 this is only possible if for any $u \in \{0,1\}^m$ there exists a $b' \in B_0$ such that $(b'_{s_1}, \ldots, b'_{s_m}) = u$. This means that B_0 also shatters S.

Now starting from B_1, execute the same transformation, but now by flipping the second component of each vector, if necessary. Again, the cardinality of the obtained set B_2 remains unchanged, and any set shattered by B_2 is also shattered by B_1 (and therefore also by B_0). Repeat the transformation for all components, arriving at the set B_n. Clearly, B_n cannot shatter sets of cardinality larger than V, since otherwise B_0 would shatter sets of the same size. On the other hand, it is easy to see that B_n is such that for every $b \in B_n$, all vectors of form $c = (c_1, \ldots, c_n)$ with $c_i \in \{b_i, 0\}$ for $1 \leq i \leq n$, are also in B_n. Then B_n is a subset of a set of form

$$T = \{b \in \{0,1\}^n : b_i = 0 \text{ if } v_i = 0\},$$

where $v = (v_1, \ldots, v_n)$ is a fixed vector containing at most V 1's. This implies that

$$\mathbb{S}_{\mathcal{A}}(n) = |B_0| = |B_n| \leq |T| = \sum_{i=0}^{V} \binom{n}{i},$$

concluding the proof. □

The following corollary makes the meaning of Sauer's lemma more transparent:

COROLLARY 1.3. *Let $\mathcal{A}$ be a class of sets with* VC *dimension $V < \infty$. Then for all n,*

$$\mathbb{S}_{\mathcal{A}}(n) \le (n+1)^V,$$

and for all $n \ge V$,

$$\mathbb{S}_{\mathcal{A}}(n) \le \left(\frac{ne}{V}\right)^V.$$

PROOF. By the binomial theorem,

$$(n+1)^V = \sum_{i=0}^{V} n^i \binom{V}{i} = \sum_{i=0}^{V} \frac{n^i V!}{i!(V-i)!} \ge \sum_{i=0}^{V} \frac{n^i}{i!} \ge \sum_{i=0}^{V} \binom{n}{i}.$$

On the other hand, if $V/n \le 1$, then

$$\left(\frac{V}{n}\right)^V \sum_{i=0}^{V} \binom{n}{i} \le \sum_{i=0}^{V} \left(\frac{V}{n}\right)^i \binom{n}{i} \le \sum_{i=0}^{n} \left(\frac{V}{n}\right)^i \binom{n}{i} = \left(1+\frac{V}{n}\right)^n \le e^V,$$

where again we used the binomial theorem. □

Recalling the Vapnik-Chervonenkis inequality, we see that if $\mathcal{A}$ is any class of sets with VC dimension V, then

$$\mathbb{E}\left\{\sup_{A\in\mathcal{A}} |\mu_n(A) - \mu(A)|\right\} \le 2\sqrt{\frac{V\log(n+1)+\log 2}{n}},$$

that is, whenever $\mathcal{A}$ has a finite VC dimension, the expected largest deviation over $\mathcal{A}$ converges to zero at a rate $O(\sqrt{\log n/n})$.

Next we calculate the VC dimension of some simple classes.

Lemma 1.5. *If $\mathcal{A}$ is the class of all rectangles in $\mathcal{R}^d$, then $V = 2d$.*

PROOF. To see that there are $2d$ points that can be shattered by $\mathcal{A}$, just consider the $2d$ vectors with $d-1$ zero components, and one non-zero component which is either 1 or -1. On the other hand, for any given set of $2d+1$ points we can choose a subset of at most $2d$ points with the property that it contains a point with largest first coordinate, a point with smallest first coordinate, a point with largest second coordinate, and so forth. Clearly, there is no set in $\mathcal{A}$ which contains these points, but not the rest. □

Lemma 1.6. *Let $\mathcal{G}$ be an m-dimensional vector space of real-valued functions defined on $\mathcal{R}^d$. The class of sets*

$$\mathcal{A} = \{\{x : g(x) \ge 0\} : g \in \mathcal{G}\}$$

has VC *dimension $V \le m$.*

PROOF. It suffices to show that no set of size $m+1$ can be shattered by sets of the form $\{x : g(x) \geq 0\}$. Fix $m+1$ arbitrary points $x_1, \ldots, x_{m+1}$, and define the linear mapping $L : \mathcal{G} \to \mathcal{R}^{m+1}$ as

$$L(g) = (g(x_1), \ldots, g(x_{m+1})) .$$

Then the image of $\mathcal{G}$, $L(\mathcal{G})$, is a linear subspace of $\mathcal{R}^{m+1}$ of dimension not exceeding m. This implies the existence of a nonzero vector $\gamma = (\gamma_1, \ldots, \gamma_{m+1}) \in \mathcal{R}^{m+1}$ orthogonal to $L(\mathcal{G})$, that is, for every $g \in \mathcal{G}$,

$$\gamma_1 g(x_1) + \ldots + \gamma_{m+1} g(x_{m+1}) = 0 .$$

We may assume that at least one of the γ_i's is negative. Rearranging this equality so that all terms with nonnegative γ_i stay on the left-hand side, we get

$$\sum_{i:\gamma_i \geq 0} \gamma_i g(x_i) = \sum_{i:\gamma_i < 0} -\gamma_i g(x_i) .$$

Now suppose that there exists a $g \in \mathcal{G}$ such that the set $\{x : g(x) \geq 0\}$ picks exactly the x_i's on the left-hand side. Then all terms on the left-hand side are nonnegative, while the terms on the right-hand side must be negative, which is a contradiction, so $x_1, \ldots, x_{m+1}$ cannot be shattered, which implies the statement. □

Generalizing a result of Schläffli (1950), Cover (1965) showed that if $\mathcal{G}$ is defined as the linear space of functions spanned by functions $\psi_1, \ldots, \psi_m : \mathcal{R}^d \to \mathcal{R}$, and the vectors $\Psi(x_i) = (\psi_1(x_i), \ldots, \psi_m(x_i))$, $i = 1, 2, \ldots, n$ are linearly independent, then for the class of sets $\mathcal{A} = \{\{x : g(x) \geq 0\} : g \in \mathcal{G}\}$ we have

$$|\mathcal{A}(x_1^n)| = 2 \sum_{i=0}^{m-1} \binom{n-1}{i} ,$$

which often gives a slightly sharper estimate than Sauer's lemma. The proof is left as an exercise. Now we may immediately deduce the following:

COROLLARY 1.4. *(1) If $\mathcal{A}$ is the class of all linear halfspaces, that is, subsets of $\mathcal{R}^d$ of the form $\{x : a^T x \geq b\}$, where $a \in \mathcal{R}^d, b \in \mathcal{R}$ take all possible values, then $V \leq d+1$.*

(2) If $\mathcal{A}$ is the class of all closed balls in $\mathcal{R}^d$, that is, sets of the form

$$\left\{ x = (x^{(1)}, \ldots, x^{(d)}) : \sum_{i=1}^{d} |x^{(i)} - a_i|^2 \leq b \right\}, \quad a_1, \ldots, a_d, b \in \mathcal{R} ,$$

then $V \leq d+2$.

(3) If $\mathcal{A}$ is the class of all ellipsoids in $\mathcal{R}^d$, that is, sets of form $\{x : x^T \Sigma^{-1} x \leq 1\}$, where Σ is a positive definite symmetric matrix, then $V \leq d(d+1)/2+1$.

Note that the above-mentioned result implies that the VC dimension of the class of all linear halfspaces actually equals $d+1$. Dudley (1979) proved that in the case of the class of all closed balls the above inequality is not tight, and the VC dimension equals $d+1$ (see exercise 5).

1.4.4 Applications to empirical risk minimization

In this section we apply the main results of the previous sections to obtain upper bounds for the performance of empirical risk minimization.

Recall the scenario set up in Chapter 2: $\mathcal{C}$ is a class of classifiers containing decision functions of the form $g : \mathcal{R}^d \to \{0,1\}$. The data $(X_1,Y_1),\ldots,(X_n,Y_n)$ may be used to calculate the empirical error $\widehat{L}_n(g)$ for any $g \in \mathcal{C}$. g_n^* denotes a classifier minimizing $\widehat{L}_n(g)$ over the class, that is,

$$\widehat{L}_n(g_n^*) \le \widehat{L}_n(g) \quad \text{for all } g \in \mathcal{C}.$$

Denote the probability of error of the optimal classifier in the class by $L_{\mathcal{C}}$, that is,

$$L_{\mathcal{C}} = \inf_{g \in \mathcal{C}} L(g).$$

(Here we implicitely assume that the infimum is achieved. This assumption is motivated by convenience in the notation, it is not essential.)

The basic Lemma 1.1 shows that

$$L(g_n^*) - L_{\mathcal{C}} \le 2 \sup_{g \in \mathcal{C}} \left| \widehat{L}_n(g) - L(g) \right|.$$

Thus, the quantity of interest is the maximal deviation between empirical probabilities of error and their expectation over the class. Such quantities are estimated by the Vapnik-Chervonenkis inequality. Indeed, the random variable $\sup_{g \in \mathcal{C}} \left| \widehat{L}_n(g) - L(g) \right|$ is of the form of

$$\sup_{A \in \mathcal{A}} |\mu_n(A) - \mu(A)|,$$

where the role of the class of sets $\mathcal{A}$ is now played by the class of error sets

$$\left\{ (x,y) \in \mathcal{R}^d \times \{0,1\} : g(x) \neq y \right\}; \quad g \in \mathcal{C}.$$

Denote the class of these error sets by $\bar{\mathcal{A}}$. Thus, the Vapnik-Chervonenkis inequality immediately bounds the expected maximal deviation in terms of the xshatter coefficients (or VC dimension) of the class of error sets.

Instead of error sets, it is more convenient to work with classes of sets of the form

$$\left\{ x \in \mathcal{R}^d : g(x) = 1 \right\}; \quad g \in \mathcal{C}.$$

We denote the class of sets above by $\mathcal{A}$. The next simple fact shows that the classes $\bar{\mathcal{A}}$ and $\mathcal{A}$ are equivalent from a combinatorial point of view:

Lemma 1.7. *For every n we have $\mathbb{S}_{\bar{\mathcal{A}}}(n) = \mathbb{S}_{\mathcal{A}}(n)$, and therefore the corresponding* VC *dimensions are also equal: $V_{\bar{\mathcal{A}}} = V_{\mathcal{A}}$.*

PROOF. Let N be a positive integer. We show that for any n pairs from $\mathcal{R}^d \times \{0,1\}$, if N sets from $\bar{\mathcal{A}}$ pick N different subsets of the n pairs, then there are N corresponding sets in $\mathcal{A}$ that pick N different subsets of n points in $\mathcal{R}^d$, and vice versa. Fix n pairs $(x_1,0),\ldots,(x_m,0),(x_{m+1},1),\ldots,(x_n,1)$. Note that since ordering does not matter, we may arrange any n pairs in this manner. Assume that for a certain set $A \in \mathcal{A}$, the corresponding set $\bar{A} = A \times \{0\} \bigcup A^c \times \{1\} \in \bar{\mathcal{A}}$ picks out the pairs $(x_1,0),\ldots,(x_k,0),(x_{m+1},1),\ldots, (x_{m+l},1)$, that is, the set of these pairs is the intersection of $\bar{\mathcal{A}}$ and the n pairs. Again, we can assume without loss of generality that the pairs are ordered in this way. This means that A picks from the set $\{x_1,\ldots,x_n\}$ the subset $\{x_1,\ldots,x_k,x_{m+l+1},\ldots,x_n\}$, and the two subsets uniquely determine each other. This proves $\mathbb{S}_{\bar{\mathcal{A}}}(n) \le \mathbb{S}_{\mathcal{A}}(n)$. To prove the other direction, notice that if A picks a subset of k points $x_1,\ldots,x_k$, then the corresponding set $\bar{A} \in \bar{\mathcal{A}}$ picks the pairs with the same indices from $\{(x_1,0),\ldots, (x_k,0)\}$. Equality of the VC dimensions follows from the equality of the shatter coefficients. □

From this point on, we will denote the common value of $\mathbb{S}_{\bar{\mathcal{A}}}(n)$ and $\mathbb{S}_{\mathcal{A}}(n)$ by $\mathbb{S}_{\mathcal{C}}(n)$, and refer to is as the n-th shatter coefficient of the class $\mathcal{C}$. It is simply the maximum number of different ways n points can be classified by classifiers in the class $\mathcal{C}$. Similarly, $V_{\bar{\mathcal{A}}} = V_{\mathcal{A}}$ will be referred to as the VC dimension of the class $\mathcal{C}$, and will be denoted by $V_{\mathcal{C}}$.

Now we are prepared to summarize our main performance bound for empirical risk minimization:

COROLLARY 1.5.

$$\mathbb{E}L(g_n^*) - L_{\mathcal{C}} \le 4\sqrt{\frac{\log 2\mathbb{S}_{\mathcal{C}}(n)}{n}} \le 4\sqrt{\frac{V_{\mathcal{C}}\log(n+1)+\log 2}{n}}$$

Bounds for $\mathbb{P}\{L(g_n^*) - L_{\mathcal{C}} > \varepsilon\}$ may now be easily obtained by combining the corollary above with the bounded difference inequality.

The inequality above may be improved in various different ways. In the appendix of this chapter we show that the factor of $\log n$ in the upper bound is unnecessary, it may be replaced by a suitable constant. In practice, however, often the sample size is so small that the inequality above provides smaller numerical values.

On the other hand, the main performance may be improved in another direction. To understand the reason, consider first an extreme situation when $L_{\mathcal{C}} = 0$, that is, there exists a classifier in $\mathcal{C}$ which classifies without error. (This also means that for som $g' \in \mathcal{C}$, $Y = g'(X)$ with probability one, a very restrictive assumption. Nevertheless, the assumption that $L_{\mathcal{C}} = 0$ is common in computational learning theory, see Blumer, Ehrenfeucht, Haussler, and Warmuth (1989). In such a case, clearly $\widehat{L}_n(g^*) = 0$, and the second statement of Corollary 1.2 implies that

$$\mathbb{P}\{L(g_n^*) - L_C > \varepsilon\} = \mathbb{P}\{L(g_n^*) > \varepsilon\} \leq 4\mathbb{S}_C(2n)e^{-n\varepsilon/4},$$

and therefore

$$\mathbb{E}L(g_n^*) - L_C = \mathbb{E}L(g_n^*) \leq \frac{4\ln(4\mathbb{S}_C(2n))}{n}.$$

(The bound on the expected value may be obtained by the following simple bounding argument: assume that for some nonnegative random variable Z, for all $\varepsilon > 0$, $\mathbb{P}\{Z > \varepsilon\} \leq Ce^{-K\varepsilon}$ for some positive constants. Then $\mathbb{E}Z = \int_0^\infty \mathbb{P}\{Z > \varepsilon\}d\varepsilon \leq u + \int_u^\infty Ce^{-K\varepsilon}$ for any $u > 0$. Integrating, and choosing u to minimize the upper bound, we obtain $\mathbb{E}Z \leq \ln C/K$.)

The main point here is that the upper bound obtained in this special case is of smaller order of magnitude than in the general case ($O(V_C \ln n/n)$ as opposed to $O\left(\sqrt{V_C \ln n/n}\right)$.) Intuition suggests that if L_C is nonzero but very small, the general bound of Corollary 1.5 should be improvable. In fact, the argument below shows that it is possible interpolate between the special case $L_C = 0$ and the fully distribution-free bound of Corollary 1.5:

Theorem 1.14.

$$\mathbb{E}L(g_n^*) - L_C \leq \sqrt{\frac{8L_C \ln(5\mathbb{S}_C(2n)) + 2}{n}} + \frac{8\ln(10\mathbb{S}_C(2n)) + 4}{n}.$$

Also, for every $\varepsilon > 0$,

$$\mathbb{P}\{L(g_n^*) - L_C > \varepsilon\} \leq 5\mathbb{S}_C(2n)e^{-n\varepsilon^2/16(L_C+\varepsilon)}.$$

PROOF. For any $\varepsilon > 0$, if

$$\sup_{g \in C} \frac{L(g) - \widehat{L}_n(g)}{\sqrt{L(g)}} \leq \frac{\varepsilon}{\sqrt{L_C + 2\varepsilon}},$$

then for each $g \in C$

$$\widehat{L}_n(g) \geq L(g) - \varepsilon\sqrt{\frac{L(g)}{L_C + 2\varepsilon}}.$$

If, in addition, g is such that $L(g) > L_C + 2\varepsilon$, then by the monotonicity of the function $x - c\sqrt{x}$ (for $c > 0$ and $x > c^2/4$),

$$\widehat{L}_n(g) \geq L_C + 2\varepsilon - \varepsilon\sqrt{\frac{L_C + 2\varepsilon}{L_C + 2\varepsilon}} = L_C + \varepsilon.$$

Therefore,

$$\mathbb{P}\left\{\inf_{g: L(g) > L_C + 2\varepsilon} \widehat{L}_n(g) < L_C + \varepsilon\right\} \leq \mathbb{P}\left\{\sup_{g \in C} \frac{L(g) - \widehat{L}_n(g)}{\sqrt{L(g)}} > \frac{\varepsilon}{\sqrt{L_C + 2\varepsilon}}\right\}.$$

But if $L(g_n^*) - L_C > 2\varepsilon$, then, denoting by g' a classifier in C such that $L(g') = L_C$, there exists an $g \in C$ such that $L(g) > L_C + 2\varepsilon$ and $\widehat{L}_n(g) \leq \widehat{L}_n(g')$. Thus,

$$\begin{aligned}
&\mathbb{P}\{L(g_n^*) - L_C > 2\varepsilon\} \\
&\leq \mathbb{P}\left\{\inf_{g:L(g)>L_C+2\varepsilon} \widehat{L}_n(g) < \widehat{L}_n(g')\right\} \\
&\leq \mathbb{P}\left\{\inf_{g:L(g)>L_C+2\varepsilon} \widehat{L}_n(g) < L_C+\varepsilon\right\} + \mathbb{P}\{\widehat{L}_n(g') > L_C+\varepsilon\} \\
&\leq \mathbb{P}\left\{\sup_{g\in C} \frac{L(g) - \widehat{L}_n(g)}{\sqrt{L(g)}} > \frac{\varepsilon}{\sqrt{L_C+2\varepsilon}}\right\} + \mathbb{P}\{\widehat{L}_n(g') - L_C > \varepsilon\}.
\end{aligned}$$

Bounding the last two probabilities by Theorem 1.11 and Bernstein's inequality, respectively, we obtain the probability bound of the statement.

The upper bound for the expected value may now be derived by some straightforward calculations which we sketch here: let $u \leq L_C$ be a positive number. Then, using the tail inequality obtained above,

$$\begin{aligned}
&\mathbb{E}L(g_n^*) - L_C \\
&= \int_0^\infty \mathbb{P}\{L(g_n^*) - L_C > \varepsilon\} d\varepsilon \\
&\leq u + \int_u^\infty 5\mathbb{S}_C(2n) \max\left(e^{-n\varepsilon^2/8L_C}, e^{-n\varepsilon/8}\right) d\varepsilon \\
&\leq \left(u/2 + \int_u^\infty 5\mathbb{S}_C(2n) e^{-n\varepsilon^2/8L_C} d\varepsilon\right) \\
&\quad + \left(u/2 + \int_u^\infty 5\mathbb{S}_C(2n) e^{-n\varepsilon/8} d\varepsilon\right).
\end{aligned}$$

The second term may be bounded as in the argument given fot the case $L_C = 0$, while the first term may be calculated similarly, using the additional observation that

$$\begin{aligned}
\int_u^\infty e^{-n\varepsilon^2} d\varepsilon &\leq \frac{1}{2}\int_u^\infty \left(2 + \frac{1}{n\varepsilon^2}\right) e^{-n\varepsilon^2} d\varepsilon \\
&= \frac{1}{2}\left[\frac{1}{n\varepsilon} e^{-n\varepsilon^2}\right]_u^\infty.
\end{aligned}$$

The details are omitted. □

1.4.5 *Convex combinations of classifiers*

Several important classification methods form a classifier as a convex combination of simple functions. To describe such a situation, consider a class C of classifiers $g : \mathcal{R}^d \to \{0,1\}$. Think of $\mathcal{C}$ as a small class of "base" classifiers such as the class of all linear splits of $\mathcal{R}^d$. In general we assume that the VC dimension $V_{\mathcal{C}}$ of $\mathcal{C}$ is finite. Define the class $\mathcal{F}$ as the class of functions $f : \mathcal{R}^d \to [0,1]$ of the form

$$f(x) = \sum_{j=1}^{N} w_j g_j(x)$$

where N is any positive integer, $w_1, \ldots, w_N$ are nonnegative weights with $\sum_{j=1}^N w_j = 1$, and $g_1, \ldots, g_N \in \mathcal{C}$. Thus, $\mathcal{F}$ may be considered as the convex hull of $\mathcal{C}$. Each function $f \in \mathcal{F}$ defines a classifier g_f, in a natural way, by

$$g_f(x) = \begin{cases} 1 & \text{if } f(x) > 1/2 \\ 0 & \text{otherwise.} \end{cases}$$

A large variety of "boosting" and "bagging" methods, based mostly on the work of Schapire (1990), Freund (1995) and Breiman (1996), construct classifiers as convex combinations of very simple functions. Typically the class of classifiers defined this way is too large in the sense that it is impossible to obtain meaningful distribution-free upper bounds for $\sup_{f\in\mathcal{F}} \left(L(g_f) - \widehat{L}_n(g_f)\right)$. Indeed, even in the simple case when $d = 1$ and $\mathcal{C}$ is the class of all linear splits of the real line, the class of all g_f is easily seen to have an infinite VC dimension.

Surprisingly, however, meaningful bounds may be obtained if we replace the empirical probability of error $\widehat{L}_n(g_f)$ by a slightly larger quantity. To this end, let $\gamma > 0$ be a fixed parameter, and define the *margin error* by

$$L_n^\gamma(g_f) = \frac{1}{n}\sum_{i=1}^{n} \mathbb{I}_{[f(X_i)(1-2Y_i)<\gamma]}.$$

Notice that for all $\gamma > 0$, $L_n^\gamma(g_f) \ge \widehat{L}_n(g_f)$ and the $L_n^\gamma(g_f)$ is increasing in γ. An interpretation of the margin error $L_n^\gamma(g_f)$ is that it counts, apart from the number of misclassified pairs (X_i, Y_i), also those which are well classified but only with a small "confidence" (or "margin") by g_f.

The purpose of this section is to present a result of Freund, Schapire, Bartlett, and Lee (1998) which states that the margin error is always a good approximate upper bound for the probability of error, at least if γ is not too small. The elegant proof shown here is due to Koltchinskii and Panchenko (2002).

Theorem 1.15. *For every* $\varepsilon > 0$,

$$\mathbb{P}\left\{\sup_{f\in\mathcal{F}} (L(g_f) - L_n^\gamma(g_f)) > \frac{2\sqrt{2}}{\gamma}\sqrt{\frac{V_{\mathcal{C}}\log(n+1)}{n}} + \varepsilon\right\} \le e^{-2n\varepsilon^2}.$$

Thus, with very high probability, the probability of error of any classifier g_f, $f \in \mathcal{F}$, may be simultaneously upper bounded by the sum

$$L_n^\gamma(g_f) + \frac{2\sqrt{2}}{\gamma}\sqrt{\frac{V_{\mathcal{C}}\log(n+1)}{n}}$$

plus a term of the order $n^{-1/2}$. Notice that, as γ grows, the first term of the sum increases, while the second decreases. The bound can be very useful whenever a classifier has a small margin error for a relatively large γ (i.e., if the classifier classifies the training data well with high "confidence") since the second term only depends on the VC dimension of the small base class $\mathcal{C}$. As shown in the next section, the second term in the above sum may be replaced by $(c/\gamma)\sqrt{V_{\mathcal{C}}/n}$ for some universal constant c.

The proof of the theorem crucially uses the following simple lemma, called the "contraction principle". Here we cite a version tailored for our needs. For the proof, see Ledoux and Talagrand (1991), pages 112–113.

Lemma 1.8. *Let $Z_1(f),\ldots,Z_n(f)$ be arbitrary real-valued bounded random variables indexed by an abstract parameter f and let $\sigma_1,\ldots,\sigma_n$ be independent symmetric sign variables, independent of the $Z_i(f)$'s (i.e., $\mathbb{P}\{\sigma_i=-1\}=\mathbb{P}\{\sigma_i=1\}=1/2$). If $\phi:\mathcal{R}\to\mathcal{R}$ is a Lipschitz function such that $|\phi(x)-\phi(y)|\le|x-y|$ with $\phi(0)=0$, then*

$$\mathbb{E}\sup_f\sum_{i=1}^n\sigma_i\phi(Z_i(f))\le\mathbb{E}\sup_f\sum_{i=1}^n\sigma_iZ_i(f).$$

PROOF OF THEOREM 1.15. For any $\gamma>0$, introduce the function

$$\phi_\gamma(x)=\begin{cases}1 & \text{if } x\le 0\\ 0 & \text{if } x\ge\gamma\\ 1-x/\gamma & \text{if } x\in(0,\gamma)\end{cases}$$

Observe that $\mathbb{I}_{[x\le 0]}\le\phi_\gamma(x)\le\mathbb{I}_{[x\le\gamma]}$. Thus,

$$\sup_{f\in\mathcal{F}}(L(g_f)-L_n^\gamma(g_f))\le\sup_{f\in\mathcal{F}}\left(\mathbb{E}\phi_\gamma((1-2Y)f(X))-\frac{1}{n}\sum_{i=1}^n\phi_\gamma((1-2Y_i)f(X))\right).$$

Introduce the notation $Z(f)=(1-2Y)f(X)$ and $Z_i(f)=(1-2Y_i)f(X_i)$. Clearly, by the bounded difference inequality,

$$\mathbb{P}\left\{\sup_{f\in\mathcal{F}}\left(\mathbb{E}\phi_\gamma(Z(f))-\frac{1}{n}\sum_{i=1}^n\phi_\gamma(Z_i(f))\right)\right.$$
$$\left.>\mathbb{E}\sup_{f\in\mathcal{F}}\left(\mathbb{E}\phi_\gamma(Z(f))-\frac{1}{n}\sum_{i=1}^n\phi_\gamma(Z_i(f))\right)+\varepsilon\right\}\le e^{-2n\varepsilon^2}$$

and therefore it suffices to prove that the expected value of the supremum is bounded by

$$\frac{2\sqrt{2}}{\gamma}\sqrt{\frac{V_{\mathcal{C}}\log(n+1)}{n}}.$$

As a first step, we proceed by a symmetrization argument just like in the proof of Theorem 1.9 to obtain

$$\begin{aligned}\mathbb{E}\sup_{f\in\mathcal{F}}\left(\mathbb{E}\phi_\gamma(Z(f))-\frac{1}{n}\sum_{i=1}^{n}\phi_\gamma(Z_i(f))\right) &\leq \mathbb{E}\sup_{f\in\mathcal{F}}\left(\frac{1}{n}\sum_{i=1}^{n}\sigma_i\left(\phi_\gamma(Z_i'(f))-\phi_\gamma(Z_i(f))\right)\right)\\ &\leq 2\mathbb{E}\sup_{f\in\mathcal{F}}\left(\frac{1}{n}\sum_{i=1}^{n}\sigma_i\left(\phi_\gamma(Z_i(f))-\phi_\gamma(0)\right)\right)\end{aligned}$$

where $\sigma_1,\ldots,\sigma_n$ are i.i.d. symmetric sign variables and $Z_i'(f)=(1-2Y_i')f(X_i')$ where the (X_i',Y_i') are independent of the (X_i,Y_i) and have the same distribution as that of the pairs (X_i,Y_i).

Observe that the function $\phi(x)=\gamma(\phi_\gamma(x)-\phi_\gamma(0))$ is Lipschitz and $\phi(0)=0$, therefore, by the contraction principle (Lemma 1.8),

$$\mathbb{E}\sup_{f\in\mathcal{F}}\frac{1}{n}\sum_{i=1}^{n}\sigma_i\left(\phi_\gamma(Z_i(f))-\phi_\gamma(0)\right)\leq\frac{1}{\gamma}\mathbb{E}\sup_{f\in\mathcal{F}}\frac{1}{n}\sum_{i=1}^{n}\sigma_iZ_i(f)=\frac{1}{\gamma}\mathbb{E}\sup_{f\in\mathcal{F}}\frac{1}{n}\sum_{i=1}^{n}\sigma_if(X_i)$$

where at the last step we used the fact that $\sigma_i(1-2Y_i)$ is a symmetric sign variable, independent of the X_i and therefore $\sigma_i(1-2Y_i)f(X_i)$ has the same distribution as that of $\sigma_if(X_i)$. The last expectation may be rewritten as

$$\mathbb{E}\sup_{f\in\mathcal{F}}\frac{1}{n}\sum_{i=1}^{n}\sigma_if(X_i)=\frac{1}{n}\mathbb{E}\sup_{N\geq1}\sup_{g_1,\ldots,g_N\in C}\sup_{w_1,\ldots,w_N}\sum_{i=1}^{n}\sum_{j=1}^{N}w_j\sigma_ig_j(X_i).$$

The key observation is that for any N and base classifiers $g_i,\ldots,g_N$, the supremum in

$$\sup_{w_1,\ldots,w_N}\sum_{i=1}^{n}\sum_{j=1}^{N}w_j\sigma_ig_j(X_i)$$

is achieved for a weight vector which puts all the mess in one index, that is, when $w_j=1$ for some j. (This may be seen by observing that a linear function over a convex polygon achieves its maximum at one of the vertices of the polygon.) Thus,

$$\mathbb{E}\sup_{f\in\mathcal{F}}\frac{1}{n}\sum_{i=1}^{n}\sigma_if(X_i)=\frac{1}{n}\mathbb{E}\sup_{g\in C}\sum_{i=1}^{n}\sigma_ig(X_i).$$

However, repeating the argument in the proof of Theorem 1.9 with the necessary adjustments, we obtain

$$\frac{1}{n}\mathbb{E}\sup_{g\in C}\left|\sum_{i=1}^{n}\sigma_ig(X_i)\right|\leq\sqrt{\frac{2\log\mathbb{S}_C(n)}{n}}\leq\sqrt{\frac{2V_C\log(n+1)}{n}}$$

which completes the proof of the desired inequality. □

1.4.6 *Appendix: sharper bounds via chaining*

In this section we present an improvement of the Vapnik-Chervonenkis inequality stating that for any class $\mathcal{A}$ of sets of VC dimension V,

$$\mathbb{E}\sup_{A\in\mathcal{A}}|\mu_n(A)-\mu(A)|\leq c\sqrt{\frac{V}{n}},$$

where c is a universal constant. This in turn implies for empirical risk minimization that

$$\mathbb{E}L(g_n^*)-L_C\leq 2c\sqrt{\frac{V_C}{n}}.$$

The new bound involves some geometric and combinatorial quantities related to the class $\mathcal{A}$. Consider a pair of bit vectors $b=(b_1,\ldots,b_n)$ and $c=(c_1,\ldots,c_n)$ from $\{0,1\}^n$, and define their distance by

$$\rho(b,c)=\sqrt{\frac{1}{n}\sum_{i=1}^{n}\mathbb{I}_{[b_i\neq c_i]}}.$$

Thus, $\rho(b,c)$ is just the square root of the normalized Hamming distance between b and c. Observe that ρ may also be considered as the normalized euclidean distance between the corners of the hypercube $[0,1]^n\subset\mathcal{R}^n$, and therefore it is indeed a distance.

Now let $B\subset\{0,1\}^n$ be any set of bit vectors, and define a *cover* of radius $r>0$ as a set $B_r\subset\{0,1\}^n$ such that for any $b\in B$ there exists a $c\in B_r$ such that $\rho(b,c)\leq r$. The *covering number* $N(r,B)$ is the cardinality of the smallest cover of radius r.

A class $\mathcal{A}$ of subsets of $\mathcal{R}^d$ and a set of n points $x_1^n=\{x_1,\ldots,x_n\}\subset\mathcal{R}^d$ define a set of bit vectors by

$$\mathcal{A}(x_1^n)=\left\{b=(b_1,\ldots,b_n)\in\{0,1\}^n : b_i=\mathbb{I}_{[x_i\in A]},\ i=1,\ldots,n \text{ for some } A\in\mathcal{A}\right\}.$$

That is, every bit vector $b\in\mathcal{A}(x_1^n)$ describes the intersection of $\{x_1,\ldots,x_n\}$ with a set A in $\mathcal{A}$. We have the following:

Theorem 1.16.

$$\mathbb{E}\left\{\sup_{A\in\mathcal{A}}|\mu_n(A)-\mu(A)|\right\}\leq\frac{24}{\sqrt{n}}\max_{x_1,\ldots,x_n\in\mathcal{R}^d}\int_0^1\sqrt{\log 2N(r,\mathcal{A}(x_1^n))}\,dr\,.$$

The theorem implies that $\mathbb{E}\{\sup_{A\in\mathcal{A}}|\mu_n(A)-\mu(A)|\}=O(1/\sqrt{n})$ whenever the integral in the bound is uniformly bounded over all $x_1,\ldots,x_n$ and all n. Note that the bound of Theorem 1.9 is always of larger order of magnitude, trivial cases excepted. The main additional idea is Dudley's *chaining* trick.

PROOF. As in the proof of Theorem 1.9, we see that

$$\begin{aligned}
&\mathbb{E}\left\{\sup_{A\in\mathcal{A}}|\mu_n(A)-\mu(A)|\right\}\\
&\le \frac{1}{n}\mathbb{E}\left\{\sup_{A\in\mathcal{A}}\left|\sum_{i=1}^n\sigma_i\left(\mathbb{I}_{[X_i\in A]}-\mathbb{I}_{[X_i'\in A]}\right)\right|\right\}\\
&\le \frac{1}{n}\mathbb{E}\left\{\sup_{A\in\mathcal{A}}\left|\sum_{i=1}^n\sigma_i\mathbb{I}_{[X_i\in A]}\right|\right\}+\frac{1}{n}\mathbb{E}\left\{\sup_{A\in\mathcal{A}}\left|\sum_{i=1}^n\sigma_i\mathbb{I}_{[X_i'\in A]}\right|\right\}\\
&= \frac{2}{n}\mathbb{E}\left\{\sup_{A\in\mathcal{A}}\left|\sum_{i=1}^n\sigma_i\mathbb{I}_{[X_i\in A]}\right|\right\}\\
&= \frac{2}{n}\mathbb{E}\mathbb{E}\left\{\sup_{A\in\mathcal{A}}\left|\sum_{i=1}^n\sigma_i\mathbb{I}_{[X_i\in A]}\right|\,\middle|\,X_1,\dots,X_n\right\}.
\end{aligned}$$

Just as in the proof of theorem 1, we fix the values $X_1=x_1,\dots,X_n=x_n$ and study

$$\mathbb{E}\left\{\sup_{A\in\mathcal{A}}\left|\sum_{i=1}^n\sigma_i\mathbb{I}_{[x_i\in A]}\right|\right\}=\mathbb{E}\left\{\max_{b\in\mathcal{A}(x_1^n)}\left|\sum_{i=1}^n\sigma_i b_i\right|\right\}.$$

Now let $B_0\stackrel{\text{def}}{=}\{b^{(0)}\}$ be the singleton set containing the all-zero vector $b^{(0)}=(0,\dots,0)$, and let $B_1,B_2,\dots,B_M$ be subsets of $\{0,1\}^n$ such that each B_k is a minimal cover of $\mathcal{A}(x_1^n)$ of radius 2^{-k}, and $M=\lfloor\log_2\sqrt{n}\rfloor+1$. Note that B_0 is also a cover of radius 2^0, and that $B_M=\mathcal{A}(x_1^n)$. Now denote the (random) vector reaching the maximum by $b^*=(b_1^*,\dots,b_n^*)\in\mathcal{A}(x_1^n)$, that is,

$$\left|\sum_{i=1}^n\sigma_i b_i^*\right|=\max_{b\in\mathcal{A}(x_1^n)}\left|\sum_{i=1}^n\sigma_i b_i\right|,$$

and, for each $k\le M$, let $b^{(k)}\in B_k$ be a nearest neighbor of b^* in the k-th cover, that is,

$$\rho(b^{(k)},b^*)\le\rho(b,b^*)\quad\text{for all } b\in B_k.$$

Note that $\rho(b^{(k)},b^*)\le 2^{-k}$, and therefore

$$\rho(b^{(k)},b^{(k-1)})\le\rho(b^{(k)},b^*)+\rho(b^{(k-1)},b^*)\le 3\cdot 2^{-k}.$$

Now clearly,

$$\begin{aligned}
\sum_{i=1}^n\sigma_i b_i^* &= \sum_{i=1}^n\sigma_i b_i^{(0)}+\sum_{k=1}^M\sum_{i=1}^n\sigma_i\left(b_i^{(k)}-b_i^{(k-1)}\right)\\
&= \sum_{k=1}^M\sum_{i=1}^n\sigma_i\left(b_i^{(k)}-b_i^{(k-1)}\right),
\end{aligned}$$

so

$$\begin{aligned}
\mathbb{E}\left\{\max_{b\in\mathcal{A}(x_1^n)}\left|\sum_{i=1}^n \sigma_i b_i\right|\right\} &= \mathbb{E}\left|\sum_{k=1}^M\sum_{i=1}^n \sigma_i\left(b_i^{(k)}-b_i^{(k-1)}\right)\right| \\
&\le \sum_{k=1}^M \mathbb{E}\left|\sum_{i=1}^n \sigma_i\left(b_i^{(k)}-b_i^{(k-1)}\right)\right| \\
&\le \sum_{k=1}^M \mathbb{E}\max_{b\in B_k, c\in B_{k-1}:\rho(b,c)\le 3\cdot 2^{-k}}\left|\sum_{i=1}^n \sigma_i(b_i-c_i)\right| .
\end{aligned}$$

Now it follows from Lemma 1.2 that for each pair $b\in B_k, c\in B_{k-1}$ with $\rho(b,c)\le 3\cdot 2^{-k}$, and for all $s>0$,

$$e^{s\sum_{i=1}^n \sigma_i(b_i-c_i)} \le e^{s^2 n\left(3\cdot 2^{-k}\right)^2/2}.$$

On the other hand, the number of such pairs is bounded by $|B_k|\cdot|B_{k-1}|\le|B_k|^2 = N(2^{-k},\mathcal{A}(x_1^n))^2$. Then Lemma 1.3 implies that for each $1\le k\le M$,

$$\mathbb{E}\max_{b\in B_k, c\in B_{k-1}:\rho(b,c)\le 3\cdot 2^{-k}}\left|\sum_{i=1}^n \sigma_i(b_i-c_i)\right| \le 3\sqrt{n}2^{-k}\sqrt{2\log 2N(2^{-k},\mathcal{A}(x_1^n))^2} .$$

Summarizing, we obtain

$$\begin{aligned}
\mathbb{E}\left\{\max_{b\in\mathcal{A}(x_1^n)}\left|\sum_{i=1}^n \sigma_i b_i\right|\right\} &\le 3\sqrt{n}\sum_{k=1}^M 2^{-k}\sqrt{2\log 2N(2^{-k},\mathcal{A}(x_1^n))^2} \\
&\le 12\sqrt{n}\sum_{k=1}^\infty 2^{-(k+1)}\sqrt{\log 2N(2^{-k},\mathcal{A}(x_1^n))} \\
&\le 12\sqrt{n}\int_0^1 \sqrt{\log 2N(r,\mathcal{A}(x_1^n))}\,dr ,
\end{aligned}$$

where at the last step we used the fact that $N(r,\mathcal{A}(x_1^n))$ is a monotonically decreasing function of r. The proof is finished. □

To complete our argument, we need to relate the VC dimension of a class of sets $\mathcal{A}$ to the covering numbers $N(r,\mathcal{A}(x_1^n))$ appearing in Theorem 3.10.

Theorem 1.17. *Let $\mathcal{A}$ be a class of sets with* VC *dimension $V<\infty$. For every $x_1,\dots,x_n\in\mathcal{R}^d$ and $0\le r\le 1$,*

$$N(r,\mathcal{A}(x_1^n)) \le \left(\frac{4e}{r^2}\right)^{V/(1-1/e)}$$

Theorem 1.17 is due to Dudley (1978). Haussler (1995) refined Dudley's probabilistic argument and showed that the stronger bound

$$N(r,\mathcal{A}(x_1^n)) \le e(V+1)\left(\frac{2e}{r^2}\right)^V .$$

also holds.

PROOF. Fix $x_1,\ldots,x_n$, and consider the set $B_0 = \mathcal{A}(x_1^n) \in \{0,1\}^n$. Fix $r \in (0,1)$, and let $B_r \subset \{0,1\}^n$ be a minimal cover of B_0 of radius r with respect to the metric

$$\rho(b,c) = \sqrt{\frac{1}{n}\sum_{i=1}^{n} \mathbb{I}_{[b_i \neq c_i]}}.$$

We need to show that $|B_r| \leq \left(4e/r^2\right)^{V/(1-1/e)}$.

First note that there exists a "packing set" $C_r \subset B_0$ such that $|B_r| \leq |C_r|$ and any two elements $b,c \in C_r$ are *r-separated*, that is, $\rho(b,c) > r$. To see this, suppose that C_r is such an r-separated set of maximal cardinality. Then for any $b \in B_0$, there exists a $c \in C_r$ with $\rho(b,c) \leq r$, since otherwise adding b to the set C_r would increase its cardinality, and it would still be r-separated. Thus, C_r is a cover of radius r, which implies that $|B_r| \leq |C_r|$. Denote the elements of C_r by $c^{(1)},\ldots,c^{(M)}$, where $M = |C_r|$. For any $i,j \leq M$, define $A_{i,j}$ as the set of indices where the binary vectors $c^{(i)}$ and $c^{(j)}$ disagree:

$$A_{i,j} = \left\{1 \leq m \leq n : c_m^{(i)} \neq c_m^{(j)}\right\} .$$

Note that any two elements of C_r differ in at least nr^2 components. Next define K independent random variables $Y_1,\ldots,Y_K$, distributed uniformly over the set $\{1,2,\ldots,n\}$, where K will be specified later. Then for any $i,j \leq M$, $i \neq j$, and $k \leq K$,

$$\mathbb{P}\{Y_k \in A_{i,j}\} \geq r^2 ,$$

and therefore the probability that no one of $Y_1,\ldots,Y_K$ falls in the set $A_{i,j}$ is less than $\left(1-r^2\right)^K$. Observing that there are less than M^2 sets $A_{i,j}$, and applying the union bound, we obtain that

$$\begin{aligned}&\mathbb{P}\left\{\text{for all } i \neq j, i,j \leq M, \text{at least one } Y_k \text{ falls in } A_{i,j}\right\}\\ &\qquad \geq 1 - M^2(1-r^2)^K \geq 1 - M^2 e^{-Kr^2} .\end{aligned}$$

If we choose $K = \lceil 2\log M/r^2 \rceil + 1$, then the above probability is strictly positive. This implies that there exist $K = \lceil 2\log M/r^2 \rceil + 1$ indices $y_1,\ldots,y_K \in \{1,2,\ldots,n\}$ such that at least one y_k falls in each set $A_{i,j}$. Therefore, restricted to the K components $y_1,\ldots,y_K$, the elements of C_r are all different, and since $C_r \subset B_0$, C_r does not shatter any set of size larger than V. Therefore, by Sauer's lemma we obtain

$$|C_r| = M \leq \left(\frac{eK}{V}\right)^V$$

for $K \leq V$. Thus, if $\log M \geq V$, then

$$\begin{aligned}
\log M &\leq V \log \frac{e\left(\lceil 2\log M/r^2 \rceil + 1\right)}{V} \\
&\leq V\left(\log \frac{4e}{r^2} + \log \frac{\log M}{V}\right) \\
&\leq V \log \frac{4e}{r^2} + \frac{1}{e}\log M \quad (\text{since } \log x \leq x/e \text{ for } x > 0)\,.
\end{aligned}$$

Therefore,

$$\log M \leq \frac{V}{1-1/e} \log \frac{4e}{r^2}\,.$$

If $\log M < V$, then the above inequality holds trivially. This concludes the proof. □

Combining this result with Theorem 3.10 we obtain that for any class $\mathcal{A}$ with VC dimension V,

$$\mathbb{E}\left\{\sup_{A\in\mathcal{A}} |\mu_n(A) - \mu(A)|\right\} \leq c\sqrt{\frac{V}{n}}\,,$$

where c is a universal constant.

1.5 Minimax lower bounds

The purpose of this section is to investigate how good the bounds obtained in the previous chapter for empirical risk minimization are. We have seen that for any class C of classifiers with VC dimension V, a classifier g_n^* minimizing the empirical risk satisfies

$$\mathbb{E}L(g_n^*) - L_C \leq O\left(\sqrt{\frac{L_C V_C \log n}{n}} + \frac{V_C \log n}{n}\right),$$

and also

$$\mathbb{E}L(g_n^*) - L_C \leq O\left(\sqrt{\frac{V_C}{n}}\right).$$

In this section we seek answers for the following questions: Are these upper bounds (at least up to the order of magnitude) tight? Is there a much better way of selecting a classifier than minimizing the empirical error?

Let us formulate exactly what we are interested in. Let C be a class of decision functions $g: \mathcal{R}^d \to \{0,1\}$. The training sequence $D_n = ((X_1,Y_1),\ldots,(X_n,Y_n))$ is used to select the classifier $g_n(X) = g_n(X,D_n)$ from C, where the selection is based on the data D_n. We emphasize here that g_n can be an arbitrary function of the data, we do not restrict our attention to empirical error minimization, where g_n is a classifier in $\mathcal{C}$ that minimizes the number errors committed on the data D_n.

As before, we measure the performance of the selected classifier by the difference between the error probability $L(g_n) = \mathbb{P}\{g_n(X) \neq Y|D_n\}$ of the selected classifier and that of the best in the class, $L_{\mathcal{C}}$. In particular, we seek lower bounds for

$$\sup \mathbb{E}L(g_n) - L_{\mathcal{C}},$$

where the supremum is taken over all possible distributions of the pair (X,Y). A lower bound for this quantities means that no matter what our method of picking a rule from $\mathcal{C}$ is, we may face a distribution such that our method performs worse than the bound.

Actually, we investigate a stronger problem, in that the supremum is taken over all distributions with $L_{\mathcal{C}}$ kept at a fixed value between zero and $1/2$. We will see that the bounds depend on n, $V_{\mathcal{C}}$, and $L_{\mathcal{C}}$ jointly. As it turns out, the situations for $L_{\mathcal{C}} > 0$ and $L_{\mathcal{C}} = 0$ are quite different. Because of its simplicity, we first treat the case $L_{\mathcal{C}} = 0$. All the proofs are based on a technique called "the probabilistic method." The basic idea here is that the existence of a "bad" distribution is proved by considering a large class of distributions, and bounding the average behavior over the class.

1.5.1 *The zero-error case*

Here we obtain lower bounds under the assumption that the best classifier in the class has zero error probability. Recall that by Corollary 1.2 the expected probability of error of an empirical risk minimizer is bounded by $O(V_{\mathcal{C}} \log n/n)$. Next we obtain minimax lower bounds close to the upper bounds.

Theorem 1.18. *Let $\mathcal{C}$ be a class of discrimination functions with* VC *dimension V. Let $\mathcal{X}$ be the set of all random variables (X,Y) for which $L_{\mathcal{C}} = 0$. Then, for every discrimination rule g_n based upon $X_1, Y_1, \ldots, X_n, Y_n$, and $n \geq V-1$,*

$$\sup_{(X,Y)\in\mathcal{X}} \mathbb{E}L(g_n) \geq \frac{V-1}{2en}\left(1 - \frac{1}{n}\right).$$

PROOF. The idea is to construct a family $\mathcal{F}$ of 2^{V-1} distributions within the distributions with $L_{\mathcal{C}} = 0$ as follows: first find points $x_1, \ldots, x_V$ that are shattered by $\mathcal{C}$. Each distribution in $\mathcal{F}$ is concentrated on the set of these points. A member in $\mathcal{F}$ is described by $V-1$ bits, $b_1, \ldots, b_{V-1}$. For convenience, this is represented as a bit vector b. Assume $V-1 \leq n$. For a particular bit vector, we let $X = x_i$ $(i < V)$ with probability $1/n$ each, while $X = x_V$ with probability $1-(V-$

$1)/n$. Then set $Y = f_b(X)$, where f_b is defined as follows:

$$f_b(x) = \begin{cases} b_i & \text{if } x = x_i, i < V \\ 0 & \text{if } x = x_V. \end{cases}$$

Note that since Y is a function of X, we must have $L^* = 0$. Also, $L_C = 0$, as the set $\{x_1, \ldots, x_V\}$ is shattered by $\mathcal{C}$, i.e., there is a $g \in \mathcal{C}$ with $g(x_i) = f_b(x_i)$ for $1 \le i \le V$. Clearly,

$$\begin{aligned}
&\sup_{(X,Y):L_C=0} \mathbb{E}\{L(g_n) - L_C\} \\
&\quad\ge \sup_{(X,Y)\in\mathcal{F}} \mathbb{E}\{L(g_n) - L_C\} \\
&\quad= \sup_b \mathbb{E}\{L(g_n) - L_C\} \\
&\quad\ge \mathbb{E}\{L(g_n) - L_C\} \\
&\quad\quad \text{(where } b \text{ is replaced by } B\text{, uniformly distributed over } \{0,1\}^{V-1}\text{)} \\
&\quad= \mathbb{E}\{L(g_n)\}, \\
&\quad= \mathbb{P}\{g_n(X, X_1, Y_1, \ldots, X_n, Y_n) \ne f_B(X)\}.
\end{aligned}$$

The last probability may be viewed as the error probability of the decision function $g_n : \mathcal{R}^d \times (\mathcal{R}^d \times \{0,1\})^n \to \{0,1\}$ in predicting the value of the random variable $f_B(X)$ based on the observation $Z_n = (X, X_1, Y_1, \ldots, X_n, Y_n)$. Naturally, this probability is bounded from below by the Bayes probability of error

$$L^*(Z_n, f_B(X)) = \inf_{g_n} \mathbb{P}\{g_n(Z_n) \ne f_B(X)\}$$

corresponding to the decision problem $(Z_n, f_B(X))$. By the results of Chapter 1,

$$L^*(Z_n, f_B(X)) = \mathbb{E}\{\min(\eta^*(Z_n), 1 - \eta^*(Z_n))\},$$

where $\eta^*(Z_n) = \mathbb{P}\{f_B(X) = 1 | Z_n\}$. Observe that

$$\eta^*(Z_n) = \begin{cases} 1/2 & \text{if } X \ne X_1, \ldots, X \ne X_n, X \ne x_V \\ 0 \text{ or } 1 & \text{otherwise.} \end{cases}$$

Thus, we see that

$$\begin{aligned}\sup_{(X,Y):L_C=0} \mathbb{E}\{L(g_n)-L_C\} &\geq L^*(Z_n, f_B(X)) \\ &= \frac{1}{2}\mathbb{P}\{X \neq X_1,\ldots,X \neq X_n, X \neq x_V\} \\ &= \frac{1}{2}\sum_{i=1}^{V-1} \mathbb{P}\{X = x_i\}(1-\mathbb{P}\{X = x_i\})^n \\ &= \frac{V-1}{2n}(1-1/n)^n \\ &\geq \frac{V-1}{2en}\left(1-\frac{1}{n}\right) \quad \text{(since } (1-1/n)^{n-1} \downarrow 1/e\text{).}\end{aligned}$$

This concludes the proof. □

1.5.2 *The general case*

In the more general case, when the best decision in the class C has positive error probability, the upper bounds derived in Chapter 2 for the expected error probability of the classifier obtained by minimizing the empirical risk are much larger than when $L_C = 0$. Theorem 1.19 below gives a lower bound for $\sup_{(X,Y):L_C \text{ fixed}} \mathbb{E}L(g_n) - L_C$. As a function of n and V_C, the bound decreases basically as in the upper bound obtained from Theorem 1.11. Interestingly, the lower bound becomes smaller as L_C decreases, as should be expected. The bound is largest when L_C is close to $1/2$.

Theorem 1.19. *Let C be a class of discrimination functions with* VC *dimension $V \geq 2$. Let X be the set of all random variables (X,Y) for which for fixed $L \in (0,1/2)$,*

$$L = \inf_{g \in C} \mathbb{P}\{g(X) \neq Y\}.$$

Then, for every discrimination rule g_n based upon $X_1, Y_1, \ldots, X_n, Y_n$,

$$\sup_{(X,Y)\in X} \mathbb{E}(L(g_n)-L) \geq \sqrt{\frac{L(V-1)}{24n}}e^{-8} \quad \textit{if } n \geq \tfrac{V-1}{2L}\max(9, 1/(1-2L)^2).$$

PROOF. Again we consider the finite family $\mathcal{F}$ from the previous section. The notation b and B is also as above. X now puts mass p at x_i, $i < V$, and mass $1-(V-1)p$ at x_V. This imposes the condition $(V-1)p \leq 1$, which will be satisfied. Next introduce the constant $c \in (0,1/2)$. We no longer have Y as a function of X. Instead, we have a uniform $[0,1]$ random variable U

independent of X and define

$$Y = \begin{cases} 1 & \text{if } U \leq \frac{1}{2} - c + 2cb_i, X = x_i, i < V \\ 0 & \text{otherwise.} \end{cases}$$

Thus, when $X = x_i$, $i < V$, Y is 1 with probability $1/2 - c$ or $1/2 + c$. A simple argument shows that the best rule for b is the one which sets

$$f_b(x) = \begin{cases} 1 & \text{if } x = x_i \,, i < V, b_i = 1 \\ 0 & \text{otherwise.} \end{cases}$$

Also, observe that

$$L = (V-1)p(1/2 - c) \,.$$

Noting that $|2\eta(x_i) - 1| = c$ for $i < V$, for fixed b, we may write

$$L(g_n) - L \geq \sum_{i=1}^{V-1} 2pcI_{\{g_n(x_i,X_1,Y_1,\ldots,X_n,Y_n)=1-f_b(x_i)\}} \,.$$

It is sometimes convenient to make the dependence of g_n upon b explicit by considering $g_n(x_i)$ as a function of $x_i, X_1, \ldots, X_n, U_1, \ldots, U_n$ (an i.i.d. sequence of uniform $[0,1]$ random variables), and b_i. We replace b by a uniformly distributed random B over $\{0,1\}^{V-1}$. After this randomization, denote $Z_n = (X, X_1, Y_1, \ldots, X_n, Y_n)$. Thus,

$$\begin{aligned} \sup_{(X,Y)\in\mathcal{F}} \mathbb{E}\{L(g_n) - L\} &= \sup_b \mathbb{E}\{L(g_n) - L\} \\ &\geq \mathbb{E}\{L(g_n) - L\} \quad \text{(with random } B\text{)} \\ &\geq \sum_{i=1}^{V-1} 2pc\mathbb{E}I_{\{g_n(x_i,X_1,\ldots,Y_n)=1-f_B(x_i)\}} \\ &= 2c\mathbb{P}\{g_n(Z_n) \neq f_B(X)\} \\ &\geq 2cL^*(Z_n, f_B(X)), \end{aligned}$$

where, as before, $L^*(Z_n, f_B(X))$ denotes the Bayes probability of error of predicting the value of $f_B(X)$ based on observing Z_n. All we have to do is to find a suitable lower bound for

$$L^*(Z_n, f_B(X)) = \mathbb{E}\{\min(\eta^*(Z_n), 1 - \eta^*(Z_n))\},$$

where $\eta^*(Z_n) = \mathbb{P}\{f_B(X) = 1 | Z_n\}$. Observe that

$$\eta^*(Z_n) = \begin{cases} 1/2 & \text{if } X \neq X_1, \ldots, X \neq X_n \text{ and } X \neq x_V \\ \mathbb{P}\{B_i = 1 | Y_{i_1}, \ldots, Y_{i_k}\} & \text{if } X = X_{i_1} = \cdots = X_{i_k} = x_i, \, i < V. \end{cases}$$

Next we compute $\mathbb{P}\{B_i = 1 | Y_{i_1} = y_1, \ldots, Y_{i_k} = y_k\}$ for $y_1, \ldots, y_k \in \{0,1\}$. Denoting the numbers of zeros and ones by $k_0 = |\{j \le k : y_j = 0\}|$ and $k_1 = |\{j \le k : y_j = 1\}|$, we see that

$$\begin{aligned}
&\mathbb{P}\{B_i = 1 | Y_{i_1} = y_1, \ldots, Y_{i_k} = y_k\} \\
&= \frac{(1-2c)^{k_1}(1+2c)^{k_0}}{(1-2c)^{k_1}(1+2c)^{k_0} + (1+2c)^{k_1}(1-2c)^{k_0}}.
\end{aligned}$$

Therefore, if $X = X_{i_1} = \cdots = X_{i_k} = x_i$, $i < V$, then

$$\begin{aligned}
&\min(\eta^*(Z_n), 1 - \eta^*(Z_n)) \\
&= \frac{\min\left((1-2c)^{k_1}(1+2c)^{k_0}, (1+2c)^{k_1}(1-2c)^{k_0}\right)}{(1-2c)^{k_1}(1+2c)^{k_0} + (1+2c)^{k_1}(1-2c)^{k_0}} \\
&= \frac{\min\left(1, \left(\frac{1+2c}{1-2c}\right)^{k_1-k_0}\right)}{1 + \left(\frac{1+2c}{1-2c}\right)^{k_1-k_0}} \\
&= \frac{1}{1 + \left(\frac{1+2c}{1-2c}\right)^{|k_1-k_0|}}.
\end{aligned}$$

In summary, denoting $a = (1+2c)/(1-2c)$, we have

$$\begin{aligned}
L^*(Z_n, f_B(X)) &= \mathbb{E}\left\{\frac{1}{1 + a^{\left|\sum_{j:X_j=X}(2Y_j-1)\right|}}\right\} \\
&\ge \mathbb{E}\left\{\frac{1}{2a^{\left|\sum_{j:X_j=X}(2Y_j-1)\right|}}\right\} \\
&\ge \frac{1}{2}\sum_{i=1}^{V-1}\mathbb{P}\{X = x_i\}\mathbb{E}\left\{a^{-\left|\sum_{j:X_j=x_i}(2Y_j-1)\right|}\right\} \\
&\ge \frac{1}{2}(V-1)pa^{-\mathbb{E}\left\{\left|\sum_{j:X_j=x_i}(2Y_j-1)\right|\right\}} \\
&\quad \text{(by Jensen's inequality).}
\end{aligned}$$

Next we bound $\mathbb{E}\left\{\left|\sum_{j:X_j=x_i}(2Y_j-1)\right|\right\}$. Clearly, if $B(k,q)$ denotes a binomial random variable with parameters k and q,

$$\mathbb{E}\left\{\left|\sum_{j:X_j=x_i}(2Y_j-1)\right|\right\} = \sum_{k=0}^{n}\binom{n}{k}p^k(1-p)^{n-k}\mathbb{E}\{|2B(k,1/2-c)-k|\}.$$

However, by straightforward calculation we see that

$$\begin{aligned}\mathbb{E}\{|2B(k,1/2-c)-k|\} &\leq \sqrt{\mathbb{E}\{(2B(k,1/2-c)-k)^2\}} \\ &= \sqrt{k(1-4c^2)+4k^2c^2} \\ &\leq 2kc+\sqrt{k}.\end{aligned}$$

Therefore, applying Jensen's inequality once again, we get

$$\sum_{k=0}^{n}\binom{n}{k}p^k(1-p)^{n-k}\mathbb{E}\{|2B(k,1/2-c)-k|\} \leq 2npc+\sqrt{np}.$$

Summarizing what we have obtained so far, we have

$$\begin{aligned}\sup_b \mathbb{E}\{L(g_n)-L\} &\geq 2cL^*(Z_n,f_B(X)) \\ &\geq 2c\frac{1}{2}(V-1)pa^{-2npc-\sqrt{np}} \\ &\geq c(V-1)pe^{-2npc(a-1)-(a-1)\sqrt{np}} \\ &\quad \text{(by the inequality } 1+x\leq e^x\text{)} \\ &= c(V-1)pe^{-8npc^2/(1-2c)-4c\sqrt{np}/(1-2c)}.\end{aligned}$$

A rough asymptotic analysis shows that the best asymptotic choice for c is given by

$$c=\frac{1}{\sqrt{4np}}\,.$$

Then the constraint $L=(V-1)p(1/2-c)$ leaves us with a quadratic equation in c. Instead of solving this equation, it is more convenient to take $c=\sqrt{(V-1)/(8nL)}$. If $2nL/(V-1)\geq 9$, then $c\leq 1/6$. With this choice for c, using $L=(V-1)p(1/2-c)$, straightforward calculation provides

$$\sup_{(X,Y)\in\mathcal{F}} \mathbb{E}(L(g_n)-L)\geq\sqrt{\frac{(V-1)L}{24n}}e^{-8}.$$

The condition $p(V-1)\leq 1$ implies that we need to ask that $n\geq (V-1)/(2L(1-2L)^2)$. This concludes the proof of Theorem 1.19. □

1.6 Complexity regularization

This section deals with the problem of automatic model selection. Our goal is to develop some data-based methods to find the class C of classifiers in a way that approximately minimizes the probability of error of the empirical risk minimizer.

1.6.1 Model selection by penalization

In empirical risk minimization one selects a classifier from a given class $\mathcal{C}$ by minimizing the error estimate $\widehat{L}_n(g)$ over all $g \in \mathcal{C}$. This provides an estimate whose loss is close to the optimal loss L^* if the class $\mathcal{C}$ is (i) sufficiently large so that the loss of the best function in $\mathcal{C}$ is close to L^* and (ii) is sufficiently small so that finding the best candidate in $\mathcal{C}$ based on the data is still possible. These two requirements are clearly in conflict. The trade-off is best understood by writing

$$\mathbb{E}L(g_n^*) - L^* = \left(\mathbb{E}L(g_n^*) - \inf_{g\in\mathcal{C}} L(g)\right) + \left(\inf_{g\in\mathcal{C}} L(g) - L^*\right).$$

The first term is often called *estimation error*, while the second is the *approximation error*.

It is common to fix in advance a sequence of model classes $\mathcal{C}_1, \mathcal{C}_2, \ldots$, which, typically, become richer for larger indices. Given the data D_n, one wishes to select a good model from *one* of these classes. This is the problem of model selection.

Denote by $\widehat{g}_k$ a function in $\mathcal{C}_k$ having minimal empirical risk. One hopes to select a model class $\mathcal{C}_K$ such that the excess error $\mathbb{E}L(\widehat{g}_K) - L^*$ is close to

$$\min_k \mathbb{E}L(\widehat{g}_k) - L^* = \min_k \left[\left(\mathbb{E}L(\widehat{g}_k) - \inf_{g\in\mathcal{C}_k} L(g)\right) + \left(\inf_{g\in\mathcal{C}_k} L(g) - L^*\right)\right].$$

The idea of *structural risk minimization*, (also known as *complexity regularization*, is to add a complexity penalty to each of the $\widehat{L}_n(\widehat{g}_k)$'s to compensate for the overfitting effect. This penalty is usually closely related to a distribution-free upper bound for $\sup_{g\in\mathcal{C}_k} |\widehat{L}_n(g) - L(g)|$ so that the penalty eliminates the effect of overfitting.

The first general result shows that any approximate upper bound on error can be used to define a (possibly data-dependent) complexity penalty $C_n(k)$ and a model selection algorithm for which the excess error is close to

$$\min_k \left[\mathbb{E}C_n(k) + \left(\inf_{g\in\mathcal{C}_k} L(g) - L^*\right)\right].$$

Our goal is to select, among the classifiers $\widehat{g}_k$ one which has approximately minimal loss. The key assumption for our analysis is that the true loss of $\widehat{g}_k$ can be estimated for all k.

Assumption 1 *There are positive numbers c and m such that for each k an estimate $R_{n,k}$ on $L(\widehat{g}_k)$ is available which satisfies*

$$\mathbb{P}[L(\widehat{g}_k) > R_{n,k} + \varepsilon] \le c e^{-2m\varepsilon^2}$$

for all $\varepsilon > 0$.

Now define the complexity penalty by

$$C_n(k) = R_{n,k} - \widehat{L}_n(\widehat{g}_k) + \sqrt{\frac{\log k}{m}}.$$

The last term is required because of technical reasons that will become apparent shortly. It is typically small. The difference $R_{n,k} - \widehat{L}_n(\widehat{g}_k)$ is simply an estimate of the 'right' amount of penalization $L(\widehat{g}_k) - \widehat{L}_n(\widehat{g}_k)$. Finally, define the prediction rule:

$$g_n^* = \arg\min_{k=1,2,\ldots} \tilde{L}_n(\widehat{g}_k),$$

where

$$\tilde{L}_n(\widehat{g}_k) = \widehat{L}_n(\widehat{g}_k) + C_n(k) = R_{n,k} + \sqrt{\frac{\log k}{m}}.$$

The following theorem summarizes the main performance bound for g_n^*.

Theorem 1.20. *Assume that the error estimates $R_{n,k}$ satisfy Assumption 1 for some positive constants c and m. Then*

$$\mathbb{E}L(g_n^*) - L^* \leq \min_k \left[\mathbb{E}C_n(k) + \left(\inf_{g\in C_k} L(g) - L^*\right)\right] + \sqrt{\frac{\log(ce)}{2m}}.$$

Theorem 1.20 shows that the prediction rule minimizing the penalized empirical loss achieves an almost optimal trade-off between the approximation error and the expected complexity, provided that the estimate $R_{n,k}$ on which the complexity is based is an approximate upper bound on the loss. In particular, if we knew in advance which of the classes C_k contained the optimal prediction rule, we could use the error estimates $R_{n,k}$ to obtain an upper bound on $\mathbb{E}L(\widehat{g}_k) - L^*$, and this upper bound would not improve on the bound of Theorem 1.20 by more than $O\left(\sqrt{\log k/m}\right)$.

PROOF. For brevity, introduce the notation

$$L_k^* = \inf_{g\in C_k} L(g).$$

Then for any $\varepsilon > 0$,

$$\begin{aligned}
\mathbb{P}\left[L(g_n^*) - \tilde{L}_n(g_n^*) > \varepsilon\right] &\le \mathbb{P}\left[\sup_{j=1,2,\dots} \left(L(\widehat{g}_j) - \tilde{L}_n(\widehat{g}_j)\right) > \varepsilon\right] \\
&\le \sum_{j=1}^{\infty} \mathbb{P}\left[L(\widehat{g}_j) - \tilde{L}_n(\widehat{g}_j) > \varepsilon\right] \\
&\quad \text{(by the union bound)} \\
&= \sum_{j=1}^{\infty} \mathbb{P}\left[L(\widehat{g}_j) - R_{n,j} > \varepsilon + \sqrt{\frac{\log j}{m}}\right] \\
&\quad \text{(by definition)} \\
&\le \sum_{j=1}^{\infty} c e^{-2m\left(\varepsilon + \sqrt{\frac{\log j}{m}}\right)^2} \quad \text{(by Assumption 1)} \\
&\le \sum_{j=1}^{\infty} c e^{-2m\left(\varepsilon^2 + \frac{\log j}{m}\right)} \\
&< 2ce^{-2m\varepsilon^2} \quad (\text{since } \textstyle\sum_{j=1}^{\infty} j^{-2} < 2).
\end{aligned}$$

To prove the theorem, for each k, we decompose $L(g_n^*) - L_k^*$ as

$$L(g_n^*) - L_k^* = \left(L(g_n^*) - \inf_j \tilde{L}_n(\widehat{g}_j)\right) + \left(\inf_j \tilde{L}_n(\widehat{g}_j) - L_k^*\right).$$

The first term may be bounded, by standard integration of the tail inequality shown above, as $\mathbb{E}\left[L(g_n^*) - \inf_j \tilde{L}_n(\widehat{g}_j)\right] \le \sqrt{\log(ce)/(2m)}$. Choosing g_k^* such that $L(g_k^*) = L_k^*$, the second term may be bounded directly by

$$\begin{aligned}
\mathbb{E}\inf_j \tilde{L}_n(\widehat{g}_j) - L_k^* &\le \mathbb{E}\tilde{L}_n(\widehat{g}_k) - L_k^* \\
&= \mathbb{E}\widehat{L}_n(\widehat{g}_k) - L_k^* + \mathbb{E}C_n(k) \\
&\quad (\text{by the definition of } \tilde{L}_n(\widehat{g}_k)) \\
&\le \mathbb{E}\widehat{L}_n(g_k^*) - L(g_k^*) + \mathbb{E}C_n(k) \\
&\quad (\text{since } \widehat{g}_k \text{ minimizes the empirical loss on } \mathcal{C}_k) \\
&= \mathbb{E}C_n(k),
\end{aligned}$$

where the last step follows from the fact that $\mathbb{E}\widehat{L}_n(g_k^*) = L(g_k^*)$. Summing the obtained bounds for both terms yields that for each k,

$$\mathbb{E}L(g_n^*) \le \mathbb{E}C_n(k) + L_k^* + \sqrt{\log(ce)/(2m)},$$

which implies the second statement of the theorem. □

1.6.2 Selection based on a test sample

In our first application of Theorem 1.20, we assume that m independent sample pairs

$$(X_1', Y_1'), \ldots, (X_m', Y_m')$$

are available. This may always be achieved by simply removing m samples from the training data. Of course, this is not very attractive, but m may be small relative to n. In this case we can estimate $L(\widehat{g}_k)$ by the hold-out error estimate

$$R_{n,k} = \frac{1}{m} \sum_{i=1}^{m} \mathbb{I}_{\widehat{g}_k(X_i') \neq Y_i'}.$$

We apply Hoeffding's inequality to show that Assumption 1 is satisfied with $c = 1$, notice that $\mathbb{E}[R_{n,k}|D_n] = L(\widehat{g}_k)$, and apply Theorem 1.20 to give the following result.

COROLLARY 1.6. *Assume that the model selection algorithm is performed with the hold-out error estimate. Then*

$$\begin{aligned} & \mathbb{E}L(g_n^*) - L^* \\ \leq \; & \min_k \left[\mathbb{E}\left[L(\widehat{g}_k) - \widehat{L}_n(\widehat{g}_k) \right] + \left(\inf_{g \in C_k} L(g) - L^* \right) + \sqrt{\frac{\log k}{m}} \right] + \frac{1}{\sqrt{2m}}. \end{aligned}$$

In other words, the estimate achieves a nearly optimal balance between the approximation error, and the quantity

$$\mathbb{E}\left[L(\widehat{g}_k) - \widehat{L}_n(\widehat{g}_k) \right],$$

which may be regarded as the amount of overfitting.

1.6.3 Penalization by the VC dimension

In the remaining examples we consider error estimates $R_{n,k}$ which avoid splitting the data. First recall that by the Vapnik-Chervonenkis inequality, $2\sqrt{(V_{C_k} \log(n+1) + \log 2)/n}$ is an upper bound for the expected maximal deviation, within class C_k, between $L(g)$ and its empirical counterpart, $\widehat{L}_n(g)$. This suggests that penalizing the empirical error by this complexity term should compensate the overfitting within class C_k. Thus, we introduce the error estimate

$$R_{n,k} = \widehat{L}_n(\widehat{g}_k) + 2\sqrt{\frac{V_C \log(n+1) + \log 2}{n}}$$

Indeed, it is easy to show that this estimate satisfies Assumption 1. Indeed,

$$\begin{aligned}
&\mathbb{P}[L(\widehat{g}_k) > R_{n,k}+\varepsilon] \\
&= \mathbb{P}\left[L(\widehat{g}_k) - \widehat{L}_n(\widehat{g}_k) > 2\sqrt{\frac{V_C \log(n+1)+\log 2}{n}} + \varepsilon\right] \\
&\leq \mathbb{P}\left[\sup_{g\in C_k}\left|L(g) - \widehat{L}_n(g)\right| > 2\sqrt{\frac{V_C \log(n+1)+\log 2}{n}} + \varepsilon\right] \\
&\leq \mathbb{P}\left[\sup_{g\in C_k}\left|L(g) - \widehat{L}_n(g)\right| > \mathbb{E}\sup_{g\in C_k}\left|L(g) - \widehat{L}_n(g)\right| + \varepsilon\right] \\
&\quad \text{(by the Vapnik-Chervonenkis inequality)} \\
&\leq e^{-2n\varepsilon^2} \quad \text{(by the bounded difference inequality).}
\end{aligned}$$

Therefore, satisfies Assumption 1 with $m = n$. Substituting this into Theorem 1.20 gives

$$\begin{aligned}
&\mathbb{E}L(g_n^*) - L^* \\
&\leq \min_k \left[2\sqrt{\frac{V_{C_k}\log(n+1)+\log 2}{n}} + \left(\inf_{g\in C_k} L(g) - L^*\right) + \sqrt{\frac{\log k}{n}}\right] + \sqrt{\frac{1}{2n}}.
\end{aligned}$$

Thus, structural risk minimization finds the best trade-off between the approximation error and a distribution-free upper bound on the estimation error.

1.6.4 Penalization by maximum discrepancy

In this section we propose a data-dependent way of computing the penalties with improved performance guarantees. Assume, for simplicity, that n is even, divide the data into two equal halves, and define, for each predictor f, the empirical loss on the two parts by

$$\widehat{L}_n^{(1)}(g) = \frac{2}{n}\sum_{i=1}^{n/2} \mathbb{I}_{g(X_i)\neq Y_i}$$

and

$$\widehat{L}_n^{(2)}(g) = \frac{2}{n}\sum_{i=n/2+1}^{n} \mathbb{I}_{g(X_i)\neq Y_i}.$$

Define the error estimate $R_{n,k}$ by

$$R_{n,k} = \widehat{L}_n(\widehat{g}_k) + \max_{g\in C_k}\left(\widehat{L}_n^{(1)}(g) - \widehat{L}_n^{(2)}(g)\right).$$

Observe that the maximum discrepancy $\max_{g\in C_k}\left(\widehat{L}_n^{(1)}(g)-\widehat{L}_n^{(2)}(g)\right)$ may be computed using the following simple trick: first flip the labels of the first half of the data, thus obtaining the modified data set $D_n' = (X_1',Y_1'),\ldots,(X_n',Y_n')$ with $(X_i',Y_i') = (X_i, 1-Y_i)$ for $i \le n/2$ and $(X_i',Y_i') = (X_i,Y_i)$ for $i > n/2$. Next find $f_k^- \in C_k$ which minimizes the empirical loss based on D_n',

$$
\begin{aligned}
\frac{1}{n}\sum_{i=1}^{n}\mathbb{I}_{g(X_i')\neq Y_i'} &= \frac{1}{2}-\frac{1}{n}\sum_{i=1}^{n/2}\mathbb{I}_{g(X_i)\neq Y_i}+\frac{1}{n}\sum_{i=n/2+1}^{n}\mathbb{I}_{g(X_i)\neq Y_i} \\
&= \frac{1-\widehat{L}_n^{(1)}(g)+\widehat{L}_n^{(2)}(g)}{2}.
\end{aligned}
$$

Clearly, the function f_k^- maximizes the discrepancy. Therefore, the same algorithm that is used to compute the empirical loss minimizer $\widehat{g}_k$ may be used to find f_k^- and compute the penalty based on maximum discrepancy. This is appealing: although empirical loss minimization is often computationally difficult, the same approximate optimization algorithm can be used for both finding prediction rules and estimating appropriate penalties. In particular, if the algorithm only approximately minimizes empirical loss over the class C_k because it minimizes over some proper subset of C_k, the theorem is still applicable.

Theorem 1.21. *If the penalties are defined using the maximum-discrepancy error estimates, and $m = n/21$, then*

$$
\begin{aligned}
\mathbb{E}L(g_n^*) - L^* \le \min_k \Bigg[& \mathbb{E}\max_{g\in C_k}\left(\widehat{L}_n^{(1)}(g)-\widehat{L}_n^{(2)}(g)\right) \\
& + \left(\inf_{g\in C_k} L(g) - L^*\right) + 4.59\sqrt{\frac{\log k}{n}}\Bigg] + \frac{4.70}{\sqrt{n}}.
\end{aligned}
$$

PROOF. Once again, we check Assumption 1 and apply Theorem 1.20. Introduce the ghost sample $(X_1',Y_1'),\ldots,(X_n',Y_n')$, which is independent of the data and has the same distribution. Denote the empirical loss based on this sample by $L_n'(g) = \frac{1}{n}\sum_{i=1}^{n}\mathbb{I}_{g(X_i')\neq Y_i'}$. The proof is based on the simple observation that for each k,

$$\begin{aligned}
\mathbb{E}\max_{g\in\mathcal{F}_k}\left(L'_n(g)-\hat{L}_n(g)\right) &= \frac{1}{n}\mathbb{E}\max_{g\in\mathcal{F}_k}\sum_{i=1}^{n}\left(\mathbb{I}_{g(X'_i)\neq Y'_i}-\mathbb{I}_{g(X_i)\neq Y_i}\right)\\
&\le \frac{1}{n}\mathbb{E}\left(\max_{g\in\mathcal{F}_k}\sum_{i=1}^{n/2}\left(\mathbb{I}_{g(X'_i)\neq Y'_i}-\mathbb{I}_{g(X_i)\neq Y_i}\right)\right.\\
&\qquad\left.+\max_{g\in\mathcal{F}_k}\sum_{i=n/2+1}^{n}\left(\mathbb{I}_{g(X'_i)\neq Y'_i}-\mathbb{I}_{g(X_i)\neq Y_i}\right)\right)\\
&= \frac{2}{n}\mathbb{E}\max_{g\in\mathcal{F}_k}\sum_{i=1}^{n/2}\left(\mathbb{I}_{g(X'_i)\neq Y'_i}-\mathbb{I}_{g(X_i)\neq Y_i}\right)\\
&= \mathbb{E}\max_{g\in\mathcal{F}_k}\left(\hat{L}_n^{(1)}(g)-\hat{L}_n^{(2)}(g)\right). \qquad (1.1)
\end{aligned}$$

The bounded difference inequality inequality (Theorem 1.8) implies

$$\mathbb{P}\left[\max_{g\in C_k}\left(L'_n(g)-\widehat{L}_n(g)\right)>\mathbb{E}\max_{g\in C_k}\left(L'_n(g)-\widehat{L}_n(g)\right)+\varepsilon\right]\le e^{-n\varepsilon^2}, \qquad (1.2)$$

$$\mathbb{P}\left[\max_{g\in C_k}\left(\widehat{L}_n^{(1)}(g)-\widehat{L}_n^{(2)}(g)\right)<\mathbb{E}\max_{g\in C_k}\left(\widehat{L}_n^{(1)}(g)-\widehat{L}_n^{(2)}(g)\right)-\varepsilon\right]\le e^{-n\varepsilon^2/2} \qquad (1.3)$$

and so for each k,

$$\begin{aligned}
&\mathbb{P}[L(\widehat{g}_k) > R_{n,k} + \varepsilon] \\
&= \mathbb{P}\left[L(\widehat{g}_k) - \widehat{L}_n(\widehat{g}_k) > \max_{g \in C_k}\left(\widehat{L}_n^{(1)}(g) - \widehat{L}_n^{(2)}(g)\right) + \varepsilon\right] \\
&\leq \mathbb{P}\left[L'_n(\widehat{g}_k) - \widehat{L}_n(\widehat{g}_k) > \max_{g \in C_k}\left(\widehat{L}_n^{(1)}(g) - \widehat{L}_n^{(2)}(g)\right) + \frac{7\varepsilon}{9}\right] \\
&\quad + \mathbb{P}\left[L(\widehat{g}_k) - L'_n(\widehat{g}_k) > \frac{2\varepsilon}{9}\right] \\
&\leq \mathbb{P}\left[L'_n(\widehat{g}_k) - \widehat{L}_n(\widehat{g}_k) > \max_{g \in C_k}\left(\widehat{L}_n^{(1)}(g) - \widehat{L}_n^{(2)}(g)\right) + \frac{7\varepsilon}{9}\right] \\
&\quad + e^{-8n\varepsilon^2/81} \quad \text{(by Hoeffding)} \\
&\leq \mathbb{P}\left[\max_{g \in C_k}\left(L'_n(g) - \widehat{L}_n(g)\right) > \max_{g \in C_k}\left(\widehat{L}_n^{(1)}(g) - \widehat{L}_n^{(2)}(g)\right) + \frac{7\varepsilon}{9}\right] \\
&\quad + e^{-8n\varepsilon^2/81} \\
&\leq \mathbb{P}\left[\max_{g \in C_k}\left(L'_n(g) - \widehat{L}_n(g)\right) > \mathbb{E}\max_{g \in C_k}\left(L'_n(g) - \widehat{L}_n(g)\right) + \frac{\varepsilon}{3}\right] \\
&\quad + \mathbb{P}\left[\max_{g \in C_k}\left(\widehat{L}_n^{(1)}(g) - \widehat{L}_n^{(2)}(g)\right) < \mathbb{E}\max_{g \in C_k}\left(\widehat{L}_n^{(1)}(g) - \widehat{L}_n^{(2)}(g)\right) - \frac{4\varepsilon}{9}\right] \\
&\quad + e^{-8n\varepsilon^2/81} \quad \text{(where we used (1.1))} \\
&\leq e^{-n\varepsilon^2/9} + e^{-8n\varepsilon^2/81} + e^{-8n\varepsilon^2/81} \quad \text{(by (1.2) and (1.3))} \\
&< 3e^{-8n\varepsilon^2/81}
\end{aligned}$$

Thus, Assumption 1 is satisfied with $m = n/21$ and $c = 3$ and the proof is finished. □

1.7 References

General [1]

M. Anthony and P. L. Bartlett, *Neural Network Learning: Theoretical Foundations*, Cambridge University Press, Cambridge, 1999.

N. Cristianini and J. Shawe-Taylor. *An Introduction to Support Vector Machines*. Cambridge University Press, Cambridge, UK, 2000.

L. Devroye, L. Györfi, and G. Lugosi. *A Probabilistic Theory of Pattern Recognition*. Springer-Verlag, New York, 1996.

V.N. Vapnik. *Estimation of Dependencies Based on Empirical Data*. Springer-Verlag, New York, 1982.

V.N. Vapnik. *The Nature of Statistical Learning Theory*. Springer-Verlag, New York, 1995.

V.N. Vapnik. *Statistical Learning Theory*. Wiley, New York, 1998.

V.N. Vapnik and A.Ya. Chervonenkis. *Theory of Pattern Recognition*. Nauka, Moscow, 1974. (in Russian); German translation: *Theorie der Zeichenerkennung*, Akademie Verlag, Berlin, 1979.

Concentration for sums of independent random variables

G. Bennett. Probability inequalities for the sum of independent random variables. *Journal of the American Statistical Association*, 57:33–45, 1962.

S.N. Bernstein. *The Theory of Probabilities*. Gastehizdat Publishing House, Moscow, 1946.

H. Chernoff. A measure of asymptotic efficiency of tests of a hypothesis based on the sum of observations. *Annals of Mathematical Statistics*, 23:493–507, 1952.

T. Hagerup and C. Rüb. A guided tour of Chernoff bounds. *Information Processing Letters*, 33:305–308, 1990.

[1] The list of references given below contains, apart from the literature cited in the text, some of the key references in each covered topics. The list is far from being complete. Its purpose it to suggest some starting points for further reading.

W. Hoeffding. Probability inequalities for sums of bounded random variables. *Journal of the American Statistical Association*, 58:13–30, 1963.

R.M. Karp. *Probabilistic Analysis of Algorithms*. Class Notes, University of California, Berkeley, 1988.

M. Okamoto. Some inequalities relating to the partial sum of binomial probabilities. *Annals of the Institute of Statistical Mathematics*, 10:29–35, 1958.

Concentration

K. Azuma. Weighted sums of certain dependent random variables. *Tohoku Mathematical Journal*, 68:357–367, 1967.

S. Boucheron, G. Lugosi, and P. Massart. A sharp concentration inequality with applications in random combinatorics and learning. *Random Structures and Algorithms*, 16:277-292, 2000.

L. Devroye. Exponential inequalities in nonparametric estimation. In G. Roussas, editor, *Nonparametric Functional Estimation and Related Topics*, pages 31–44. NATO ASI Series, Kluwer Academic Publishers, Dordrecht, 1991.

J. H. Kim. The Ramsey number $R(3,t)$ has order of magnitude $t^2/\log t$. *Random Structures and Algorithms*, 7:173–207, 1995.

M. Ledoux. On Talagrand's deviation inequalities for product measures. *ESAIM: Probability and Statistics*, **1**, 63–87, `http://www.emath.fr/ps/`, (1996).

K. Marton. A simple proof of the blowing-up lemma. *IEEE Transactions on Information Theory*, 32:445–446, 1986.

K. Marton. Bounding d-distance by informational divergence: a way to prove measure concentration. *Annals of Probability*, to appear:0–0, 1996.

K. Marton. A measure concentration inequality for contracting Markov chains. *Geometric and Functional Analysis*, 6:556–571, 1996. Erratum: 7:609–613, 1997.

P. Massart. About the constant in Talagrand's concentration inequalities from empirical processes. *Annals of Probability*, 28:863–884, 2000.

C. McDiarmid. On the method of bounded differences. In *Surveys in Combinatorics 1989*, pages 148–188. Cambridge University Press, Cambridge, 1989.

W. Rhee and M. Talagrand. Martingales, inequalities, and NP-complete problems. *Mathematics of Operations Research*, 12:177–181, 1987.

J.M. Steele. An Efron-Stein inequality for nonsymmetric statistics. *Annals of Statistics*, 14:753–758, 1986.

M. Talagrand. Concentration of measure and isoperimetric inequalities in product spaces. I.H.E.S. Publications Mathématiques, 81:73–205, 1996.

M. Talagrand. New concentration inequalities in product spaces. *Invent. Math.* 126:505-563, 1996.

M. Talagrand. A new look at independence. *Annals of Probability*, 24:0–0, 1996. special invited paper.

VC theory

K. Alexander. Probability inequalities for empirical processes and a law of the iterated logarithm. *Annals of Probability*, 4:1041–1067, 1984.

M. Anthony and J. Shawe-Taylor. A result of Vapnik with applications. *Discrete Applied Mathematics*, 47:207–217, 1993.

P. Bartlett and G. Lugosi. An inequality for uniform deviations of sample averages from their means. *Statistics and Probability Letters*, 44:55–62, 1999.

L. Breiman. Bagging predictors. *Machine Learning*, 24:123–140, 1996.

Devroye, L. Bounds for the uniform deviation of empirical measures. *Journal of Multivariate Analysis*, 12:72–79, 1982.

A. Ehrenfeucht, D. Haussler, M. Kearns, and L. Valiant. A general lower bound on the number of examples needed for learning. *Information and Computation*, 82:247–261, 1989.

Y. Freund. Boosting a weak learning algorithm by majority. *Information and Computation*, 121:256–285, 1995.

E. Giné and J. Zinn. Some limit theorems for empirical processes. *Annals of Probability*, 12:929–989, 1984.

D. Haussler. Decision theoretic generalizations of the PAC model for neural net and other learning applications. *Information and Computation*, 100:78–150, 1992.

V. Koltchinskii and D. Panchenko. Empirical margin distributions and bounding the generalization error of combined classifiers, *Annals of Statistics*, 30, 2002.

M. Ledoux and M. Talagrand. *Probability in Banach Space*, Springer-Verlag, New York, 1991.

G. Lugosi. Improved upper bounds for probabilities of uniform deviations. *Statistics and Probability Letters*, 25:71–77, 1995.

D. Pollard. *Convergence of Stochastic Processes*. Springer-Verlag, New York, 1984.

R.E. Schapire, Y. Freund, P. Bartlett, and W.S. Lee. Boosting the margin: a new explanation for the effectiveness of voting methods, *Annals of Statistics*, 26:1651–1686, 1998.

R.E. Schapire. The strength of weak learnability. *Machine Learning*, 5:197–227, 1990.

M. Talagrand. Sharper bounds for Gaussian and empirical processes. *Annals of Probability*, 22:28–76, 1994.

S. Van de Geer. Estimating a regression function. *Annals of Statistics*, 18:907–924, 1990.

V.N. Vapnik. *Estimation of Dependencies Based on Empirical Data*. Springer-Verlag, New York, 1982.

V.N. Vapnik. *The Nature of Statistical Learning Theory*. Springer-Verlag, New York, 1995.

V.N. Vapnik. *Statistical Learning Theory*. Wiley, New York, 1998.

V.N. Vapnik and A.Ya. Chervonenkis. On the uniform convergence of relative frequencies of events to their probabilities. *Theory of Probability and its Applications*, 16:264–280, 1971.

V.N. Vapnik and A.Ya. Chervonenkis. *Theory of Pattern Recognition*. Nauka, Moscow, 1974. (in Russian); German translation: *Theorie der Zeichenerkennung*, Akademie Verlag, Berlin, 1979.

A. W. van der Vaart and J. A. Wellner. *Weak convergence and empirical processes*, Springer-Verlag, New York, 1996.

Shatter coefficients, VC dimension

P. Assouad, Sur les classes de Vapnik-Chervonenkis, *C.R. Acad. Sci. Paris*, vol. 292, Sér.I, pp. 921–924, 1981.

T. M. Cover, Geometrical and statistical properties of systems of linear inequalities with applications in pattern recognition, *IEEE Transactions on Electronic Computers*, vol. 14, pp. 326–334, 1965.

R. M. Dudley, Central limit theorems for empirical measures, *Annals of Probability*, vol. 6, pp. 899–929, 1978.

R. M. Dudley, Balls in R^k do not cut all subsets of $k+2$ points, *Advances in Mathematics*, vol. 31 (3), pp. 306–308, 1979.

P. Frankl, On the trace of finite sets, *Journal of Combinatorial Theory, Series A*, vol. 34, pp. 41–45, 1983.

D. Haussler, Sphere packing numbers for subsets of the boolean n-cube with bounded Vapnik-Chervonenkis dimension, *Journal of Combinatorial Theory, Series A*, vol. 69, pp. 217–232, 1995.

N. Sauer, On the density of families of sets, *Journal of Combinatorial Theory Series A*, vol. 13, pp. 145–147, 1972.

L. Schläffli, *Gesammelte Mathematische Abhandlungen*, Birkhäuser-Verlag, Basel, 1950.

S. Shelah, A combinatorial problem: stability and order for models and theories in infinity languages, *Pacific Journal of Mathematics*, vol. 41, pp. 247–261, 1972.

J. M. Steele, Combinatorial entropy and uniform limit laws, Ph.D. dissertation, Stanford University, Stanford, CA, 1975.

J. M. Steele, Existence of submatrices with all possible columns, *Journal of Combinatorial Theory, Series A*, vol. 28, pp. 84–88, 1978.

R. S. Wenocur and R. M. Dudley, Some special Vapnik-Chervonenkis classes, *Discrete Mathematics*, vol. 33, pp. 313–318, 1981.

Lower bounds.

A. Antos and G. Lugosi. Strong minimax lower bounds for learning. *Machine Learning*, vol.30, 31–56, 1998.

P. Assouad. Deux remarques sur l'estimation. *Comptes Rendus de l'Académie des Sciences de Paris*, 296:1021–1024,1983.

L. Birgé. Approximation dans les espaces métriques et théorie de l'estimation. *Zeitschrift für Wahrscheinlichkeitstheorie und verwandte Gebiete*, 65:181–237, 1983.

L. Birgé. On estimating a density using Hellinger distance and some other strange facts. *Probability Theory and Related Fields*, 71:271–291,1986.

A. Blumer, A. Ehrenfeucht, D. Haussler, and M.K. Warmuth. Learnability and the Vapnik-Chervonenkis dimension. *Journal of the ACM*, 36:929–965, 1989.

L. Devroye and G. Lugosi. Lower bounds in pattern recognition and learning. *Pattern Recognition*, 28:1011–1018, 1995.

A. Ehrenfeucht, D. Haussler, M. Kearns, and L. Valiant. A general lower bound on the number of examples needed for learning. *Information and Computation*, 82:247–261, 1989.

D. Haussler, N. Littlestone, and M. Warmuth. Predicting $\{0,1\}$-functions on randomly drawn points. *Information and Computation*, 115:248–292, 1994.

E. Mammen, A. B. Tsybakov. Smooth discrimination analysis. *The Annals of Statistics*, 27:1808–1829, 1999.

D. Schuurmans. Characterizing rational versus exponential learning curves. In *Computational Learning Theory: Second European Conference. EuroCOLT'95*, pages 272–286. Springer Verlag, 1995.

V.N. Vapnik and A.Ya. Chervonenkis. *Theory of Pattern Recognition*. Nauka, Moscow, 1974. (in Russian); German translation: *Theorie der Zeichenerkennung*, Akademie Verlag, Berlin, 1979.

Complexity regularization

H. Akaike. A new look at the statistical model identification. *IEEE Transactions on Automatic Control*, 19:716–723, 1974.

A.R. Barron. Logically smooth density estimation. Technical Report TR 56, Department of Statistics, Stanford University, 1985.

A.R. Barron. Complexity regularization with application to artificial neural networks. In G. Roussas, editor, *Nonparametric Functional Estimation and Related Topics*, pages 561–576. NATO ASI Series, Kluwer Academic Publishers, Dordrecht, 1991.

A.R. Barron, L. Birgé, and P. Massart. Risk bounds for model selection via penalization. *Probability Theory and Related fields*, 113:301–413, 1999.

A.R. Barron and T.M. Cover. Minimum complexity density estimation. *IEEE Transactions on Information Theory*, 37:1034–1054, 1991.

P. L. Bartlett. The sample complexity of pattern classification with neural networks: the size of the weights is more important than the size of the network. *IEEE Transactions on Information Theory*, 44(2):525–536, March 1998.

P. Bartlett, S. Boucheron, and G. Lugosi, Model selection and error estimation. *Proceedings of the 13th Annual Conference on Computational Learning Theory*, ACM Press, pp.286–297, 2000.

L. Birgé and P. Massart. From model selection to adaptive estimation. In E. Torgersen D. Pollard and G. Yang, editors, *Festschrift for Lucien Le Cam: Research papers in Probability and Statistics*, pages 55–87. Springer, New York, 1997.

L. Birgé and P. Massart. Minimum contrast estimators on sieves: exponential bounds and rates of convergence. *Bernoulli*, 4:329–375, 1998.

Y. Freund. Self bounding learning algorithms. *Proceedings of the Eleventh Annual Conference on Computational Learning Theory*, pages 247–258, 1998.

A.R. Gallant. *Nonlinear Statistical Models*. John Wiley, New York, 1987.

S. Geman and C.R. Hwang. Nonparametric maximum likelihood estimation by the method of sieves. *Annals of Statistics*, 10:401–414, 1982.

M. Kearns, Y. Mansour, A.Y. Ng, and D. Ron. An experimental and theoretical comparison of model selection methods. In *Proceedings of the Eighth Annual ACM Workshop on Computational Learning Theory*, pages 21–30. Association for Computing Machinery, New York, 1995.

A. Krzyzak and T. Linder. Radial basis function networks and complexity regularization in function learning. *IEEE Transactions on Neural Networks*, 9:247–256, 1998.

G. Lugosi and A. Nobel. Adaptive model selection using empirical complexities. *Annals of Statistics*, vol. 27, no.6, 1999.

G. Lugosi and K. Zeger. Nonparametric estimation via empirical risk minimization. *IEEE Transactions on Information Theory*, 41:677–678, 1995.

G. Lugosi and K. Zeger. Concept learning using complexity regularization. *IEEE Transactions on Information Theory*, 42:48–54, 1996.

C.L. Mallows. Some comments on c_p. *IEEE Technometrics*, 15:661–675, 1997.

P. Massart. Some applications of concentration inequalities to statistics. *Annales de la faculté des sciences de l'université de Toulouse, Mathématiques*, série 6, **IX**:245–303, 2000.

R. Meir. Performance bounds for nonlinear time series prediction. In *Proceedings of the Tenth Annual ACM Workshop on Computational Learning Theory*, page 122–129. Association for Computing Machinery, New York, 1997.

D.S. Modha and E. Masry. Minimum complexity regression estimation with weakly dependent observations. *IEEE Transactions on Information Theory*, 42:2133–2145, 1996.

J. Rissanen. A universal prior for integers and estimation by minimum description length. *Annals of Statistics*, 11:416–431, 1983.

G. Schwarz. Estimating the dimension of a model. *Annals of Statistics*, 6:461–464, 1978.

J. Shawe-Taylor, P. L. Bartlett, R. C. Williamson, and M. Anthony. Structural risk minimization over data-dependent hierarchies. *IEEE Transactions on Information Theory*, 44(5):1926–1940, 1998.

X. Shen and W.H. Wong. Convergence rate of sieve estimates. *Annals of Statistics*, 22:580–615, 1994.

Y. Yang and A.R. Barron. Information-theoretic determination of minimax rates of convergence. *Annals of Statistics*, to appear, 1997.

Y. Yang and A.R. Barron. An asymptotic property of model selection criteria. *IEEE Transactions on Information Theory*, 44:to appear, 1998.

2

NONPARAMETRIC REGRESSION ESTIMATION

L. Györfi
Budapest University of Technology and Economics, Budapest, Hungary

M. Kohler
Universität Stuttgart, Stuttgart, Germany

The basic aim of mathematical statistics is to learn a probability law or its characteristics from data.

The classical mathematical statistics consists mainly of parametric rules, when the statistician has some a priori information about the actual problem in order to assume that the unknown probability law belongs to a parametric class of distributions.

In case of nonparametric statistics the statistician has no such knowledge, therefore
(i) the construction of the rules applied cannot use any specific feature of the law, and
(ii) the basic properties of the rule should be distribution-free.

2.1 Regression problem

Let Y be a real valued random variable and let X be a d-dimensional random vector (observation). The coordinates of X may have various types of distributions, some of them may be discrete (for example binary), others may be absolutely continuous. Hence, we do not assume anything about the distribution of X. In **regression analysis** one wishes to estimate Y given X, i.e., one wants to find a function f defined on the range of X so that $f(X)$ is "close" to Y. Assume that the main aim of the analysis is to minimize the mean squared error :

$$\min_f \mathbb{E}\{(f(X)-Y)^2\}.$$

The **regression function** $m(x)$ is defined by

$$m(x) = \mathbb{E}\{Y \mid X = x\} \tag{2.1}$$

for $\mathbb{E}|Y| < \infty$. For each measurable function f one has

$$\begin{aligned}
&\mathbb{E}\{(f(X)-Y)^2\}\\
&= \mathbb{E}\{(m(X)-Y)^2\}\\
&\quad +2\mathbb{E}\{(m(X)-Y)(f(X)-m(X))\}+\mathbb{E}\{(m(X)-f(X))^2\}\\
&= \mathbb{E}\{(m(X)-Y)^2\}\\
&\quad +2\mathbb{E}\{\mathbb{E}\{(m(X)-Y)(f(X)-m(X))|X\}\}+\mathbb{E}\{(m(X)-f(X))^2\}\\
&= \mathbb{E}\{(m(X)-Y)^2\}+\mathbb{E}\{(m(X)-f(X))^2\}\\
&= \mathbb{E}\{(m(X)-Y)^2\}+\int_{\mathcal{R}^d} | m(x)-f(x) |^2 \mu(dx),
\end{aligned}$$

where μ denotes the distribution of X. The second term on the right hand side is called L_2 error, excess error or integrated squared error for the function f and is denoted by

$$J(f) = \int_{\mathcal{R}^d} | m(x)-f(x) |^2 \mu(dx), \tag{2.2}$$

i.e., the mean squared error for f is close to its minimum if and only if the excess error $J(f)$ is close to 0. Thus

$$\mathbb{E}\{(m(X)-Y)^2\} = \min_f \mathbb{E}\{(f(X)-Y)^2\}.$$

In the regression estimation problem let $(X_1,Y_1),\dots,(X_n,Y_n)$ be independent and identically distributed copies of (X,Y). The regression function estimate is denoted by

$$m_n(x) = m_n(x,(X_1,Y_1),\dots,(X_n,Y_n)).$$

We are interested in the $L_2(\mu)$ convergence of the regression estimate m_n to m.

DEFINITION 2.1. *The estimator m_n is called* **weakly universally consistent** *if*

$$\mathbb{E}\{J(m_n)\} \to 0$$

for all distributions of (X,Y) with $\mathbb{E}|Y|^2 < \infty$.

DEFINITION 2.2. *The estimator m_n is called* **strongly universally consistent** *if*

$$J(m_n) \to 0 \text{ a.s.}$$

for all distributions of (X,Y) with $\mathbb{E}|Y|^2 < \infty$.

Stone (1977) first pointed out that there exist weakly universally consistent estimators, for example the nearest neighbor estimates. He considered estimates of form

$$m_n(x) = \sum_{i=1}^{n} W_{ni}(x;X_1,\dots,X_n)Y_i$$

where the weights $W_{ni}(x) = W_{ni}(x, X_1, \ldots, X_n)$ satisfy some conditions. These estimators are the **local averaging** regression function estimates. A powerful tool for proving weak consistency for local averaging regression function estimates is the Stone's Theorem:

Theorem 2.1. (STONE (1977)). *Assume that the following conditions are satisfied for any distribution of X:*

(i) There is a constant c such that for every nonnegative measurable function f satisfying $\mathbb{E}f(X) < \infty$,

$$\mathbb{E}\left\{\sum_{i=1}^{n} |W_{ni}(X)| f(X_i)\right\} \leq c\mathbb{E}f(X).$$

(ii) There is $D \geq 1$ such that

$$\mathbb{P}\left\{\sum_{i=1}^{n} |W_{n,i}(X)| \leq D\right\} = 1,$$

for all n.

(iii) For all $a > 0$,

$$\lim_{n\to\infty} \mathbb{E}\left\{\sum_{i=1}^{n} |W_{ni}(X)| I_{\{\|X_i - X\| > a\}}\right\} = 0.$$

(iv)

$$\sum_{i=1}^{n} W_{n,i}(X) \to 1$$

in probability,

(v)

$$\lim_{n\to\infty} \mathbb{E}\left\{\sum_{i=1}^{n} W_{ni}(X)^2\right\} = 0.$$

Then the corresponding regression function estimate m_n is weakly universally consistent.

2.2 Local averaging estimates

Let $\mathcal{P}_n = \{A_{n,1}, A_{n,2}, \ldots\}$ be a partition of $\mathcal{R}^d$ and for each $x \in \mathcal{R}^d$ let $A_n(x)$ denote the cell of $\mathcal{P}_n$ containing x. The **partitioning estimate** of the regression function is defined as

$$m_n(x) = \begin{cases} \dfrac{\sum_{i=1}^{n} Y_i I_{\{X_i \in A_n(x)\}}}{\sum_{i=1}^{n} I_{\{X_i \in A_n(x)\}}} & \text{if } \sum_{i=1}^{n} I_{\{X_i \in A_n(x)\}} > 0 \\ 0 & \text{otherwise,} \end{cases}$$

where I denotes the indicator. For cubic partition $A_{n,j}$ are cubes of volume h_n^d. The partitioning estimate was introduced by Tukey (1947, 1961) (called regressogram). For these weights $\sum_{i=1}^{n} W_{ni}(x) \leq 1$.

Theorem 2.2. *If for each sphere S centered at the origin*

$$\lim_{n\to\infty} \max_{j:A_{n,j}\cap S\neq\emptyset} diam(A_{n,j}) = 0 \tag{2.3}$$

and

$$\lim_{n\to\infty} \frac{|\{j: A_{n,j}\cap S\neq\emptyset\}|}{n} = 0 \tag{2.4}$$

then the partitioning regression function estimate is weakly universally consistent.

Results on weak and strong universal consistencies are in Devroye, Györfi (1983) and Györfi (1991).

The universal consistency of some nonparametric regression estimates is usually shown such that first the consistency is proved for continuous regression function, and then it is extended to arbitrary regression function. This extension is possible because of the following lemma:

Lemma 2.1. *For any probability measure μ, the set of continuous functions of bounded support is dense in $L_2(\mu)$.*

For the proof see, e.g., Györfi et al. (2002), it is similar to the proof of Lemma 5.1 below.
PROOF OF THEOREM 2.2. We implicitely prove the Stone Theorem with constants $c = D = 1$. Put

$$W_{n,i}(x) = \frac{I_{\{X_i\in A_n(x)\}}}{\sum_{j=1}^n I_{\{X_j\in A_n(x)\}}}$$

Then

$$\begin{aligned}
\mathbb{E}\{(m_n(X)-m(X))^2\} &\leq 3\mathbb{E}\left\{\left(\sum_{i=1}^n W_{n,i}(X)(Y_i - m(X_i))\right)^2\right\}\\
&+ 3\mathbb{E}\left\{\left(\sum_{i=1}^n W_{n,i}(X)(m(X_i) - m(X))\right)^2\right\}\\
&+ 3\mathbb{E}\left\{\left(\left(\sum_{i=1}^n W_{n,i}(X) - 1\right) m(X)\right)^2\right\}\\
&= 3I_n + 3J_n + 3L_n.
\end{aligned}$$

If μ_n denotes the empirical distribution for sample size n then

$$\begin{aligned}
\mathbb{P}\left\{\sum_{i=1}^{n} W_{ni}(X) \neq 1\right\} &= \mathbb{P}\{\mu_n(A_n(X)) = 0\} \\
&= \sum_j \mathbb{P}\{X \in A_{nj}, \mu_n(A_{nj}) = 0\} \\
&= \sum_j \mu(A_{nj})(1-\mu(A_{nj}))^n \\
&\leq \sum_{j:A_{nj}\cap S=\emptyset} \mu(A_{nj}) + \sum_{j:A_{nj}\cap S\neq\emptyset} \mu(A_{nj})(1-\mu(A_{nj}))^n.
\end{aligned}$$

By the elementary inequalities

$$x(1-x)^n \leq xe^{-nx} \leq \frac{1}{en}, \qquad (0 \leq x \leq 1)$$

and by (2.4) we have that

$$\mathbb{P}\left\{\sum_{i=1}^{n} W_{ni}(X) \neq 1\right\} \leq \mu(S^c) + \frac{1}{en}\left|\{j : A_{nj}\cap S \neq \emptyset\}\right| \to \mu(S^c).$$

therefore $\sum_{i=1}^{n} W_{ni}(X) \to 1$ in probability, and because of $0 \leq \sum_{i=1}^{n} W_{ni}(X) \leq 1$ concerning the third term

$$L_n = \mathbb{E}\left\{\left(\left(\sum_{i=1}^{n} W_{n,i}(X) - 1\right) m(X)\right)^2\right\} \to 0$$

by the dominated convergence theorem. We show that for any nonnegative function f with $\mathbb{E}f(X) < \infty$,

$$\mathbb{E}\left\{\sum_{i=1}^{n} f(X_i)\frac{I_{\{X_i\in A_n(X)\}}}{\sum_{l=1}^{n} I_{\{X_l\in A_n(X)\}}}\right\} \leq \mathbb{E}f(X). \tag{2.5}$$

Let the random variable $B(n,p)$ be binomially distributed with parameters n and p. Then

$$\begin{aligned}
\mathbb{E}\left\{\frac{1}{1+B(n,p)}\right\} &= \sum_{k=0}^{n} \frac{1}{k+1}\binom{n}{k} p^k (1-p)^{n-k} \\
&= \frac{1}{(n+1)p}\sum_{k=0}^{n}\binom{n+1}{k+1} p^{k+1}(1-p)^{n-k} \\
&\leq \frac{1}{(n+1)p}\sum_{k=0}^{n+1}\binom{n+1}{k} p^{k}(1-p)^{n-k+1} \\
&= \frac{1}{(n+1)p}
\end{aligned} \tag{2.6}$$

and

$$\mathbb{E}\left\{\frac{1}{B(n,p)}I_{\{B(n,p)>0\}}\right\} \leq \mathbb{E}\left\{\frac{2}{1+B(n,p)}\right\} \leq \frac{2}{(n+1)p}. \tag{2.7}$$

Observe that

$$\begin{aligned}
&\mathbb{E}\left\{\sum_{i=1}^{n} f(X_i)\frac{I_{\{X_i\in A_n(X)\}}}{\sum_{l=1}^{n} I_{\{X_l\in A_n(X)\}}}\right\}\\
&= \sum_{i=1}^{n}\mathbb{E}\left\{f(X_i)\frac{I_{\{X_i\in A_n(X)\}}}{1+\sum_{l\neq i} I_{\{X_l\in A_n(X)\}}}\right\}\\
&= n\mathbb{E}\left\{f(X_1)I_{\{X_1\in A_n(X)\}}\frac{1}{1+\sum_{l\neq 1} I_{\{X_l\in A_n(X)\}}}\right\}\\
&= n\mathbb{E}\left\{\mathbb{E}\left\{f(X_1)I_{\{X_1\in A_n(X)\}}\frac{1}{1+\sum_{l=2}^{n} I_{\{X_l\in A_n(X)\}}}\middle| X, X_1\right\}\right\}\\
&= n\mathbb{E}\left\{\mathbb{E}\{f(X_1)I_{\{X_1\in A_n(X)\}}|X\}\,\mathbb{E}\left\{\frac{1}{1+\sum_{l=2}^{n} I_{\{X_l\in A_n(X)\}}}\middle| X\right\}\right\}
\end{aligned}$$

by the independence of the random variables $X, X_1, \ldots, X_n$, therefore (2.6) implies that the expected value above can be bounded from above by

$$\begin{aligned}
&n\mathbb{E}\left\{\mathbb{E}\{f(X_1)I_{\{X_1\in A_n(X)\}}|X\}\frac{1}{n\mu(A_n(X))}\right\}\\
&= \sum_j \mathbb{P}\{X\in A_{nj}\}\int_{A_{nj}} f(u)\mu(du)\frac{1}{\mu(A_{nj})}\\
&= \int_{\mathcal{R}^d} f(u)\mu(du) = \mathbb{E}f(X).
\end{aligned}$$

By Jensen's inequality

$$\begin{aligned}
J_n &\leq \mathbb{E}\left\{\left(\sum_{i=1}^{n} W_{n,i}(X)|m(X_i)-m(X)|\right)^2\right\}\\
&\leq \mathbb{E}\left\{\sum_{i=1}^{n} W_{n,i}(X)(m(X_i)-m(X))^2\right\}\\
&= J_n'.
\end{aligned}$$

If m is uniformly continuous with bounded support then by (2.3) $J_n'\to 0$. In order to see this, for $\varepsilon>0$ let $\delta>0$ be such that for $\|x-z\|<\delta$ we have $|m(x)-m(z)|<\varepsilon$. Choose a ball S centered at the origin such that it contains the support of m and by condition (2.3) for large n and for

$A_{n,j} \cap S \neq 0$ we have $diam(A_{n,j}) < \delta$. Then for such large n

$$\begin{aligned} J'_n &= \mathbb{E}\left\{\sum_{i=1}^{n} W_{n,i}(X)(m(X_i)-m(X))^2\right\} \\ &= \mathbb{E}\left\{\sum_{i=1}^{n} W_{n,i}(X)(m(X_i)-m(X))^2 I_{\{\|X_i-X\|\leq\delta\}}\right\} \\ &= \mathbb{E}\left\{\sum_{i=1}^{n} W_{n,i}(X)\varepsilon^2 I_{\{\|X_i-X\|\leq\delta\}}\right\} \\ &\leq \varepsilon^2. \end{aligned}$$

Because of Lemma 2.1 for general m and $\varepsilon > 0$ we can choose uniformly continuous $\tilde{m}$ with bounded support such that

$$\mathbb{E}\{(m(X)-\tilde{m}(X))^2\} < \varepsilon.$$

Then

$$\begin{aligned} J'_n &\leq 3\mathbb{E}\left\{\sum_{i=1}^{n} W_{n,i}(X)(m(X_i)-\tilde{m}(X_i))^2\right\} \\ &\quad +3\mathbb{E}\left\{\sum_{i=1}^{n} W_{n,i}(X)(\tilde{m}(X_i)-\tilde{m}(X))^2\right\} \\ &\quad +3\mathbb{E}\left\{\sum_{i=1}^{n} W_{n,i}(X)(\tilde{m}(X)-m(X))^2\right\} \\ &= 3J_{n1}+3J_{n2}+3J_{n3} \end{aligned}$$

By the choice of $\tilde{m}$

$$J_{n2} \to 0$$

and

$$J_{n3} \leq \mathbb{E}\{(\tilde{m}(X)-m(X))^2\} < \varepsilon.$$

Apply (2.5) for $f(x) = (\tilde{m}(x)-m(x))^2$, then

$$J_{n1} \leq \mathbb{E}\{(\tilde{m}(X)-m(X))^2\} \leq \varepsilon,$$

so

$$\limsup_n J'_n \leq 2\varepsilon.$$

Because of

$$\sum_{i=1}^{n} W_{ni}(x)^2 = \begin{cases} \frac{1}{n\mu_n(A_n(x))} & \text{if } \mu_n(A_n(x)) > 0 \\ 0 & \text{if } \mu_n(A_n(x)) = 0, \end{cases}$$

we have that

$$\begin{aligned}
&\mathbb{E}\left\{\sum_{i=1}^{n} W_{ni}(X)^2\right\} \\
&\le \mathbb{P}\{X \in S^c\} + \sum_{j:A_{nj}\cap S\neq\emptyset} \mathbb{E}\left\{ I_{\{X\in A_{nj}\}} \frac{1}{n\mu_n(A_{nj})} I_{\{\mu_n(A_{nj})>0\}} \right\} \\
&\le \mu(S^c) + \sum_{j:A_{nj}\cap S\neq\emptyset} \mu(A_{nj}) \frac{2}{n\mu(A_{nj})} \\
&= \mu(S^c) + \frac{2|\{j : A_{nj}\cap S\neq\emptyset\}|}{n},
\end{aligned}$$

where we used (2.7). The first term on the right-hand side can be made arbitrarily small by choice of S, while the second term goes to zero by the second condition on the partition, therefore

$$\mathbb{E}\left\{\sum_{i=1}^{n} W_{ni}(X)^2\right\} \to 0 \tag{2.8}$$

Put

$$\sigma^2(x) = \mathbb{E}\{(Y - m(X))^2 | X = x\}$$

then $\mathbb{E}Y^2 < \infty$ implies that $\mathbb{E}\sigma^2(X) < \infty$, and

$$\begin{aligned}
I_n &= \mathbb{E}\left\{ \left(\sum_{i=1}^{n} W_{n,i}(X)(Y_i - m(X_i)) \right)^2 \right\} \\
&= \mathbb{E}\left\{ \sum_{i=1}^{n}\sum_{j=1}^{n} W_{n,i}(X)W_{n,j}(X)(Y_i - m(X_i))(Y_j - m(X_j)) \right\} \\
&= \mathbb{E}\left\{ \sum_{i=1}^{n} W_{n,i}(X)^2 (Y_i - m(X_i))^2 \right\} \\
&= \mathbb{E}\left\{ \sum_{i=1}^{n} W_{n,i}(X)^2 \sigma^2(X_i) \right\}.
\end{aligned}$$

If $\sigma^2(x)$ is bounded then (2.8) implies that $I_n \to 0$. For general $\sigma^2(x)$ and $\varepsilon > 0$ choose bounded $\tilde{\sigma}^2(x) \le L$ such that

$$\mathbb{E}\{|\tilde{\sigma}^2(X) - \sigma^2(X)|\} < \varepsilon.$$

Then

$$\begin{aligned}
I_n &\le \mathbb{E}\left\{ \sum_{i=1}^{n} W_{n,i}(X)^2 \tilde{\sigma}^2(X_i) \right\} + \mathbb{E}\left\{ \sum_{i=1}^{n} W_{n,i}(X)^2 |\sigma^2(X_i) - \tilde{\sigma}^2(X_i)| \right\} \\
&\le L\mathbb{E}\left\{ \sum_{i=1}^{n} W_{n,i}(X)^2 \right\} + \mathbb{E}\left\{ \sum_{i=1}^{n} W_{n,i}(X) |\sigma^2(X_i) - \tilde{\sigma}^2(X_i)| \right\},
\end{aligned}$$

therefore because of (2.5)

$$\limsup_n I_n \leq \mathbb{E}\{|\tilde{\sigma}^2(X) - \sigma^2(X)|\} < \varepsilon.$$

□

There are strong consistency results for bounded Y (Devroye, Györfi (1983)).

For the **k-nearest neighbor estimate**, $W_{ni}(x;X_1,\ldots,X_n)$ is $1/k$ if X_i is one of the k nearest neighbors of x among $X_1,\ldots,X_n$, and W_{ni} is zero otherwise (cf. Cover and Hart (1967), Cover (1968)). Note in particular that $\sum_{i=1}^n W_{ni}(x) = 1$.

Theorem 2.3. (STONE (1977)) *If*

$$k_n \to \infty, \quad k_n/n \to 0 \tag{2.9}$$

then the k_n-nearest neighbor estimate is weakly universally consistent.

Theorem 2.4. (DEVROYE, GYÖRFI, KRZYŻAK AND LUGOSI (1994)) *If*

$$\lim_{n\to\infty} k_n/\log(n) = \infty, \quad \lim_{n\to\infty} k_n/n = 0$$

then the k_n-nearest neighbor estimate is strongly universally consistent.

The regression function can be estimated by the **kernel estimate**

$$m_n(x) = \frac{\sum_{i=1}^n Y_i K_{h_n}(x - X_i)}{\sum_{i=1}^n K_{h_n}(x - X_i)},$$

where $h_n > 0$ is a smoothing factor, K is an absolutely integrable function (the kernel), and $K_h(x) = K(x/h)$ (cf. Nadaraya (1964, 1970) and Watson (1964)). For these weights $\sum_{i=1}^n W_{ni}(x) \leq 1$.

Theorem 2.5. (DEVROYE AND WAGNER (1980), SPIEGELMAN AND SACKS (1980)) *Assume that*

$$b I_{x \in S_{0,r}} \leq K(x) \leq I_{x \in S_{0,R}} \tag{2.10}$$

and

$$\lim_{n\to\infty} h_n = 0, \quad \lim_{n\to\infty} n h_n^d = \infty. \tag{2.11}$$

Then the kernel estimate is weakly universally consistent.

The kernel estimate is strongly consistent if Y is uniformly bounded (Devroye and Krzyżak, 1989). The strong universal consistency of the kernel regression function estimate has been proved recently by Walk (2000).

2.3 Consequences in pattern recognition

In pattern recognition or discrimination one has to make a decision on a $\{1,2,\dots M\}$ valued label (class) Y given an observation (feature) vector X. A **classifier** or a **decision function** is formulated by a function

$$g : \mathcal{R}^d \to \{1,2,\dots M\}.$$

The quality of the classifier is measured by the probability of error:

$$L_g = \mathbb{P}(g(X) \neq Y).$$

For any $x \in \mathcal{R}^d$, introduce the **a posteriori probability** functions:

$$P_i(x) = \mathbb{P}\{Y = i | X = x\}.$$

Thus, $P_i(x)$ is the conditional probability that Y is i given $X = x$.

Of particular interest is the **Bayes decision** function

$$g^*(x) = \arg\max_i P_i(x).$$

This decision function minimizes the error probability (cf. Theorem 1.1).

The smallest error probability $L^* = \mathbb{P}\{g^*(X) \neq Y\}$ is called **Bayes error** or **Bayes risk**.

In order to create the Bayes decision we should know the conditional distributions of Y given X. In **pattern recognition** these are unknown, but we are given data $(X_1,Y_1),\dots,(X_n,Y_n)$ that are independent and identically distributed copies of (X,Y). A classifier depends on the data:

$$g_n(x) = g_n((X_1,Y_1),\dots,(X_n,Y_n),x).$$

Again we have two types of universal consistencies:

DEFINITION 2.3. *The classifier g_n is called* **weakly universally consistent** *if*

$$\mathbb{P}(g_n(X) \neq Y) \to L^*$$

for all distributions of (X,Y).

DEFINITION 2.4. *The classifier g_n is called* **strongly universally consistent** *if*

$$\mathbb{P}(g_n(X) \neq Y | (X_1,Y_1),\dots,(X_n,Y_n)) \to L^* \ a.s.$$

for all distributions of (X,Y).

Assume that the functions $\tilde{P}_i(x)$ are some approximations of the a posteriori probability $P_i(x)$, then plug-in decision g is defined as

$$g(x) = \arg\max_i \tilde{P}_i(x).$$

The next well-known inequality states that if $\tilde{P}_i(x)$ is close to the real a posteriori probability $P_i(x)$ in L_1-sense, then the error probability of plug-in decision g is near the optimal decision g^*.

Lemma 2.2. *For the error probability of the plug-in decision g defined above, we have*

$$\mathbb{P}\{g(X) \neq Y\} - L^* \leq \sum_{i=1}^{M} \int_{\mathcal{R}^d} |P_i(x) - \tilde{P}_i(x)|\mu(dx).$$

PROOF. Given $X = x$, the conditional error probability of any decision g may be expressed as

$$\begin{aligned}
\mathbb{P}\{g(X) \neq Y|X = x\} &= 1 - \mathbb{P}\{Y = g(X)|X = x\} \\
&= 1 - \sum_{i=1}^{M} \mathbb{P}\{Y = i, g(X) = i|X = x\} \\
&= 1 - \sum_{i=1}^{M} I_{\{g(x)=i\}} \mathbb{P}\{Y = i|X = x\} \\
&= 1 - \sum_{i=1}^{M} I_{\{g(x)=i\}} P_i(x),
\end{aligned}$$

therefore

$$\mathbb{P}\{g(X) \neq Y|X = x\} - \mathbb{P}\{g^*(X) \neq Y|X = x\} = \sum_{i=1}^{M} (I_{\{g^*(x)=i\}} - I_{\{g(x)=i\}}) P_i(x).$$

If $g^*(x) = g(x)$ then it is zero, otherwise

$$\begin{aligned}
&\mathbb{P}\{g(X) \neq Y|X = x\} - \mathbb{P}\{g^*(X) \neq Y|X = x\} \\
= \; & \sum_{i=1}^{M} I_{\{g^*(x)=i\}} P_i(x) - \sum_{i=1}^{M} I_{\{g(x)=i\}} P_i(x) \\
= \; & P_{\arg\max_i P_i(x)}(x) - P_{\arg\max_i \tilde{P}_i(x)}(x) \\
= \; & P_{\arg\max_i P_i(x)}(x) - \tilde{P}_{\arg\max_i P_i(x)}(x) \\
& + \tilde{P}_{\arg\max_i P_i(x)}(x) - \tilde{P}_{\arg\max_i \tilde{P}_i(x)}(x) \\
& + \tilde{P}_{\arg\max_i \tilde{P}_i(x)}(x) - P_{\arg\max_i \tilde{P}_i(x)}(x) \\
\leq \; & P_{\arg\max_i P_i(x)}(x) - \tilde{P}_{\arg\max_i P_i(x)}(x) \\
& + \tilde{P}_{\arg\max_i \tilde{P}_i(x)}(x) - P_{\arg\max_i \tilde{P}_i(x)}(x) \\
\leq \; & |P_{\arg\max_i P_i(x)}(x) - \tilde{P}_{\arg\max_i P_i(x)}(x)| \\
& + |\tilde{P}_{\arg\max_i \tilde{P}_i(x)}(x) - P_{\arg\max_i \tilde{P}_i(x)}(x)| \\
\leq \; & \sum_{i=1}^{M} |P_i(x) - \tilde{P}_i(x)|.
\end{aligned}$$

The statement follows by integrating both sides with respect to μ. □

Similarly to the consistent regression estimates we can consider consistent classifiers. The local averaging regression estimates imply pattern recognition rules of **local majority voting** type. The **partitioning rule** is

$$g_n(x) = \arg\max_j \sum_{i=1}^{n} I_{\{X_i \in A_n(x)\}} I_{\{Y_i = j\}},$$

where $A_n(x)$ is the cell from $\mathcal{P}_n$ into which x falls. **The kernel rule rule** is

$$g_n(x) = \arg\max_j \sum_{i=1}^{n} K_{h_n}(X_i - x) I_{\{Y_i = j\}}.$$

The k-**nearest neighbor rule** is defined by

$$g_n(x) = \arg\max_j \sum_{i=1}^{n} W_{n,i}(x) I_{\{Y_i = j\}},$$

where $W_{n,i}$ are the nearest neighbor weights defined in the previous section.

Theorem 2.6. (SEE DEVROYE, GYÖRFI AND LUGOSI (1996)) *Under the conditions of Theorem 2.4 the k-NN rule, or under the conditions of Theorem 2.2 the partitioning rule, or under the conditions of Theorem 2.5 the kernel rule are strongly universally consistent.*

2.4 Definition of (penalized) least squares estimates

The other important concept for estimating regression function is the **(penalized) least squares principle**. It is based on the simple idea to estimate the L_2 risk of f

$$\mathbb{E}\{(f(X)-Y)^2\}$$

by the empirical L_2 risk

$$\frac{1}{n}\sum_{j=1}^{n}|f(X_j)-Y_j|^2 \tag{2.12}$$

and to choose as a regression function estimate a function which minimizes the empirical L_2 risk.

If $X_1,\ldots,X_n$ are all distinct (which happens with probability 1 if X has a density) then minimizing (2.12) leads to a function interpolating the data $(X_1,Y_1),\ldots,(X_n,Y_n)$ and having empirical L_2 risk 0. Obviously, such a function will not be consistent in general.

There are two different strategies to avoid such a "overfitting". For **least squares estimates** one first chooses a "suitable" class of functions $\mathcal{F}_n$ (maybe depending on the data, but at least depending on the sample size n) and then selects a function from this class which minimizes the empirical L_2 risk, i.e. one defines the estimate m_n by

$$m_n \in \mathcal{F}_n \text{ and } \frac{1}{n}\sum_{j=1}^{n}|m_n(X_j)-Y_j|^2 = \min_{f\in\mathcal{F}_n}\frac{1}{n}\sum_{j=1}^{n}|f(X_j)-Y_j|^2. \tag{2.13}$$

Here we assumed the existence of minimizing functions, though not necessarily their uniqueness. In cases where the minima do not exist, the same analysis can be carried out with functions whose empirical L_2 risk is arbitrarily close to the infimum, but for the sake of simplicity we maintain the assumption of existence throughout this notes. The class of candidate functions grows as the sample-size n grows.

Examples of possible choices of the set $\mathcal{F}_n$ are sets of piecewise polynomials with respect to a partition $\mathcal{P}_n$ of $\mathcal{R}^d$, or sets of smooth piecewise polynomials (splines). The use of spline spaces ensures that the estimate is a smooth function.

For **penalized least squares estimates** one adds a penalty term

$$J_n(f)\geq 0$$

to the empirical L_2 risk which penalizes the "roughness" of the function f:

$$m_n = \arg\min_f\left\{\frac{1}{n}\sum_{i=1}^{n}|f(X_i)-Y_i|^2+J_n(f)\right\}, \tag{2.14}$$

where one minimizes over all measurable functions f.

A popular choice for $J_n(f)$ in the case $d = 1$ is

$$J_n(f) = \lambda_n \int |f''(t)|^2 dt, \tag{2.15}$$

where f'' denotes the second derivative of f and λ_n is some positive constant. One can show that for this penalty term the minimum in (2.14) is achieved by a cubic spline with knots at the X_i's, i.e., by a twice differential function which is equal to a polynomial of degree 3 (or less) between adjacent values of the X_i's.

A generalization of (2.15) is

$$J_{n,k}(f) = \lambda_n \int |f^{(k)}(t)|^2 dt,$$

where $f^{(k)}$ denotes the k–th derivative of f. In this case one can show that m_n is equal to a polynomial of degree $2k-1$ (or less) between adjacent X_i's and $2k-2$ times continuously differentiable on $\mathcal{R}$, and that m_n can be computed by solving a linear equation system.

For multivariate X one can use

$$J_{n,k}(f) = \lambda_n \int \sum_{i_1,\dots,i_k \in \{1,\dots,d\}} \left| \frac{\partial^k f}{\partial x_{i_1} \dots \partial x_{i_k}} \right|^2 dx,$$

which leads to thin plate spline estimates.

2.5 Consistency of least squares estimates

We consider the least squares estimate

$$m_n \in \mathcal{F}_n \text{ and } \frac{1}{n} \sum_{j=1}^{n} |m_n(X_j) - Y_j|^2 = \min_{f \in \mathcal{F}_n} \frac{1}{n} \sum_{j=1}^{n} |f(X_j) - Y_j|^2, \tag{2.16}$$

where $\mathcal{F}_n$ is a set of functions $f : \mathcal{R}^d \to \mathcal{R}$.

The choice of $\mathcal{F}_n$ has two effects on the error of the estimate. On one hand, if $\mathcal{F}_n$ is not too "massive" then the empirical L_2 risk will be close to the L_2 risk uniformly over $\mathcal{F}_n$. Thus the error introduced by minimizing the empirical L_2 risk instead of the L_2 risk will be small. On the other hand, because of the requirement that our estimate is contained in $\mathcal{F}_n$ it cannot be better (with respect to the L_2 error) than the best function in $\mathcal{F}_n$. This is formulated in the following lemma.

Lemma 2.3. *Let $\mathcal{F}_n$ be a class of functions $f : \mathcal{R}^d \to \mathcal{R}$. If m_n satisfies (2.16) then*

$$\int |m_n(x) - m(x)|^2 \mu(dx)$$

$$\leq 2 \sup_{f\in\mathcal{F}_n} \left| \frac{1}{n}\sum_{j=1}^{n} |f(X_j)-Y_j|^2 - \mathbb{E}\{(f(X)-Y)^2\} \right|$$
$$+ \inf_{f\in\mathcal{F}_n} \int |f(x)-m(x)|^2 \mu(dx)$$

PROOF. It follows that

$$\int |m_n(x)-m(x)|^2\mu(dx) = \mathbb{E}\{|m_n(X)-Y|^2|D_n\} - \mathbb{E}\{|m(X)-Y|^2\}$$
$$= \left(\mathbb{E}\{|m_n(X)-Y|^2|D_n\} - \inf_{f\in\mathcal{F}_n} \mathbb{E}\{|f(X)-Y|^2\}\right)$$
$$+ \left(\inf_{f\in\mathcal{F}_n} \mathbb{E}\{|f(X)-Y|^2\} - \mathbb{E}\{|m(X)-Y|^2\}\right). \tag{2.17}$$

Then

$$\inf_{f\in\mathcal{F}_n} \mathbb{E}\{|f(X)-Y|^2\} - \mathbb{E}\{|m(X)-Y|^2\} = \inf_{f\in\mathcal{F}_n} \int |f(x)-m(x)|^2\mu(dx).$$

Thus all we need is an upper bound for the first term. By (2.16) one gets

$$\mathbb{E}\left\{|m_n(X)-Y|^2 \middle| D_n\right\} - \inf_{f\in\mathcal{F}_n} \mathbb{E}\{|f(X)-Y|^2\}$$
$$= \sup_{f\in\mathcal{F}_n} (\mathbb{E}\left\{|m_n(X)-Y|^2 \middle| D_n\right\} - \frac{1}{n}\sum_{j=1}^{n} |m_n(X_j)-Y_j|^2$$
$$+\frac{1}{n}\sum_{j=1}^{n} |m_n(X_j)-Y_j|^2 - \frac{1}{n}\sum_{j=1}^{n} |f(X_j)-Y_j|^2$$
$$+\frac{1}{n}\sum_{j=1}^{n} |f(X_j)-Y_j|^2 - \mathbb{E}\{|f(X)-Y|^2\})$$
$$\leq \sup_{f\in\mathcal{F}_n} (\mathbb{E}\left\{|m_n(X)-Y|^2 \middle| D_n\right\} - \frac{1}{n}\sum_{j=1}^{n} |m_n(X_j)-Y_j|^2$$
$$+\frac{1}{n}\sum_{j=1}^{n} |f(X_j)-Y_j|^2 - \mathbb{E}\{|f(X)-Y|^2\})$$
$$\leq 2\sup_{f\in\mathcal{F}_n} \left| \frac{1}{n}\sum_{j=1}^{n} |f(X_j)-Y_j|^2 - \mathbb{E}\{|f(X)-Y|^2\} \right|.$$

□

We call the first term in the right hand side of (2.17) the estimation error, and the second term the approximation error of the estimator. The estimation error measures the distance between the L_2 risk of the estimate and the L_2 risk of the best function in $\mathcal{F}_n$. The approximation error measures how well the regression function can be approximated by functions of $\mathcal{F}_n$ in L_2.

In order to get universally consistent estimates it suffices to show that both terms converge to 0 for all distributions of (X,Y) with $\mathbb{E}Y^2 < \infty$.

For the approximation error this is often quite simple: if, for example, $\mathcal{F}_n$ is nested, that is $\mathcal{F}_n \subset \mathcal{F}_{n+1}$ for all n then

$$\lim_{n\to\infty} \inf_{f\in\mathcal{F}_n} \int |f(x)-m(x)|^2\mu(dx) = 0$$

for all measures μ and all $m \in L_2$ simply means that $\bigcup_{n=1}^{\infty} \mathcal{F}_n$ is dense in $L_2 = L_2(\mu)$ for all distributions μ. This is true, for example, if $\bigcup_{n=1}^{\infty} \mathcal{F}_n$ is dense in $C_0^\infty(\mathcal{R}^d)$ with respect to the sup norm $\|.\|_\infty$ since $C_0^\infty(\mathcal{R}^d)$ is dense in $L_2(\mu)$ for all distributions μ and $\int |f(x)-m(x)|^2\mu(dx) \leq \|f-m\|_\infty^2$.

For the estimation error it suffices to show

$$\sup_{f\in\mathcal{F}_n} \left| \frac{1}{n}\sum_{i=1}^n |f(X_i)-Y_i|^2 - \mathbb{E}\{|f(X)-Y|^2\} \right| \to 0 \quad (n\to\infty) \quad a.s. \tag{2.18}$$

Set $Z=(X,Y)$, $Z_i=(X_i,Y_i)$ $(i=1,\ldots,n)$, $g_f(x,y)=|f(x)-y|^2$ for $f\in\mathcal{F}_n$ and $\mathcal{G}_n=\{g_f : f\in\mathcal{F}_n\}$.

Then (2.18) can be written as

$$\sup_{g\in\mathcal{G}_n} \left| \frac{1}{n}\sum_{i=1}^n g(Z_i) - \mathbb{E}\{g(Z)\} \right| \to 0 \quad (n\to\infty) \quad a.s.$$

Thus we are interested in bounding the distance between an average and its expectation uniformly over a set of functions. In the sequel we discuss techniques for doing this.

Let $Z, Z_1, Z_2, \ldots$ be independent identically distributed random variables with values in $\mathcal{R}^d$, and for $n\in\mathcal{N}$ let $\mathcal{G}_n$ be a class of functions $g:\mathcal{R}^d\to\mathcal{R}$. We derive sufficient conditions for

$$\sup_{g\in\mathcal{G}_n} \left| \frac{1}{n}\sum_{i=1}^n g(Z_i) - \mathbb{E}\{g(Z)\} \right| \to 0 \quad (n\to\infty) \quad a.s. \tag{2.19}$$

For each fixed function g with $\mathbb{E}|g(Z)| < \infty$ the strong law of large numbers implies

$$\lim_{n\to\infty} \left| \frac{1}{n}\sum_{j=1}^n g(Z_j) - \mathbb{E}\{g(Z)\} \right| = 0 \quad a.s.$$

We will use Hoeffding's inequality to extend it to sets of functions. Recall that if g is a function $g:\mathcal{R}^d\to[0,B]$ then by Hoeffding's inequality

$$\mathbb{P}\left\{ \left| \frac{1}{n}\sum_{j=1}^n g(Z_j) - \mathbb{E}\{g(Z)\} \right| > \varepsilon \right\} \leq 2e^{-\frac{2n\varepsilon^2}{B^2}},$$

which together with the union bound implies

$$\mathbb{P}\left\{\sup_{g\in\mathcal{G}_n}\left|\frac{1}{n}\sum_{j=1}^{n}g(Z_j)-\mathbb{E}\{g(Z)\}\right|>\varepsilon\right\}\le 2|\mathcal{G}_n|e^{-\frac{2n\varepsilon^2}{B^2}}. \tag{2.20}$$

Thus, for finite classes $\mathcal{G}_n$ satisfying

$$\sum_{n=1}^{\infty}|\mathcal{G}_n|e^{-\frac{2n\varepsilon^2}{B^2}}<\infty \tag{2.21}$$

for all $\varepsilon > 0$, (2.19) follows from (2.20) and the Borel-Cantelli Lemma (see proof of Lemma 2.4 below for details).

In our applications (2.21) is never satisfied because the cardinality of $\mathcal{G}_n$ is always infinite. But sometimes it is possible to choose finite sets $\mathcal{G}_{n,\varepsilon}$ which satisfy (2.21) and

$$\left\{\sup_{g\in\mathcal{G}_n}\left|\frac{1}{n}\sum_{j=1}^{n}g(Z_j)-\mathbb{E}\{g(Z)\}\right|>\varepsilon\right\}$$
$$\subset\left\{\sup_{g\in\mathcal{G}_{n,\varepsilon}}\left|\frac{1}{n}\sum_{j=1}^{n}g(Z_j)-\mathbb{E}\{g(Z)\}\right|>\varepsilon'\right\} \tag{2.22}$$

for some ε' depending on ε (but not on n). Clearly, then (2.19) follows from (2.22) and (2.21) with the above argument applied to $\mathcal{G}_{n,\varepsilon}$.

To construct classes $\mathcal{G}_{n,\varepsilon}$ which satisfy (2.22) one can use covers with respect to the supremum norm:

DEFINITION 2.5. *Let $\varepsilon > 0$ and let $\mathcal{G}$ be a set of functions $\mathcal{R}^d \to \mathcal{R}$. Every finite collection of functions $g_1,\dots,g_N : \mathcal{R}^d \to \mathcal{R}$ with the property, that for every $g \in \mathcal{G}$ there is a $j = j(g) \in \{1,\dots,N\}$ such that*

$$\|g-g_j\|_\infty := \sup_z |g(z)-g_j(z)| < \varepsilon,$$

is called an **ε-cover of** *$\mathcal{G}$ with respect to $\|\cdot\|_\infty$.*

If $\{g_1,\dots,g_N\}$ is an ε-cover of $\mathcal{G}$ w.r.t. $\|\cdot\|_\infty$, then $\mathcal{G}$ is a subset of the union of all $\|\cdot\|_\infty$-balls with center g_i and radius ε $(i = 1,\dots,n)$. The fewer balls are needed to cover $\mathcal{G}$, the smaller $\mathcal{G}$ is in some sense.

DEFINITION 2.6. *Let $\varepsilon > 0$ and let $\mathcal{G}$ be a set of functions $\mathcal{R}^d \to \mathcal{R}$. Let $\mathcal{N}(\varepsilon,\mathcal{G},\|\cdot\|_\infty)$ be the size of the smallest ε-cover of $\mathcal{G}$ w.r.t. $\|\cdot\|_\infty$. Take $\mathcal{N}(\varepsilon,\mathcal{G},\|\cdot\|_\infty)=\infty$ if no finite ε-cover exists. Then $\mathcal{N}(\varepsilon,\mathcal{G},\|\cdot\|_\infty)$ is called* **ε-covering number of** *$\mathcal{G}$ w.r.t. $\|\cdot\|_\infty$ and will be abbreviated to $\mathcal{N}_\infty(\varepsilon,\mathcal{G})$.*

Lemma 2.4. *For $n \in \mathcal{N}$ let $\mathcal{G}_n$ be a set of functions $g : \mathcal{R}^d \to [0, B]$ and let $\varepsilon > 0$. Then*

$$\mathbb{P}\left\{ \sup_{g\in\mathcal{G}_n} \left| \frac{1}{n}\sum_{j=1}^n g(Z_j) - \mathbb{E}\{g(Z)\} \right| > \varepsilon \right\} \leq 2\mathcal{N}_\infty(\varepsilon/3, \mathcal{G}_n)\, e^{-\frac{2n\varepsilon^2}{9B^2}}.$$

Furthermore, if

$$\sum_{n=1}^\infty \mathcal{N}_\infty(\varepsilon/3, \mathcal{G}_n)\, e^{-\frac{2n\varepsilon^2}{9B^2}} < \infty \tag{2.23}$$

for all $\varepsilon > 0$, then (2.19) holds.

Since in the above probability the supremum is taken over a possible uncountable set, there may be some measurability problems. In the book of van de Vaart and Wellner (1996) this issue is handled very elegantly using the notion of outer probability. In most of our applications it will suffice to consider countable sets of functions, therefore here and in the sequel we shall completely ignore this problem.

PROOF. Let $\mathcal{G}_{n,\frac{\varepsilon}{3}}$ be an $\frac{\varepsilon}{3}$–cover of $\mathcal{G}_n$ w.r.t. $\|\cdot\|_\infty$ of minimal cardinality. Fix $g \in \mathcal{G}_n$. Then there exists $\bar{g} \in \mathcal{G}_{n,\frac{\varepsilon}{3}}$ such that $\|g - \bar{g}\|_\infty < \frac{\varepsilon}{3}$. Using this one gets

$$\begin{aligned}
&\left| \frac{1}{n}\sum_{i=1}^n g(Z_i) - \mathbb{E}\{g(Z)\} \right| \\
&\leq \left| \frac{1}{n}\sum_{i=1}^n g(Z_i) - \frac{1}{n}\sum_{i=1}^n \bar{g}(Z_i) \right| + \left| \frac{1}{n}\sum_{i=1}^n \bar{g}(Z_i) - \mathbb{E}\{\bar{g}(Z)\} \right| + |\mathbb{E}\{\bar{g}(Z)\} - \mathbb{E}\{g(Z)\}| \\
&\leq \|g - \bar{g}\|_\infty + \left| \frac{1}{n}\sum_{i=1}^n \bar{g}(Z_i) - \mathbb{E}\{\bar{g}(Z)\} \right| + \|g - \bar{g}\|_\infty \\
&\leq \frac{2}{3}\varepsilon + \left| \frac{1}{n}\sum_{i=1}^n \bar{g}(Z_i) - \mathbb{E}\{\bar{g}(Z)\} \right|.
\end{aligned}$$

Hence,

$$\begin{aligned}
&\mathbb{P}\left\{ \sup_{g\in\mathcal{G}_n} \left| \frac{1}{n}\sum_{i=1}^n g(Z_i) - \mathbb{E}\{g(Z)\} \right| > \varepsilon \right\} \\
&\leq \mathbb{P}\left\{ \sup_{g\in\mathcal{G}_{n,\frac{\varepsilon}{3}}} \left| \frac{1}{n}\sum_{i=1}^n g(Z_i) - \mathbb{E}\{g(Z)\} \right| > \frac{\varepsilon}{3} \right\} \\
&\overset{(2.20)}{\leq} 2\left| \mathcal{G}_{n,\frac{\varepsilon}{3}} \right| e^{-\frac{2n(\frac{\varepsilon}{3})^2}{B^2}} = 2\mathcal{N}_\infty\left(\frac{\varepsilon}{3}, \mathcal{G}_n\right) e^{-\frac{2n\varepsilon^2}{9B^2}}.
\end{aligned}$$

If (2.23) holds, then this implies

$$\sum_{n=1}^\infty \mathbb{P}\left\{ \sup_{g\in\mathcal{G}_n} \left| \frac{1}{n}\sum_{i=1}^n g(Z_i) - \mathbb{E}\{g(Z)\} \right| > \frac{1}{k} \right\} < \infty$$

for each $k \in \mathcal{N}$. Using the Borel–Cantelli Lemma one concludes that for each $k \in \mathcal{N}$

$$\limsup_{n\to\infty} \sup_{g\in\mathcal{G}_n} \left| \frac{1}{n}\sum_{i=1}^{n} g(Z_i) - \mathbf{E}\{g(Z)\} \right| \le \frac{1}{k} \quad a.s.,$$

hence also with probability one

$$\limsup_{n\to\infty} \sup_{g\in\mathcal{G}_n} \left| \frac{1}{n}\sum_{i=1}^{n} g(Z_i) - \mathbf{E}\{g(Z)\} \right| \le \frac{1}{k} \quad \text{for all } k \in \mathcal{N},$$

which implies (2.19). □

Supremum norm covers are often too large to satisfy (2.23). But clearly, (2.23) is not necessary for the proof of (2.19). To motivate a weaker condition let us again consider (2.22). For the sake of simplicity ignore the expected value $\mathbf{E}g(Z)$ (think of it as an average $\frac{1}{n}\sum_{j=n+1}^{2n} g(Z_j)$, where $Z_1,\ldots,Z_{2n}$ are i.i.d.). So instead of a data independent set $\tilde{\mathcal{G}}_{n,\varepsilon}$ we now try to construct a set $\bar{\mathcal{G}}_{n,\varepsilon}$ depending on $Z_1,\ldots,Z_{2n}$ which satisfies

$$\left\{ \sup_{g\in\mathcal{G}_n} \left| \frac{1}{n}\sum_{j=1}^{n} g(Z_j) - \frac{1}{n}\sum_{j=n+1}^{2n} g(Z_j) \right| > \varepsilon \right\}$$

$$\subset \left\{ \sup_{g\in\bar{\mathcal{G}}_{n,\varepsilon}} \left| \frac{1}{n}\sum_{j=1}^{n} g(Z_j) - \frac{1}{n}\sum_{j=n+1}^{2n} g(Z_j) \right| > \varepsilon' \right\}$$

for some ε' depending on ε (but not on n). Then it is clear that all we need is a data dependent cover which can approximate each $g \in \mathcal{G}_n$ with respect to

$$\|g-h\|_{2n} = \frac{1}{2n}\sum_{j=1}^{2n} |g(Z_j) - h(Z_j)|.$$

To formulate this idea introduce the following covering numbers:

DEFINITION 2.7. *Let $\varepsilon > 0$, $\mathcal{G}$ be a set of functions $\mathcal{R}^d \to \mathcal{R}$, $1 \le p < \infty$ and ν be a probability measure on $\mathcal{R}^d$. For a function $f : \mathcal{R}^d \to \mathcal{R}$ set*

$$\|f\|_{L_p(\nu)} := \left\{ \int |f(z)|^p d\nu \right\}^{\frac{1}{p}}.$$

a) *Every finite collection of functions $g_1,\ldots,g_N : \mathcal{R}^d \to \mathcal{R}$ with the property, that for every $g \in \mathcal{G}$ there is a $j = j(g) \in \{1,\ldots,N\}$ such that*

$$\|g - g_j\|_{L_p(\nu)} < \varepsilon,$$

is called an **ε-cover of** *$\mathcal{G}$ with respect to $\|\cdot\|_{L_p(\nu)}$.*

b) *Let* $\mathcal{N}(\varepsilon, \mathcal{G}, \|\cdot\|_{L_p(\nu)})$ *be the size of the smallest* ε*-cover of* $\mathcal{G}$ *w.r.t.* $\|\cdot\|_{L_p(\nu)}$. *Take* $\mathcal{N}(\varepsilon, \mathcal{G}, \|\cdot\|_{L_p(\nu)}) = \infty$ *if no finite* ε*-cover exists. Then* $\mathcal{N}(\varepsilon, \mathcal{G}, \|\cdot\|_{L_p(\nu)})$ *is called* **ε-covering number of $\mathcal{G}$** *w.r.t.* $\|\cdot\|_{L_p(\nu)}$.

c) *Let* $z_1^n = (z_1, \dots, z_n)$ *be n fixed points in* $\mathcal{R}^d$. *Let* ν_n *be the corresponding empirical measure, i.e.,*

$$\nu_n(A) = \frac{1}{n}\sum_{i=1}^{n} I_A(z_i) \quad (A \subseteq \mathcal{R}^d).$$

Then

$$\|f\|_{L_p(\nu_n)} = \left\{\frac{1}{n}\sum_{i=1}^{n} |f(z_i)|^p\right\}^{\frac{1}{p}}$$

and any ε*-cover of* $\mathcal{G}$ *w.r.t.* $\|\cdot\|_{L_p(\nu_n)}$ *will be called* **L_p ε-cover of $\mathcal{G}$ on z_1^n** *and the* ε*-covering number of* $\mathcal{G}$ *w.r.t.* $\|\cdot\|_{L_p(\nu_n)}$ *will be denoted by*

$$\mathcal{N}_p(\varepsilon, \mathcal{G}, z_1^n).$$

In other words, $\mathcal{N}_p(\varepsilon, \mathcal{G}, z_1^n)$ *is the minimal* $N \in \mathcal{N}$ *such that there exist functions* $g_1, \dots, g_N : \mathcal{R}^d \to \mathcal{R}$ *with the property, that for every* $g \in \mathcal{G}$ *there is a* $j = j(g) \in \{1, \dots, N\}$ *such that*

$$\left\{\frac{1}{n}\sum_{i=1}^{n} |g(z_i) - g_j(z_i)|^p\right\}^{\frac{1}{p}} < \varepsilon.$$

If $Z_1^n = (Z_1, \dots, Z_n)$ is a sequence of i.i.d. random variables, then $\mathcal{N}_1(\varepsilon, \mathcal{G}, Z_1^n)$ is a random variable, whose expected value plays a central role in our problem. With these notations one can extend the first half of Lemma 2.4 from fixed supremum norm covers to L_1 covers on a random set of points.

Theorem 2.7. (POLLARD (1984)). *Let* $\mathcal{G}$ *be a set of functions* $g : \mathcal{R}^d \to [0, B]$. *For any n and any* $\varepsilon > 0$,

$$\mathbb{P}\left\{\sup_{g \in \mathcal{G}} \left|\frac{1}{n}\sum_{i=1}^{n} g(Z_i) - \mathbb{E}\{g(Z_1)\}\right| > \varepsilon\right\} \le 8\mathbb{E}\{\mathcal{N}_1(\varepsilon/8, \mathcal{G}, Z_1^n)\}\, e^{-n\varepsilon^2/(128B^2)}.$$

PROOF. See Györfi et al. (2002), Theorem 9.1 □

From Theorem 2.7 one can derive (2.19) by an application of the Borel-Cantelli lemma, if one has suitable upper bounds for the L_1 covering numbers. These upper bounds will be described in the sequel. Therefore we need the following definitions (cf. Chapter 1, Section 1.4.3).

DEFINITION 2.8. *Let* $\mathcal{A}$ *be a class of subsets of* $\mathcal{R}^d$ *and let* $n \in \mathcal{N}$.

a) *For* $z_1, \dots, z_n \in \mathcal{R}^d$ *define*

$$s(\mathcal{A}, \{z_1, \dots, z_n\}) = |\{A \cap \{z_1, \dots, z_n\} : A \in \mathcal{A}\}|,$$

that is, $s(\mathcal{A},\{z_1,\dots,z_n\})$ is the number of different subsets of $\{z_1,\dots,z_n\}$ of the form $A\cap\{z_1,\dots,z_n\}$, $A\in\mathcal{A}$.

b) *Let G be a subset of $\mathcal{R}^d$ of size n. One says that $\mathcal{A}$* **shatters** *G if $s(\mathcal{A},G)=2^n$, i.e., if each subset of G can be represented in the form $A\cap G$ for some $A\in\mathcal{A}$.*

c) *The n-th* **shatter coefficient** *of $\mathcal{A}$ is*

$$S(\mathcal{A},n)=\max_{\{z_1,\dots,z_n\}\subseteq\mathcal{R}^d} s(\mathcal{A},\{z_1,\dots,z_n\}).$$

That is, the shatter coefficient is the maximal number of different subsets of n points that can be picked out by sets from $\mathcal{A}$.

Clearly, $s(\mathcal{A},\{z_1,\dots,z_n\})\le 2^n$ and $S(\mathcal{A},n)\le 2^n$. If $S(\mathcal{A},n)<2^n$ then $s(\mathcal{A},\{z_1,\dots,z_n\})<2^n$ for all $z_1,\dots,z_n\in\mathcal{R}^d$. If $s(\mathcal{A},\{z_1,\dots,z_n\})<2^n$, then $\{z_1,\dots,z_n\}$ has a subset such that there is no set in $\mathcal{A}$ that contains exactly that subset of $\{z_1,\dots,z_n\}$.

It is easy to see that $S(\mathcal{A},k)<2^k$ implies $S(\mathcal{A},n)<2^n$ for all $n>k$. The last time when $S(\mathcal{A},k)=2^k$ is important:

DEFINITION 2.9. *Let $\mathcal{A}$ be a class of subsets of $\mathcal{R}^d$ with $\mathcal{A}\neq\emptyset$. The* **VC dimension** *(or Vapnik-Chervonenkis dimension) $V_{\mathcal{A}}$ of $\mathcal{A}$ is defined by*

$$V_{\mathcal{A}}=\sup\left\{n\in\mathcal{N}\ :\ S(\mathcal{A},n)=2^n\right\},$$

i.e., the VC dimension $V_{\mathcal{A}}$ is the largest integer n such that there exists a set of n points in $\mathcal{R}^d$ which can be shattered by $\mathcal{A}$.

Let $\mathcal{G}$ be a class of functions on $\mathcal{R}^d$ taking their values in $[0,B]$. To bound the L_p covering of $\mathcal{G}$ on z_1^n we will use the VC dimension $V_{\mathcal{G}^+}$ of the set

$$\mathcal{G}^+:=\left\{\{(z,t)\in\mathcal{R}^d\times\mathcal{R}\,;\,t\le g(z)\}\,;\,g\in\mathcal{G}\right\}$$

of all subgraphs of functions of $\mathcal{G}$.

Theorem 2.8. (HAUSSLER (1992)). *Let $\mathcal{G}$ be a class of functions $g:\mathcal{R}^d\to[0,B]$ with $V_{\mathcal{G}^+}\ge 2$, let $p\ge 1$, let $z_1^n\in\mathcal{R}^{d\cdot n}$ and let $0<\varepsilon<\frac{B}{4}$. Then*

$$\mathcal{N}_p(\varepsilon,\mathcal{G},z_1^n)\le 3\left(\frac{2eB^p}{\varepsilon^p}\log\frac{3eB^p}{\varepsilon^p}\right)^{V_{\mathcal{G}^+}}$$

PROOF. See Györfi et al. (2002), Theorem 9.4 □

In order to bound the VC dimension one can use

Theorem 2.9. *(Steele (1975), Dudley (1978)) Let $\mathcal{F}$ be a k–dimensional vector space of functions $f:\mathcal{R}^d\to\mathcal{R}$. Then the class of sets of the form $\{x\in\mathcal{R}^d\ :\ f(x)\ge 0\}$, $f\in\mathcal{F}$, has VC dimension less than or equal to k.*

PROOF. See Lemma 1.6 in Chapter 1 □

Next we use the above cited results to prove consistency of least squares estimates using piecewise polynomials. For simplicity we assume $X \in [0,1]$ *a.s.* and $|Y| \leq L$ *a.s.* for some $L > 0$.

Let $M \in \mathcal{N}_0$ and for $n \in \mathcal{N}$ let $K_n \in \mathcal{N}$ and let $\mathcal{P}_n = \{A_{n,1}, \ldots, A_{n,K_n}\}$ be a partition of $[0,1]$ consisting of K_n cells. Let $\mathcal{G}_M$ be the set of all polynomials of degree M (or less), and set

$$\mathcal{G}_M \circ \mathcal{P}_n = \left\{ \sum_{j=1}^{K_n} p_j I_{A_{n,j}} \ : \ p_j \in \mathcal{G}_M \ (j = 1, \ldots, K_n) \right\}.$$

$\mathcal{G}_M \circ \mathcal{P}_n$ is the set of all piecewise polynomials of degree M (or less) w.r.t. $\mathcal{P}_n$. Set

$$\mathcal{F}_n = \{ f \in \mathcal{G}_M \circ \mathcal{P}_n \ : \ \|f\|_\infty \leq L \}.$$

Define the piecewise polynomial partitioning estimate by

$$m_n(\cdot) = \arg\min_{f \in \mathcal{F}_n} \frac{1}{n} \sum_{i=1}^{n} |f(X_i) - Y_i|^2. \tag{2.24}$$

Theorem 2.10. *Let M, K_n, $\mathcal{P}_n$ be as above and define the estimate m_n by (2.24).*
If

$$\frac{K_n}{n} \to 0, \tag{2.25}$$

and

$$\max_{j=1,\ldots K_n} diam\,(A_{n,j}) \to 0 \quad (n \to \infty) \tag{2.26}$$

then

$$\mathbf{E} \int |m_n(x) - m(x)|^2 \mu(dx) \to 0 \quad (n \to \infty)$$

for all distributions of (X,Y) with $X \in [0,1]$ a.s. and $|Y| \leq L$ a.s..

PROOF. Because of Lemma 2.3 it suffices to show

$$\inf_{f \in \mathcal{F}_n} \int |f(x) - m(x)|^2 \mu(dx) \to 0 \quad (n \to \infty) \tag{2.27}$$

and

$$\mathbf{E}\left\{ \sup_{f \in \mathcal{F}_n} \left| \frac{1}{n} \sum_{i=1}^{n} |f(X_i) - Y_i|^2 - \mathbf{E}\{|f(X) - Y|^2\} \right| \right\} \to 0 \quad (n \to \infty). \tag{2.28}$$

In order to show (2.27) let $\varepsilon > 0$ be arbitrary. By Lemma 2.1 there exists a continuous function $\bar{m}$ such that

$$\int |\bar{m}(x) - m(x)|^2 \mu(dx) \leq \frac{\varepsilon}{4}.$$

W.l.o.g. we may assume that $|\bar{m}(x)| \leq L$ for all $x \in [0,1]$. Since $\bar{m}$ is uniformly continuous on the compact interval $[0,1]$ there exists a $\delta > 0$ such that $|\bar{m}(x) - \bar{m}(z)| < \sqrt{\varepsilon}/2$ for all $x, z \in [0,1], |x-z| < \delta$. Choose arbitrary points $z_{n,j} \in A_{n,j}$ $(j = 1, \ldots, K_n)$ and set

$$f_n = \sum_{j=1}^{K_n} \bar{m}(z_{n,j}) \cdot I_{A_{n,j}} \in \mathcal{G}_M \circ \mathcal{P}_n.$$

Then $z \in A_{n,j}$ and $diam(A_{n,j}) < \delta$ imply

$$|f_n(z) - \bar{m}(z)| = |\bar{m}(z_{n,j}) - \bar{m}(z)| < \sqrt{\varepsilon}/2.$$

Using this one gets for n sufficiently large (i.e., for n so large that $\max_{j=1,\ldots,K_n} diam(A_{n,j}) < \delta$):

$$\begin{aligned}
\inf_{f \in \mathcal{F}_n} \int |f(x) - m(x)|^2 \mu(dx) &\leq 2 \inf_{f \in \mathcal{F}_n} \int |f(x) - \bar{m}(x)|^2 \mu(dx) + 2 \cdot \frac{\varepsilon}{4} \\
&\leq 2 \int |f_n(x) - \bar{m}(x)|^2 \mu(dx) + \frac{\varepsilon}{2} \\
&\leq 2 \sup_{x \in [0,1]} |f_n(x) - \bar{m}(x)|^2 + \frac{\varepsilon}{2} \\
&\leq 2 \cdot \frac{\varepsilon}{4} + \frac{\varepsilon}{2} = \varepsilon.
\end{aligned}$$

With $\varepsilon \to 0$ this implies (2.27).

Next we prove (2.28). Set

$$Z = (X,Y),\ Z_1 = (X_1, Y_1), \ldots,\ Z_n = (X_n, Y_n)$$

and

$$\mathcal{H}_n = \left\{ h : \mathcal{R}^d \times \mathcal{R} \to \mathcal{R} \ : \ \exists f \in \mathcal{F}_n \text{ such that } h(x,y) = |f(x) - y|^2 \right\}.$$

Observe, that the functions in $\mathcal{H}_n$ satisfy

$$0 \leq h(x,y) \leq 2L^2 + 2L^2 \leq 4L^2 \quad ((x,y) \in \mathcal{R}^d \times [-L, L]).$$

By Theorem 2.7 one has for arbitrary $\varepsilon > 0$:

$$\begin{aligned}
&\mathbb{P}\left[\sup_{f \in \mathcal{F}_n} \left| \frac{1}{n} \sum_{i=1}^n |f(X_i) - Y_i|^2 - \mathbb{E}\{|f(X) - Y|^2\} \right| > \varepsilon \right] \\
&= \mathbb{P}\left[\sup_{h \in \mathcal{H}_n} \left| \frac{1}{n} \sum_{i=1}^n h(Z_i) - \mathbb{E}\{h(Z)\} \right| > \varepsilon \right] \\
&\leq 8 \mathbb{E} \mathcal{N}_1 \left(\frac{\varepsilon}{8}, \mathcal{H}_n, Z_1^n \right) e^{-\frac{n\varepsilon^2}{128(4L^2)^2}}. \qquad (2.29)
\end{aligned}$$

Next we bound the covering number in (2.29). Let $h_i(x,y) = |f_i(x) - y|^2$ $((x,y) \in \mathcal{R}^d \times \mathcal{R})$ for some $f_i \in \mathcal{F}_n$. Then

$$\begin{aligned}
&\frac{1}{n}\sum_{i=1}^{n} |h_1(Z_i) - h_2(Z_i)| \\
&= \frac{1}{n}\sum_{i=1}^{n} \left| |f_1(X_i) - Y_i|^2 - |f_2(X_i) - Y_i|^2 \right| \\
&= \frac{1}{n}\sum_{i=1}^{n} |f_1(X_i) - f_2(X_i)| \cdot |f_1(X_i) - Y_i + f_2(X_i) - Y_i| \\
&\le 4L\frac{1}{n}\sum_{i=1}^{n} |f_1(X_i) - f_2(X_i)|,
\end{aligned}$$

which implies

$$\mathcal{N}_1\left(\frac{\varepsilon}{8}, \mathcal{H}_n, Z_1^n\right) \le \mathcal{N}_1\left(\frac{\varepsilon}{32L}, \mathcal{F}_n, X_1^n\right). \tag{2.30}$$

By Theorem 2.8 we can bound the latter term:

$$\begin{aligned}
\mathcal{N}_1\left(\frac{\varepsilon}{32L}, \mathcal{F}_n, X_1^n\right) &\le 3\left(\frac{2e(2L)}{\frac{\varepsilon}{32L}} \log\left(\frac{3e(2L)}{\frac{\varepsilon}{32L}}\right)\right)^{V_{\mathcal{F}_n^+}} \\
&= 3\left(\frac{128eL^2}{\varepsilon} \log\left(\frac{192eL^2}{\varepsilon}\right)\right)^{V_{\mathcal{F}_n^+}}
\end{aligned} \tag{2.31}$$

$V_{\mathcal{F}_n^+}$ can be bounded by Theorem 2.9. Observe, that

$$\begin{aligned}
\mathcal{F}_n^+ &= \{\{(x,t) : t \le f(x)\} \quad : \quad f \in \mathcal{F}_n\} \\
&\subseteq \{\{(x,t) : f(x) + a_0 t \ge 0\} \quad : \quad f \in \mathcal{F}_n,\ a_0 \in \mathcal{R}\}.
\end{aligned}$$

Now

$$\begin{aligned}
&\left\{ f(x) + a_0 t \,:\, f \in \mathcal{F}_n, a_0 \in \mathcal{R} \right\} \\
&\subseteq \left\{ \sum_{j=1}^{K_n}\sum_{l=0}^{M} a_{j,l} x^l I_{A_{j,n}}(x) + a_0 t \,:\, a_0, a_{j,l} \in \mathcal{R} \right\}
\end{aligned}$$

and the last set is a linear vector space of dimension $K_n(M+1)+1$, thus Theorem 2.9 implies

$$V_{\mathcal{F}_n^+} \le K_n(M+1)+1. \tag{2.32}$$

(2.29) - (2.32) imply

$$\begin{aligned}
&\mathbb{P}\left\{ \sup_{f \in \mathcal{F}_n} \left| \frac{1}{n}\sum_{i=1}^{n} |f(X_i) - Y_i|^2 - \mathbb{E}\{|f(X) - Y|^2\} \right| > \varepsilon \right\} \\
&\le 24\left(\frac{192eL^2}{\varepsilon}\right)^{2(K_n(M+1)+1)} e^{-\frac{n\varepsilon^2}{2048L^4}},
\end{aligned} \tag{2.33}$$

where we have used $\log(x) \le x-1 \le x$ $(x \in \mathcal{R}_+)$.

Let $\tilde{Z}$ be a non–negative random variable and let $\varepsilon > 0$. Then

$$\mathbf{E}\{\tilde{Z}\} = \int_0^\infty \mathbf{P}\{\tilde{Z} > t\}\,dt \le \varepsilon + \int_\varepsilon^\infty \mathbf{P}\{\tilde{Z} > t\}\,dt.$$

Using this and (2.33) one gets

$$\begin{aligned}
&\mathbf{E}\left\{\sup_{f\in\mathcal{F}_n}\left|\frac{1}{n}\sum_{i=1}^n |f(X_i)-Y_i|^2 - \mathbf{E}\{|f(X)-Y|^2\}\right|\right\}\\
&\le \varepsilon + \int_\varepsilon^\infty 24\cdot\left(\frac{192eL^2}{t}\right)^{2(K_n(M+1)+1)}\cdot\exp\left(-\frac{n\cdot t^2}{2048L^4}\right)dt\\
&\le \varepsilon + 24\cdot\left(\frac{192eL^2}{\varepsilon}\right)^{2(K_n(M+1)+1)}\cdot\left[-\frac{2048L^4}{n\cdot\varepsilon}\cdot\exp\left(-\frac{n\cdot\varepsilon\cdot t}{2048L^4}\right)\right]_{t=\varepsilon}^{\infty}\\
&= \varepsilon + 24\cdot\left(\frac{192eL^2}{\varepsilon}\right)^{2(K_n(M+1)+1)}\cdot\frac{2048L^4}{n\cdot\varepsilon}\cdot\exp\left(-\frac{n\cdot\varepsilon^2}{2048L^4}\right)\\
&\to \varepsilon \quad (n\to\infty),
\end{aligned}$$

if (2.25) holds. With $\varepsilon \to 0$ one gets (2.28).

□

It is possible to modify the estimate such that it is weakly and strongly universally consistent. Multivariate piecewise polynomial partitioning estimates can be defined by using piecewise multivariate polynomials. For details see Györfi et al. (2002).

2.6 Consistency of penalized least squares estimates

In order to simplify the notation we will assume in the sequel that $X \in (0,1)$ *a.s.* and $|Y| \le L$ *a.s.* for some $L \in \mathcal{R}_+$. It is possible to show that after a minor modification of the definition of the estimate this assumption is no longer necessary and the resulting estimate is universally consistent (cf. Kohler and Krzyżak (2001)).

Let $k \in \mathcal{N}$ and let the estimate $\tilde{m}_n$ be defined via

$$\tilde{m}_n(\cdot) = \arg\min_{f\in C^k(\mathcal{R})}\left\{\frac{1}{n}\sum_{i=1}^n |f(X_i)-Y_i|^2 + \lambda_n J_k^2(f)\right\}, \tag{2.34}$$

where $\lambda_n > 0$ and

$$J_k^2(f) = \int_0^1 |f^{(k)}(x)|^2\,dx.$$

In order to show consistency of the estimate $\tilde{m}_n$, we will use the following error decomposition:

$$\begin{aligned}
&\int |\tilde{m}_n(x) - m(x)|^2 \mu(dx) \\
&= \mathbf{E}\{|\tilde{m}_n(X) - Y|^2 | D_n\} - \mathbf{E}\{|m(X) - Y|^2\} \\
&= \mathbf{E}\{|\tilde{m}_n(X) - Y|^2 | D_n\} - \frac{1}{n}\sum_{i=1}^n |\tilde{m}_n(X_i) - Y_i|^2 \\
&\quad + \frac{1}{n}\sum_{i=1}^n |\tilde{m}_n(X_i) - Y_i|^2 - \frac{1}{n}\sum_{i=1}^n |m(X_i) - Y_i|^2 \\
&\quad + \frac{1}{n}\sum_{i=1}^n |m(X_i) - Y_i|^2 - \mathbf{E}\{|m(X) - Y|^2\} \\
&=: T_{1,n} + T_{2,n} + T_{3,n}.
\end{aligned}$$

By the strong law of large numbers,

$$T_{3,n} = \frac{1}{n}\sum_{i=1}^n |m(X_i) - Y_i|^2 - \mathbf{E}\{|m(X) - Y|^2\} \to 0 \quad (n \to \infty) \quad a.s.$$

Furthermore, if $m \in C^k(\mathcal{R})$, then the definition of $\tilde{m}_n$ implies

$$\begin{aligned}
T_{2,n} &\leq \frac{1}{n}\sum_{i=1}^n |\tilde{m}_n(X_i) - Y_i|^2 + \lambda_n J_k^2(\tilde{m}_n) - \frac{1}{n}\sum_{i=1}^n |m(X_i) - Y_i|^2 \\
&\leq \frac{1}{n}\sum_{i=1}^n |m(X_i) - Y_i|^2 + \lambda_n J_k^2(m) - \frac{1}{n}\sum_{i=1}^n |m(X_i) - Y_i|^2 \\
&= \lambda_n J_k^2(m).
\end{aligned}$$

Hence, if $m \in C^k(\mathcal{R})$ and if we choose λ_n such that

$$\lambda_n \to 0 \quad (n \to \infty),$$

then

$$\limsup_{n\to\infty} T_{2,n} \leq 0.$$

Here the assumption $m \in C^k(\mathcal{R})$ can be avoided by approximating m by a smooth function (which we will do in the proof of Theorem 2.11 below).

Thus in order to obtain strong consistency of the estimate (2.34) we basically have to show

$$T_{1,n} = \mathbf{E}\{|\tilde{m}_n(X) - Y|^2 | D_n\} - \frac{1}{n}\sum_{i=1}^n |\tilde{m}_n(X_i) - Y_i|^2 \to 0 \quad (n \to \infty) \quad a.s. \tag{2.35}$$

By definition of the estimate,

$$\frac{1}{n}\sum_{i=1}^{n}|\tilde{m}_n(X_i)-Y_i|^2+\lambda_n J_k^2(\tilde{m}_n)$$
$$\le \frac{1}{n}\sum_{i=1}^{n}|0-Y_i|^2+\lambda_n\cdot 0$$
$$=\frac{1}{n}\sum_{i=1}^{n}|Y_i|^2 \to \mathbf{E}\{Y^2\} \quad (n\to\infty) \quad a.s.,$$

which implies that with probability one

$$J_k^2(\tilde{m}_n)\le \frac{2}{\lambda_n}\mathbf{E}\{Y^2\} \tag{2.36}$$

for n sufficiently large. Set

$$\mathcal{F}_n := \left\{ f\in C^k(\mathcal{R}) \,:\, J_k^2(f)\le \frac{2}{\lambda_n}\mathbf{E}\{Y^2\}\right\}.$$

Then $\tilde{m}_n \in \mathcal{F}_n$ for n sufficiently large and therefore (2.35) follows from

$$\sup_{f\in\mathcal{F}_n}\left|\mathbf{E}\{|f(X)-Y|^2\}-\frac{1}{n}\sum_{i=1}^{n}|f(X_i)-Y_i|^2\right| \to 0 \quad (n\to\infty) \quad a.s. \tag{2.37}$$

To show (2.37) we will use the results of Section 2.5. Recall that these results require that the random variables $|f(X)-Y|^2$ $(f\in\mathcal{F}_n)$ are bounded uniformly by some constant, which may depend on n. To ensure this, we will truncate our estimate, i.e., we will set

$$m_n(x)=T_L\tilde{m}_n(x)=\begin{cases} L & \text{if} \quad \tilde{m}_n(x)>L, \\ \tilde{m}_n(x) & \text{if} \quad -L\le \tilde{m}_n(x)\le L, \\ -L & \text{if} \quad \tilde{m}_n(x)<-L. \end{cases} \tag{2.38}$$

Because of $|Y|\le L$ *a.s.* we have $|m(x)|\le L$ $(x\in(0,1))$ which implies

$$|m_n(x)-m(x)|\le|\tilde{m}_n(x)-m(x)| \text{ for all } x\in(0,1).$$

All what we need then is an upper bound on the covering number of

$$\left\{T_L f \,:\, f\in C^k(\mathcal{R}) \text{ and } J_k^2(f)\le\frac{2}{\lambda_n}\mathbf{E}\{Y^2\}\right\}.$$

This bound will be given in the next lemma.

Lemma 2.5. *Let $L,c>0$ and set*

$$\mathcal{F}=\left\{T_L f \,:\, f\in C^k(\mathcal{R}) \text{ and } J_k^2(f)\le c\right\}.$$

Then for any $1 \le p < \infty$, $0 < \delta < 4L$ *and* $x_1, \ldots, x_n \in [0,1]$

$$\mathcal{N}_p(\delta, \mathcal{F}, x_1^n) \le \left(\frac{6e4^pL^pn}{\delta^p}\right)^{4(k+2)\cdot\left((\sqrt{c}/\delta)^{\frac{1}{k}}+1\right)}$$

PROOF. We proceed similarily as in the proof of Lemma 3.3.1 in van de Geer (1987). In the first step of the proof we approximate in the supremum norm the functions of $\mathcal{F}$ by piecewise polynomials. In the second step we bound the covering number of these piecewise polynomials.

Fix $g = T_L f \in \mathcal{F}$ where $f \in C^k(\mathcal{R})$, $J_k^2(f) \le c$. Choose

$$0 = u_0 < u_1 < \cdots < u_{\tilde{K}} = 1$$

such that

$$\int_{u_i}^{u_{i+1}} |f^{(k)}(x)|^2\,dx = c \cdot \left(\frac{\delta}{2\sqrt{c}}\right)^{\frac{1}{k}} \quad (i = 0, 1, \ldots, \tilde{K}-2),$$

and

$$\int_{u_{\tilde{K}-1}}^{u_{\tilde{K}}} |f^{(k)}(x)|^2\,dx \le c \cdot \left(\frac{\delta}{2\sqrt{c}}\right)^{\frac{1}{k}}.$$

Then

$$c \ge \int_0^1 |f^{(k)}(x)|^2\,dx = \sum_{j=0}^{\tilde{K}-1} \int_{u_i}^{u_{i+1}} |f^{(k)}(x)|^2\,dx \ge (\tilde{K}-1) \cdot c \cdot \left(\frac{\delta}{2\sqrt{c}}\right)^{\frac{1}{k}},$$

which implies

$$\tilde{K} \le \left(\frac{2\sqrt{c}}{\delta}\right)^{\frac{1}{k}} + 1 \le 2\left(\frac{\sqrt{c}}{\delta}\right)^{\frac{1}{k}} + 1.$$

By a refinement of the partition $\{[u_0, u_1), \ldots, [u_{\tilde{K}-1}, u_{\tilde{K}}]\}$ one can construct points

$$0 = v_0 < v_1 < \cdots < v_K = 1$$

such that

$$\int_{v_i}^{v_{i+1}} |f^{(k)}(x)|^2\,dx \le c \cdot \left(\frac{\delta}{2\sqrt{c}}\right)^{\frac{1}{k}} \quad (i = 0, 1, \ldots, K-1),$$

$$|v_{i+1} - v_i| \le \left(\frac{\delta}{2\sqrt{c}}\right)^{\frac{1}{k}} \quad (i = 0, 1, \ldots, K-1)$$

and

$$K \le 4\left(\frac{\sqrt{c}}{\delta}\right)^{\frac{1}{k}} + 2.$$

Let p_i be the Taylor polynomial of degree $k-1$ of f around v_i. For $v_i \le x \le v_{i+1}$ one gets

$$
\begin{aligned}
&|f(x)-p_i(x)|^2 \\
&= \left| \frac{1}{(k-1)!} \int_{v_i}^{x} (t-v_i)^{k-1} f^{(k)}(t)\,dt \right|^2 \\
&\le \frac{1}{(k-1)!^2} \int_{v_i}^{x} (t-v_i)^{2k-2}\,dt \cdot \int_{v_i}^{x} |f^{(k)}(t)|^2\,dt \\
&\le \frac{1}{(k-1)!^2} \frac{(v_{i+1}-v_i)^{2k-1}}{2k-1} \cdot \int_{v_i}^{v_{i+1}} |f^{(k)}(t)|^2\,dt \\
&\le \frac{1}{(k-1)!^2(2k-1)} \left(\frac{\delta}{2\sqrt{c}}\right)^{\frac{2k-1}{k}} \cdot c \cdot \left(\frac{\delta}{2\sqrt{c}}\right)^{\frac{1}{k}} \\
&\le \frac{\delta^2}{4}.
\end{aligned}
$$

Let $\mathcal{G}_{k-1}$ be the set of all polynomials of degree less than or equal to $k-1$ and let Π be the family of all partitions of $[0,1]$ into $K \le 4\left(\frac{\sqrt{c}}{\delta}\right)^{\frac{1}{k}} + 2$ intervals. We have shown that for each $g \in \mathcal{F}$ there exists $p \in \mathcal{G}_{k-1} \circ \Pi$ such that

$$\sup_{x\in[0,1]} |g(x) - T_L p(x)| \le \frac{\delta}{2}.$$

This implies

$$\mathcal{N}_p(\delta, \mathcal{F}, x_1^n) \le \mathcal{N}_p\left(\frac{\delta}{2}, T_L \mathcal{G}_{k-1} \circ \Pi, x_1^n\right). \tag{2.39}$$

By using some standard arguments for bounding covering numbers of piecewise defined functions (see Lemma 13.1 in Györfi et al. (2002)) we get

$$
\begin{aligned}
&\mathcal{N}_p\left(\frac{\delta}{2}, T_L \mathcal{G}_{k-1} \circ \Pi, x_1^n\right) \\
&\le n^{2(4(\sqrt{c}/\delta)^{\frac{1}{k}}+2)} \left\{ \sup_{z_1,\dots,z_l \in x_1^n, l \le n} \mathcal{N}_p\left(\frac{\delta}{2}, T_L \mathcal{G}_{k-1}, z_1^l\right) \right\}^{4(\sqrt{c}/\delta)^{\frac{1}{k}}+2} \\
&\le n^{2(4(\sqrt{c}/\delta)^{\frac{1}{k}}+2)} \cdot \left(3 \cdot \left(\frac{2e(2L)^p}{(\delta/2)^p}\right)^{2(k+1)}\right)^{4(\sqrt{c}/\delta)^{\frac{1}{k}}+2}
\end{aligned}
$$

The assertion follows from this and (2.39). □

Using this lemma and the results of the previous section we show next:

Lemma 2.6. *Let $k \in \mathcal{N}$ and for $n \in \mathcal{N}$ choose $\lambda_n > 0$ such that*

$$\lambda_n \to 0 \quad (n \to \infty) \tag{2.40}$$

and

$$\lambda_n \left(\frac{n}{\log(n)} \right)^{2k} \to \infty \quad (n \to \infty). \tag{2.41}$$

Let the estimate m_n be defined by (2.34) and (2.38). Then

$$\mathbf{E}\{|m_n(X) - Y|^2 | D_n\} - \frac{1}{n} \sum_{i=1}^{n} |m_n(X_i) - Y_i|^2 \to 0 \quad (n \to \infty) \quad a.s.$$

for every distribution of (X,Y) with $X \in (0,1)$ a.s. and $|Y| \leq L$ a.s.

PROOF. By definition of the estimate and strong law of large numbers

$$\frac{1}{n} \sum_{i=1}^{n} |\tilde{m}_n(X_i) - Y_i|^2 + \lambda_n J_k^2(\tilde{m}_n) \leq \frac{1}{n} \sum_{i=1}^{n} |0 - Y_i|^2 + \lambda_n \cdot 0 \to \mathbf{E}Y^2 \quad (n \to \infty) \quad a.s.,$$

which implies that with probability one we have for n sufficiently large

$$m_n \in \mathcal{F}_n = \left\{ T_L f \, : \, f \in C^k(\mathcal{R}) \text{ and } J_k^2(f) \leq \frac{2\mathbf{E}Y^2}{\lambda_n} \right\}.$$

Hence it suffices to show

$$\sup_{g \in \mathcal{G}_n} \left| \mathbf{E}\{g(X,Y)\} - \frac{1}{n} \sum_{i=1}^{n} g(X_i, Y_i) \right| \to 0 \quad (n \to \infty) \quad a.s., \tag{2.42}$$

where $\mathcal{G}_n = \left\{ g : \mathcal{R}^d \times \mathcal{R} \to \mathcal{R} \, : \, g(x,y) = |f(x) - T_L y|^2 \text{ for some } f \in \mathcal{F}_n \right\}$.

If $g_j(x,y) = |f_j(x) - T_L y|^2$ $((x,y) \in \mathcal{R}^d \times \mathcal{R})$ for some function f_j bounded in absolute value by L $(j = 1,2)$, then

$$\frac{1}{n} \sum_{i=1}^{n} |g_1(X_i, Y_i) - g_2(X_i, Y_i)| \leq 4L \frac{1}{n} \sum_{i=1}^{n} |f_1(X_i) - f_2(X_i)|,$$

which implies

$$\mathcal{N}_1 \left(\frac{\varepsilon}{8}, \mathcal{G}_n, (X,Y)_1^n \right) \leq \mathcal{N}_1 \left(\frac{\varepsilon}{32L}, \mathcal{F}_n, X_1^n \right).$$

Using this, Theorem 2.7 and Lemma 2.5, one gets for $\varepsilon > 0$ arbitrary

$$\mathbf{P} \left\{ \sup_{g \in \mathcal{G}_n} \left| \mathbf{E}\{g(X,Y)\} - \frac{1}{n} \sum_{i=1}^{n} g(X_i, Y_i) \right| > \varepsilon \right\}$$

$$\leq 8 \left(\frac{c_1 L n}{\frac{\varepsilon}{32L}} \right)^{c_2 \left(\frac{\sqrt{2 \cdot \mathbf{E}Y^2} 32L}{\sqrt{\lambda_n} \varepsilon} \right)^{\frac{1}{k}} + c_3} \exp \left(- \frac{n\varepsilon^2}{128(4L^2)^2} \right)$$

$$= 8 \exp \left(\left(c_2 \left(\frac{\sqrt{2 \cdot \mathbf{E}Y^2} 32}{\varepsilon} \frac{L}{\sqrt{\lambda_n}} \right)^{\frac{1}{k}} + c_3 \right) \cdot \log \left(\frac{32 c_1 L^2 n}{\varepsilon} \right) - \frac{n\varepsilon^2}{2048 L^4} \right)$$

$$\leq 8 \exp \left(- \frac{1}{2} \cdot \frac{n\varepsilon^2}{2048 L^4} \right)$$

for n sufficiently large, where we have used that (2.41) implies

$$\frac{(1/\sqrt{\lambda_n})^{\frac{1}{k}}\log(n)}{n} = \left(\left(\frac{1}{\lambda_n}\right)\cdot\left(\frac{\log(n)}{n}\right)^{2k}\right)^{\frac{1}{2k}} \to 0 \quad (n\to\infty).$$

From this one gets the assertion by an application of the Borel–Cantelli lemma. □

We are now ready to formulate and proof our main result.

Theorem 2.11. (KOHLER AND KRZYŻAK (2001)) *Let $k \in \mathcal{N}$ and for $n \in \mathcal{N}$ choose $\lambda_n > 0$ such that (2.40) and (2.41) hold. Let the estimate m_n be defined by (2.34) and (2.38). Then*

$$\int |m_n(x) - m(x)|^2 \mu(dx) \to 0 \quad (n\to\infty) \quad a.s.$$

for every distribution of (X,Y) with $X \in (0,1)$ a.s. and $|Y| \le L$ a.s..

PROOF. Let $\varepsilon > 0$ be arbitray. Choose $g_\varepsilon \in C^k(\mathcal{R})$ such that

$$\int |m(x) - g_\varepsilon(x)|^2 \mu(dx) < \varepsilon \text{ and } J_k^2(g_\varepsilon) < \infty.$$

We use the following error decomposition:

$$\begin{aligned}
\int |m_n(x) - m(x)|^2 \mu(dx) &= \mathbb{E}\{|m_n(X) - Y|^2|D_n\} - \mathbb{E}\{|m(X)-Y|^2\} \\
&= \mathbb{E}\{|m_n(X) - Y|^2|D_n\} - \frac{1}{n}\sum_{i=1}^n |m_n(X_i) - Y_i|^2 \\
&\quad + \frac{1}{n}\sum_{i=1}^n |m_n(X_i) - Y_i|^2 - \frac{1}{n}\sum_{i=1}^n |\tilde{m}_n(X_i) - Y_i|^2 \\
&\quad + \frac{1}{n}\sum_{i=1}^n |\tilde{m}_n(X_i) - Y_i|^2 - \frac{1}{n}\sum_{i=1}^n |g_\varepsilon(X_i) - Y_i|^2 \\
&\quad + \frac{1}{n}\sum_{i=1}^n |g_\varepsilon(X_i) - Y_i|^2 - \mathbb{E}\{|g_\varepsilon(X) - Y|^2\} \\
&\quad + \mathbb{E}\{|g_\varepsilon(X) - Y|^2\} - \mathbb{E}\{|m(X)-Y|^2\} \\
&= \sum_{j=1}^5 T_{j,n}.
\end{aligned}$$

By Lemma 2.6,

$$T_{1,n} \to 0 \quad (n\to\infty) \quad a.s.$$

Furthermore, if $x, y \in \mathcal{R}$ with $|y| \le L$ and $z = T_L x$, then $|z - y| \le |x - y|$, which implies

$$T_{2,n} \le 0.$$

It follows from the definition of the estimate and (2.40) that

$$T_{3,n} \leq \lambda_n J_k^2(g_\varepsilon) - \lambda_n J_k^2(\tilde{m}_n) \leq \lambda_n J_k^2(g_\varepsilon) \to 0 \quad (n \to \infty).$$

By the strong law of large numbers,

$$T_{4,n} \to 0 \quad (n \to \infty) \quad a.s.$$

Finally,

$$T_{5,n} = \int |g_\varepsilon(x) - m(x)|^2 \mu(dx) \leq \varepsilon.$$

Using this one concludes

$$\limsup_{n \to \infty} \int |m_n(x) - m(x)|^2 \mu(dx) \leq \varepsilon \quad a.s.$$

With $\varepsilon \to 0$ the result follows. □

2.7 Rate of convergence of least squares estimates

Throughout this section we will assume $|Y| \leq L$ a.s. and the estimate m_n will be defined by

$$m_n(\cdot) = \arg \min_{f \in \mathcal{F}_n} \frac{1}{n} \sum_{i=1}^{n} |f(X_i) - Y_i|^2, \tag{2.43}$$

where $\mathcal{F}_n$ is some set of functions $f : \mathcal{R}^d \to [-L, L]$. Let us first try to apply the results which we have derived in order to show the consistency of the estimate: it follows from Lemma 2.3 that

$$\mathbb{E} \int |m_n(x) - m(x)|^2 \mu(dx)$$

is bounded by

$$2\mathbb{E} \sup_{f \in \mathcal{F}_n} \left| \frac{1}{n} \sum_{j=1}^{n} |f(X_j) - Y_j|^2 - \mathbb{E}\{(f(X) - Y)^2\} \right|$$
$$+ \inf_{f \in \mathcal{F}_n} \int |f(x) - m(x)|^2 \mu(dx).$$

To bound the first term we can apply Lemma 2.4 or Theorem 2.7, where we have used Hoeffding's inequality on fixed sup–norm and random L_1 norm covers, resp. In both cases we have bounded the probability

$$\mathbb{P}\left\{ \sup_{f \in \mathcal{F}_n} \left| \frac{1}{n} \sum_{j=1}^{n} |f(X_j) - Y_j|^2 - \mathbb{E}\{|f(X) - Y|^2\} \right| > \varepsilon_n \right\}$$

by some term tending to infinity as $n \to \infty$ times

$$\exp\left(-c\frac{n\varepsilon_n^2}{L^4}\right). \tag{2.44}$$

Thus, if we want these upper bounds converge to zero as $n \to \infty$ then ε_n must converge to zero not faster than $n^{-\frac{1}{2}}$. Unfortunately, this is far away from the optimal rate of convergence.

Therefore, to analyze the rate of convergence of the expected value of the L_2 error, we will use a different decomposition than in Section 2.5:

$$\begin{aligned}
&\int |m_n(x) - m(x)|^2 \mu(dx) \\
&= \left\{ \mathbb{E}\{|m_n(X) - Y|^2 | D_n\} - \mathbb{E}|m(X) - Y|^2 \right. \\
&\qquad \left. -2\frac{1}{n}\sum_{i=1}^{n}\{|m_n(X_i) - Y_i|^2 - |m(X_i) - Y_i|^2\} \right\} \\
&+2\frac{1}{n}\sum_{i=1}^{n}\{|m_n(X_i) - Y_i|^2 - |m(X_i) - Y_i|^2\}
\end{aligned} \tag{2.45}$$

Let us first observe that we can get a nice upper bound for the expectation of the second term on the right side of (2.45). For simplicity we ignore the factor 2. By using the definition of m_n one gets

$$\begin{aligned}
&\mathbb{E}\left\{\frac{1}{n}\sum_{i=1}^{n}\{|m_n(X_i) - Y_i|^2 - |m(X_i) - Y_i|^2\}\right\} \\
&= \mathbb{E}\left\{\inf_{f \in \mathcal{F}_n}\frac{1}{n}\sum_{i=1}^{n}\{|f(X_i) - Y_i|^2 - |m(X_i) - Y_i|^2\}\right\} \\
&\le \inf_{f \in \mathcal{F}_n}\mathbb{E}\left\{\frac{1}{n}\sum_{i=1}^{n}\{|f(X_i) - Y_i|^2 - |m(X_i) - Y_i|^2\}\right\} \\
&= \inf_{f \in \mathcal{F}_n}\mathbb{E}|f(X) - Y|^2 - \mathbb{E}|m(X) - Y|^2 \\
&= \inf_{f \in \mathcal{F}_n}\int |f(x) - m(x)|^2 \mu(dx).
\end{aligned} \tag{2.46}$$

Next we derive an upper bound for the first term on the right–hand side of (2.45). We have

$$\mathbb{P}\Big\{\mathbb{E}\{|m_n(X)-Y|^2|D_n\}-\mathbb{E}|m(X)-Y|^2 - 2\frac{1}{n}\sum_{i=1}^n\{|m_n(X_i)-Y_i|^2-|m(X_i)-Y_i|^2\}>\varepsilon\Big\}$$

$$=\mathbb{P}\Big\{\mathbb{E}\{|m_n(X)-Y|^2|D_n\}-\mathbb{E}|m(X)-Y|^2 -\frac{1}{n}\sum_{i=1}^n\{|m_n(X_i)-Y_i|^2-|m(X_i)-Y_i|^2\} >\frac{\varepsilon}{2}+\frac{1}{2}\{\mathbb{E}\{|m_n(X)-Y|^2|D_n\}-\mathbb{E}|m(X)-Y|^2\}\Big\}$$

$$\leq \mathbb{P}\Big\{\exists f\in\mathcal{F}_n:\mathbb{E}|f(X)-Y|^2-\mathbb{E}|m(X)-Y|^2 -\frac{1}{n}\sum_{i=1}^n\{|f(X_i)-Y_i|^2-|m(X_i)-Y_i|^2\} >\frac{\varepsilon}{2}+\frac{1}{2}\{\mathbb{E}|f(X)-Y|^2-\mathbb{E}|m(X)-Y|^2\}\Big\}. \quad (2.47)$$

For simplicity let us consider for a moment the right–hand side of (2.47) for only one fixed function $f\in\mathcal{F}_n$. Set $Z=(X,Y)$, $Z_i=(X_i,Y_i)$ $(i=1,\dots,n)$,

$$g(Z)=|f(X)-Y|^2-|m(X)-Y|^2$$

and

$$g(Z_i)=|f(X_i)-Y_i|^2-|m(X_i)-Y_i|^2 \quad (i=1,...,n).$$

Then $g(Z),g(Z_1),...,g(Z_n)$ are i.i.d random variables such that $|g(Z)|\leq 4L^2$ and we want to bound

$$\mathbb{P}\left\{\mathbb{E}g(Z)-\frac{1}{n}\sum_{i=1}^n g(Z_i)>\frac{\varepsilon}{2}+\frac{1}{2}\mathbb{E}g(Z)\right\}. \quad (2.48)$$

The main trick is that the variance of $g(Z)$ is bounded by some constant times the expectation of $g(Z)$: Indeed,

$$\begin{aligned} g(Z) &= (f(X)-Y+m(X)-Y)((f(X)-Y)-(m(X)-Y)) \\ &= (f(X)+m(X)-2Y)(f(X)-m(X)) \end{aligned}$$

and thus

$$\sigma^2=\mathbf{Var}(g(Z))\leq \mathbb{E}g(Z)^2\leq 16L^2\,\mathbb{E}|f(X)-m(X)|^2 = 16L^2\,(\mathbb{E}|f(X)-Y|^2-\mathbb{E}|m(X)-Y|^2)=16L^2\,\mathbb{E}g(Z). \quad (2.49)$$

This enables us to derive an excellent bound for (2.48) by applying Bernstein inequality:

$$\begin{aligned}
&\mathbb{P}\left\{\mathbb{E}g(Z)-\frac{1}{n}\sum_{i=1}^{n}g(Z_i)>\frac{\varepsilon}{2}+\frac{1}{2}\mathbb{E}g(Z)\right\}\\
&\leq \mathbb{P}\left\{\mathbb{E}g(Z)-\frac{1}{n}\sum_{i=1}^{n}g(Z_i)>\frac{\varepsilon}{2}+\frac{1}{2}\frac{\sigma^2}{16L^2}\right\}\\
&\leq \exp\left(-\frac{n\{\frac{\varepsilon}{2}+\frac{\sigma^2}{32L^2}\}^2}{2\sigma^2+2\frac{8L^2}{3}\left\{\frac{\varepsilon}{2}+\frac{\sigma^2}{32L^2}\right\}}\right)\\
&\leq \exp\left(-\frac{n\{\frac{\varepsilon}{2}+\frac{\sigma^2}{32L^2}\}^2}{\left(64L^2+\frac{16L^2}{3}\right)\cdot\left\{\frac{\varepsilon}{2}+\frac{\sigma^2}{32L^2}\right\}}\right)\\
&= \exp\left(-\frac{n\left\{\frac{\varepsilon}{2}+\frac{\sigma^2}{32L^2}\right\}}{64L^2+\frac{16}{3}L^2}\right)\\
&\leq \exp\left(-\frac{1}{128+\frac{32}{3}}\cdot\frac{n\varepsilon}{L^2}\right).
\end{aligned}$$

The main advantage of this upper bound is that ε appears only in linear and not squared form as in (2.44) and therefore this upper bound converges to zero whenever

$$\varepsilon n = \varepsilon_n n \to \infty \quad \text{as } n \to \infty.$$

What we now need is an extension of this upper bound from the case of a fixed function to the general case of a set of functions as in (2.47). The result is summarized in the next theorem.

Theorem 2.12. (LEE, BARTLETT AND WILLIAMSON (1996)).
Assume $|Y| \leq B$ a.s. and $B \geq 1$. Let $\mathcal{F}$ be a set of functions $f : \mathcal{R}^d \to \mathcal{R}$ and let $|f(x)| \leq B, B \geq 1$. Then for each $n \geq 1$

$$\begin{aligned}
&\mathbb{P}[\exists f \in \mathcal{F}:\\
&\qquad \frac{\mathbb{E}|f(X)-Y|^2-\mathbb{E}|m(X)-Y|^2-\frac{1}{n}\sum_{i=1}^{n}\{|f(X_i)-Y_i|^2-|m(X_i)-Y_i|^2\}}{\alpha+\beta+\mathbb{E}|f(X)-Y|^2-\mathbb{E}|m(X)-Y|^2}\geq\varepsilon\Bigg]\\
&\leq 14\sup_{x_1^n}\mathcal{N}_1\left(\frac{\beta\varepsilon}{20B},\mathcal{F},x_1^n\right)\exp\left(-\frac{\varepsilon^2(1-\varepsilon)\alpha n}{214(1+\varepsilon)B^4}\right)
\end{aligned}$$

where $\alpha, \beta > 0$ and $0 < \varepsilon \leq 1/2$.

In the proof of this theorem one uses single cover of some fixed radius to bound the differences between expectations and sample means. As it turns out, due to this one gets an additional log–factor in the rate of convergence. In the sequel we will use the so-called chaining technique, which introduces a sequence of covers of increasing cardinality. There one has to control a sum of covering numbers instead of one fixed covering number, and this sum will be bounded above by an integral of covering numbers (cf (2.51)).

Theorem 2.13. (VAN DE GEER (1990, 2000)) *Let $L \in \mathcal{R}_+$ and let $\varepsilon_1, \ldots, \varepsilon_n$ be independent random variables with expectation zero and values in $[-L, L]$. Let $z_1, \ldots, z_n \in \mathcal{X}$, let $R > 0$ and let $\mathcal{F}$ be a class of functions $f : \mathcal{X} \to \mathcal{R}$ with the property*

$$\|f\|_n^2 := \frac{1}{n}\sum_{i=1}^n |f(z_i)|^2 \le R^2 \quad (f \in \mathcal{F}). \tag{2.50}$$

Then

$$\sqrt{n}\delta \ge 48\sqrt{2}L \int_{\frac{\delta}{8L}}^{\frac{R}{2}} \left(\log \mathcal{N}_2(u, \mathcal{F}, z_1^n)\right)^{1/2} du \tag{2.51}$$

and

$$\sqrt{n}\delta \ge 36R \cdot L$$

imply

$$\mathbf{P}\left\{\sup_{f\in\mathcal{F}}\left|\frac{1}{n}\sum_{i=1}^n f(z_i)\cdot\varepsilon_i\right| \ge \delta\right\} \le 5\exp\left(-\frac{n\delta^2}{2304L^2R^2}\right).$$

PROOF. Let

$$\mathcal{F} = \{f_\theta \quad : \quad \theta \in \Theta\}.$$

For $s \in \mathcal{N}_0$ let $\{f_1^s, \ldots, f_{N_s}^s\}$ be a $\|\cdot\|_n$-cover of $\mathcal{F}$ of radius $\frac{R}{2^s}$ of size

$$N_s = \mathcal{N}_2\left(\frac{R}{2^s}, \mathcal{F}, z_1^n\right).$$

Because of (2.50) we can assume w.l.o.g. $f_1^0 = 0$ and $N_0 = 1$. For $f_\theta \in \mathcal{F}$ choose

$$f_\theta^s \in \{f_1^s, \ldots, f_{N_s}^s\}$$

such that

$$\|f_\theta - f_\theta^s\|_n \le \frac{R}{2^s}.$$

Set

$$S = \min\left\{s \ge 1 : \frac{R}{2^s} \le \frac{\delta}{2L}\right\}.$$

Because of

$$f_\theta = f_\theta - f_\theta^0 = f_\theta - f_\theta^S + \sum_{s=1}^{S} (f_\theta^s - f_\theta^{s-1})$$

we get by definition of S

$$\begin{aligned}
&\left| \frac{1}{n} \sum_{i=1}^{n} f_\theta(z_i) \cdot \varepsilon_i \right| \\
&= \left| \frac{1}{n} \sum_{i=1}^{n} (f_\theta(z_i) - f_\theta^S(z_i)) \cdot \varepsilon_i + \sum_{s=1}^{S} \frac{1}{n} \sum_{i=1}^{n} (f_\theta^s(z_i) - f_\theta^{s-1}(z_i)) \cdot \varepsilon_i \right| \\
&\leq \left| \frac{1}{n} \sum_{i=1}^{n} (f_\theta(z_i) - f_\theta^S(z_i)) \cdot \varepsilon_i \right| + \sum_{s=1}^{S} \left| \frac{1}{n} \sum_{i=1}^{n} (f_\theta^s(z_i) - f_\theta^{s-1}(z_i)) \cdot \varepsilon_i \right| \\
&\leq \|f_\theta - f_\theta^S\|_n \cdot \sqrt{\frac{1}{n} \sum_{i=1}^{n} \varepsilon_i^2} + \sum_{s=1}^{S} \left| \frac{1}{n} \sum_{i=1}^{n} (f_\theta^s(z_i) - f_\theta^{s-1}(z_i)) \cdot \varepsilon_i \right| \\
&\leq \frac{R}{2^S} \cdot L + \sum_{s=1}^{S} \left| \frac{1}{n} \sum_{i=1}^{n} (f_\theta^s(z_i) - f_\theta^{s-1}(z_i)) \cdot \varepsilon_i \right| \\
&\leq \frac{\delta}{2} + \sum_{s=1}^{S} \left| \frac{1}{n} \sum_{i=1}^{n} (f_\theta^s(z_i) - f_\theta^{s-1}(z_i)) \cdot \varepsilon_i \right|.
\end{aligned}$$

Therefore for any $\eta_1, \dots, \eta_S \geq 0$, $\eta_1 + \dots + \eta_S \leq 1$:

$$\begin{aligned}
&\mathbb{P}\left[\sup_{f \in \mathcal{F}} \left| \frac{1}{n} \sum_{i=1}^{n} f(z_i) \cdot \varepsilon_i \right| > \delta \right] \\
&\leq \mathbb{P}\left[\exists \theta \in \Theta : \frac{\delta}{2} + \sum_{s=1}^{S} \left| \frac{1}{n} \sum_{i=1}^{n} (f_\theta^s(z_i) - f_\theta^{s-1}(z_i)) \cdot \varepsilon_i \right| > \frac{\delta}{2} + \sum_{s=1}^{S} \eta_s \cdot \frac{\delta}{2} \right] \\
&\leq \sum_{s=1}^{S} \mathbb{P}\left[\exists \theta \in \Theta : \left| \frac{1}{n} \sum_{i=1}^{n} (f_\theta^s(z_i) - f_\theta^{s-1}(z_i)) \cdot \varepsilon_i \right| > \eta_s \cdot \frac{\delta}{2} \right] \\
&\leq \sum_{s=1}^{S} N_s \cdot N_{s-1} \cdot \max_{\theta \in \Theta} \mathbb{P}\left[\left| \frac{1}{n} \sum_{i=1}^{n} (f_\theta^s(z_i) - f_\theta^{s-1}(z_i)) \cdot \varepsilon_i \right| > \eta_s \cdot \frac{\delta}{2} \right].
\end{aligned}$$

Fix $s \in \{1, \dots, S\}$ and $\theta \in \Theta$. The random variables

$$(f_\theta^s(z_1) - f_\theta^{s-1}(z_1)) \cdot \varepsilon_1, \dots, (f_\theta^s(z_n) - f_\theta^{s-1}(z_n)) \cdot \varepsilon_n$$

are independent, have zero mean and take values in

$$[-L \cdot |f_\theta^s(z_i) - f_\theta^{s-1}(z_i)|, L \cdot |f_\theta^s(z_i) - f_\theta^{s-1}(z_i)|] \quad (i = 1, \dots, n).$$

Therefore

$$\begin{aligned}
&\frac{1}{n}\sum_{i=1}^{n}\left(2L\cdot|f_\theta^s(z_i)-f_\theta^{s-1}(z_i)|\right)^2\\
&=4L^2\|f_\theta^s-f_\theta^{s-1}\|_n^2\\
&\le 4L^2\left(\|f_\theta^s-f_\theta\|_n+\|f_\theta-f_\theta^{s-1}\|_n\right)^2\\
&\le 4L^2\left(\frac{R}{2^s}+\frac{R}{2^{s-1}}\right)^2=36\frac{R^2L^2}{2^{2s}}
\end{aligned}$$

together with Hoeffding inequality imply

$$\mathbb{P}\left[\left|\frac{1}{n}\sum_{i=1}^{n}(f_\theta^s(z_i)-f_\theta^{s-1}(z_i))\cdot\varepsilon_i\right|>\eta_s\cdot\frac{\delta}{2}\right]\le 2\exp\left(-\frac{2n(\frac{\eta_s\delta}{2})^2}{36\frac{R^2L^2}{2^{2s}}}\right).$$

It follows

$$\begin{aligned}
&\mathbb{P}\left[\sup_{f\in\mathcal{F}}\left|\frac{1}{n}\sum_{i=1}^{n}f(z_i)\cdot\varepsilon_i\right|>\delta\right]\\
&\le\sum_{s=1}^{S}2N_s^2\exp\left(-\frac{n\delta^2\eta_s^2 2^{2s}}{72R^2L^2}\right)\\
&=\sum_{s=1}^{S}2\exp\left(2\log N_s-\frac{n\delta^2\eta_s^2 2^{2s}}{72R^2L^2}\right).
\end{aligned}$$

In order to get rid of the covering number N_s, we choose η_s such that

$$2\log N_s\le\frac{1}{2}\cdot\frac{n\delta^2\eta_s^2 2^{2s}}{72R^2L^2}, \tag{2.52}$$

which is equivalent to

$$\eta_s\ge\bar{\eta}_s:=\frac{12\cdot\sqrt{2}\cdot R\cdot L}{2^s\delta\sqrt{n}}\cdot\{\log N_s\}^{1/2}.$$

More precisely, we set

$$\eta_s:=\max\left\{\frac{2^{-s}\sqrt{s}}{4},\bar{\eta}_s\right\}.$$

Because of

$$\sum_{s=1}^{S}\frac{2^{-s}\sqrt{s}}{4}\le\frac{1}{8}\sum_{s=1}^{\infty}s\cdot(\tfrac{1}{2})^{s-1}=\frac{1}{8}\frac{1}{(1-\frac{1}{2})^2}=\frac{1}{2}$$

and

$$\begin{aligned}
\sum_{s=1}^{S} \bar{\eta}_s &= \sum_{s=1}^{S} \frac{24\sqrt{2}L}{\delta\sqrt{n}} \cdot \frac{R}{2^{s+1}} \{\log \mathcal{N}_2(\frac{R}{2^s}, \mathcal{F}, z_1^n)\}^{1/2} \\
&\leq \sum_{s=1}^{S} \frac{24\sqrt{2}L}{\delta\sqrt{n}} \cdot \int_{R/2^{s+1}}^{R/2^s} \{\log \mathcal{N}_2(u, \mathcal{F}, z_1^n)\}^{1/2} du \\
&\leq \frac{24\sqrt{2}L}{\delta\sqrt{n}} \cdot \int_{R/2^{S+1}}^{R/2} \{\log \mathcal{N}_2(u, \mathcal{F}, z_1^n)\}^{1/2} du \\
&\leq \frac{1}{2},
\end{aligned}$$

where the last inequality follows from (2.51) and $R/2^{S+1} \geq \delta/(8L)$, we get

$$\sum_{s=1}^{S} \eta_s \leq \sum_{s=1}^{S} \frac{2^{-s}\sqrt{s}}{4} + \sum_{s=1}^{S} \bar{\eta}_s \leq 1.$$

Furthermore, $\eta_s \geq \bar{\eta}_s$ implies (2.52), from which we can conclude

$$\begin{aligned}
&\mathbb{P}\left[\sup_{f \in \mathcal{F}} \left| \frac{1}{n} \sum_{i=1}^{n} f(z_i) \cdot \varepsilon_i \right| > \delta \right] \\
&\leq \sum_{s=1}^{S} 2\exp\left(-\frac{n\delta^2 \eta_s^2 2^{2s}}{144 L^2 R^2}\right) \\
&\leq \sum_{s=1}^{S} 2\exp\left(-\frac{n\delta^2}{16 \cdot 144 L^2 R^2} \cdot s\right) \\
&\leq \frac{2}{1 - \exp\left(-\frac{n\delta^2}{16 \cdot 144 L^2 R^2}\right)} \cdot \exp\left(-\frac{n\delta^2}{16 \cdot 144 L^2 R^2}\right).
\end{aligned}$$

Now,

$$\frac{n\delta^2}{16 \cdot 144 L^2 R^2} \geq \frac{36^2 L^2 R^2}{16 \cdot 144 L^2 R^2} = \frac{9}{16}$$

implies

$$\frac{2}{1 - \exp\left(-\frac{n\delta^2}{16 \cdot 144 L^2 R^2}\right)} \leq \frac{2}{1 - \exp\left(-\frac{9}{16}\right)} \leq 5$$

which in turn implies the assertion. □

In the squel we show the following modification of Theorem 2.12.

Theorem 2.14. (KOHLER (2000)) *Let $Z, Z_1, \ldots, Z_n$ be independent identically distributed random variables with values in some set $\mathcal{X}$. Let $K_1, K_2 \geq 1$ and let $\mathcal{F}$ be a class of functions $f : \mathcal{X} \to \mathcal{R}$ with the properties*

$$|f(z)| \leq K_1 \quad (z \in \mathcal{X}) \quad \textit{and} \quad \mathbb{E}\{f(Z)^2\} \leq K_2 \mathbb{E}\{f(Z)\}.$$

Let $0 < \varepsilon < 1$ *and* $\alpha > 0$. *Assume that*

$$\sqrt{n}\varepsilon\sqrt{1-\varepsilon}\sqrt{\alpha} \geq 288\max\{2K_1, \sqrt{2K_2}\} \tag{2.53}$$

and that for all $z_1, \ldots, z_n \in \mathcal{X}$ *and all* $\delta \geq \frac{\alpha}{4}$

$$\sqrt{n}\varepsilon(1-\varepsilon)\delta \geq$$
$$288\max\{K_1, 2K_2\} \int_0^{\sqrt{\delta}} \left(\log \mathcal{N}_2\left(u, \left\{f \in \mathcal{F} : \frac{1}{n}\sum_{i=1}^n f(z_i)^2 \leq 4\delta\right\}, z_1^n\right)\right)^{1/2} du. \tag{2.54}$$

Then

$$\mathbb{P}\left\{\sup_{f\in\mathcal{F}} \frac{\left|\mathbb{E}\{f(Z)\} - \frac{1}{n}\sum_{i=1}^n f(Z_i)\right|}{\alpha + \mathbb{E}\{f(Z)\}} > \varepsilon\right\} \leq 50\exp\left(-\frac{n\alpha\varepsilon^2(1-\varepsilon)}{128\cdot 2304\max\{K_1^2, K_2\}}\right). \tag{2.55}$$

PROOF. The proof is divided into 6 steps.
STEP 1. Replace the expectation in the nominator of (2.55) by an empirical mean.
Draw a "ghost" sample $Z_1'^n = (Z_1', \ldots, Z_n')$ of i.i.d. random variables distributed as Z_1 and independent of Z_1^n. Let $f^* = f^*(Z_1^n)$ be a function $f \in \mathcal{F}$ such that

$$\left|\frac{1}{n}\sum_{i=1}^n f(Z_i) - \mathbb{E}\{f(Z)\}\right| > \varepsilon(\alpha + \mathbb{E}\{f(Z)\}),$$

if there exists any such function, and let f^* be an arbitrary other function contained in $\mathcal{F}$, if there doesn't exist such a function.

We have

$$\mathbb{P}\left\{\exists f\in\mathcal{F}:\left|\frac{1}{n}\sum_{i=1}^{n}f(Z_i)-\frac{1}{n}\sum_{i=1}^{n}f(Z_i')\right|>\frac{\varepsilon}{2}\alpha+\frac{\varepsilon}{2}\mathbb{E}\{f(Z)\}\right\}$$

$$\geq\mathbb{P}\left\{\left|\frac{1}{n}\sum_{i=1}^{n}f^*(Z_i)-\frac{1}{n}\sum_{i=1}^{n}f^*(Z_i')\right|>\frac{\varepsilon}{2}\alpha+\frac{\varepsilon}{2}\mathbb{E}\{f^*(Z)|Z_1^n\}\right\}$$

$$\geq\mathbb{P}\left\{\left|\frac{1}{n}\sum_{i=1}^{n}f^*(Z_i)-\mathbb{E}\{f^*(Z)|Z_1^n\}\right|>\varepsilon\alpha+\varepsilon\mathbb{E}\{f^*(Z)|Z_1^n\},\right.$$

$$\left.\left|\frac{1}{n}\sum_{i=1}^{n}f^*(Z_i')-\mathbb{E}\{f^*(Z)|Z_1^n\}\right|\leq\frac{\varepsilon}{2}\alpha+\frac{\varepsilon}{2}\mathbb{E}\{f^*(Z)|Z_1^n\}\right\}$$

$$=\mathbb{E}\left\{I_{\{|\frac{1}{n}\sum_{i=1}^{n}f^*(Z_i)-\mathbb{E}\{f^*(Z)|Z_1^n\}|>\varepsilon\alpha+\varepsilon\mathbb{E}\{f^*(Z)|Z_1^n\}\}}\cdot\right.$$

$$\left.\mathbb{P}\left\{\left|\frac{1}{n}\sum_{i=1}^{n}f^*(Z_i')-\mathbb{E}\{f^*(Z)|Z_1^n\}\right|\leq\frac{\varepsilon}{2}\alpha+\frac{\varepsilon}{2}\mathbb{E}\{f^*(Z)|Z_1^n\}\middle|Z_1^n\right\}\right\}.$$

Chebychev inequality together with

$$0\leq\mathbb{E}\left\{f^*(Z)^2|Z_1^n\right\}\leq K_2\mathbb{E}\{f^*(Z)|Z_1^n\}$$

imply that for $n\geq\frac{20K_2}{\varepsilon^2\alpha}$

$$\mathbb{P}\left\{\left|\frac{1}{n}\sum_{i=1}^{n}f^*(Z_i')-\mathbb{E}\{f^*(Z)|Z_1^n\}\right|>\frac{\varepsilon}{2}\alpha+\frac{\varepsilon}{2}\mathbb{E}\{f^*(Z)|Z_1^n\}\middle|Z_1^n\right\}$$

$$\leq\frac{K_2\mathbb{E}\{f^*(Z)|Z_1^n\}}{n\left(\frac{\varepsilon}{2}\alpha+\frac{\varepsilon}{2}\mathbb{E}\{f^*(Z)|Z_1^n\}\right)^2}\leq\frac{2K_2}{n\varepsilon^2\alpha}\leq\frac{1}{10}.$$

Thus for $n\geq\frac{20K_2}{\varepsilon^2\alpha}$

$$\mathbb{P}\left\{\exists f\in\mathcal{F}:\left|\frac{1}{n}\sum_{i=1}^{n}f(Z_i)-\frac{1}{n}\sum_{i=1}^{n}f(Z_i')\right|>\frac{\varepsilon}{2}\alpha+\frac{\varepsilon}{2}\mathbb{E}\{f(Z)\}\right\}$$

$$\geq\frac{9}{10}\mathbb{P}\left\{\left|\frac{1}{n}\sum_{i=1}^{n}f^*(Z_i)-\mathbb{E}\{f^*(Z)|Z_1^n\}\right|>\varepsilon\alpha+\varepsilon\mathbb{E}\{f^*(Z)|Z_1^n\}\right\}\Bigg\}$$

$$=\frac{9}{10}\mathbb{P}\left\{\exists f\in\mathcal{F}:\left|\frac{1}{n}\sum_{i=1}^{n}f(Z_i)-\mathbb{E}\{f(Z)\}\right|>\varepsilon\alpha+\varepsilon\mathbb{E}\{f(Z)\}\right\}\Bigg\},$$

which implies

$$\mathbb{P}\left\{\sup_{f\in\mathcal{F}} \frac{\left|\frac{1}{n}\sum_{i=1}^{n} f(Z_i)-\mathbb{E}\{f(Z)\}\right|}{\alpha+\mathbb{E}\{f(Z)\}}>\varepsilon\right\}$$

$$\leq \frac{10}{9}\mathbb{P}\left\{\exists f\in\mathcal{F}:\left|\frac{1}{n}\sum_{i=1}^{n} f(Z_i)-\frac{1}{n}\sum_{i=1}^{n} f(Z_i')\right|>\frac{\varepsilon}{2}\alpha+\frac{\varepsilon}{2}\mathbb{E}\{f(Z)\}\right\}.$$

STEP 2. Replace the expectation in the above probability by an empirical mean. Using $\mathbb{E}\{f(Z)\}\geq\frac{1}{K_2}\mathbb{E}\{f(Z)^2\}$ we get

$$\mathbb{P}\left\{\exists f\in\mathcal{F}:\left|\frac{1}{n}\sum_{i=1}^{n} f(Z_i)-\frac{1}{n}\sum_{i=1}^{n} f(Z_i')\right|>\frac{\varepsilon}{2}\alpha+\frac{\varepsilon}{2}\mathbb{E}\{f(Z)\}\right\}$$

$$\leq \mathbb{P}\left\{\exists f\in\mathcal{F}:\left|\frac{1}{n}\sum_{i=1}^{n} f(Z_i)-\frac{1}{n}\sum_{i=1}^{n} f(Z_i')\right|>\frac{\varepsilon}{2}\alpha+\frac{\varepsilon}{2}\frac{1}{K_2}\mathbb{E}\{f(Z)^2\}\right\}$$

$$\begin{aligned}\leq \mathbb{P}\Bigg\{&\exists f\in\mathcal{F}:\left|\frac{1}{n}\sum_{i=1}^{n} f(Z_i)-\frac{1}{n}\sum_{i=1}^{n} f(Z_i')\right|>\frac{\varepsilon}{2}\alpha+\frac{\varepsilon}{2}\frac{1}{K_2}\mathbb{E}\{f(Z)^2\},\\ &\frac{1}{n}\sum_{i=1}^{n} f(Z_i)^2-\mathbb{E}\{f(Z)^2\}\leq\varepsilon\left(\alpha+\frac{1}{n}\sum_{i=1}^{n} f(Z_i)^2+\mathbb{E}\{f(Z)^2\}\right),\\ &\frac{1}{n}\sum_{i=1}^{n} f(Z_i')^2-\mathbb{E}\{f(Z)^2\}\leq\varepsilon\left(\alpha+\frac{1}{n}\sum_{i=1}^{n} f(Z_i')^2+\mathbb{E}\{f(Z)^2\}\right)\Bigg\}\end{aligned}$$

$$+2\mathbb{P}\left\{\sup_{f\in\mathcal{F}} \frac{\left|\frac{1}{n}\sum_{i=1}^{n} f(Z_i)^2-\mathbb{E}\{f(Z)^2\}\right|}{\alpha+\frac{1}{n}\sum_{i=1}^{n} f(Z_i)^2+\mathbb{E}\{f(Z)^2\}}>\varepsilon\right\}.$$

The second inequality in the first probability on the right hand side of the above inequality is equivalent to

$$\mathbb{E}\{f(Z)^2\}\geq\frac{1}{1+\varepsilon}\left(-\varepsilon\alpha+(1-\varepsilon)\frac{1}{n}\sum_{i=1}^{n} f(Z_i)^2\right),$$

which implies

$$\frac{1}{2}\frac{\varepsilon}{2}\frac{1}{K_2}\mathbb{E}\{f(Z)^2\}\geq-\frac{\varepsilon^2}{4(1+\varepsilon)K_2}\alpha+\frac{1-\varepsilon}{4(1+\varepsilon)K_2}\varepsilon\frac{1}{n}\sum_{i=1}^{n} f(Z_i)^2.$$

This together with a similar argument applied to the third inequality and

$$\frac{\varepsilon\alpha}{2}-\frac{\varepsilon^2\alpha}{2(1+\varepsilon)K_2}\geq\frac{\varepsilon\alpha}{4}$$

(which follows from $\varepsilon/(K_2(1+\varepsilon)) \le 1/2$ for $0<\varepsilon<1$ and $K_2 \ge 1$) yields

$$\mathbb{P}\left\{\exists f\in\mathcal{F}: \left|\frac{1}{n}\sum_{i=1}^n f(Z_i)-\frac{1}{n}\sum_{i=1}^n f(Z_i')\right| > \frac{\varepsilon}{2}\alpha+\frac{\varepsilon}{2}\mathbb{E}\{f(Z)\}\right\}$$

$$\le \mathbb{P}\left\{\exists f\in\mathcal{F}: \left|\frac{1}{n}\sum_{i=1}^n f(Z_i)-\frac{1}{n}\sum_{i=1}^n f(Z_i')\right|\right.$$

$$\left. > \frac{\varepsilon\alpha}{4}+\frac{(1-\varepsilon)\varepsilon}{4(1+\varepsilon)K_2}\left(\frac{1}{n}\sum_{i=1}^n f(Z_i)^2+\frac{1}{n}\sum_{i=1}^n f(Z_i')^2\right)\right\}$$

$$+2\mathbb{P}\left\{\sup_{f\in\mathcal{F}}\frac{\left|\frac{1}{n}\sum_{i=1}^n f(Z_i)^2-\mathbb{E}\{f(Z)^2\}\right|}{\alpha+\frac{1}{n}\sum_{i=1}^n f(Z_i)^2+\mathbb{E}\{f(Z)^2\}} > \varepsilon\right\}. \qquad (2.56)$$

STEP 3. Bounding the second second probability on the right hand side of the above inequality. Using arguments similar to the ones given below one can show

$$\mathbb{P}\left\{\sup_{f\in\mathcal{F}}\frac{\left|\frac{1}{n}\sum_{i=1}^n f(Z_i)^2-\mathbb{E}\{f(Z)^2\}\right|}{\alpha+\frac{1}{n}\sum_{i=1}^n f(Z_i)^2+\mathbb{E}\{f(Z)^2\}} > \varepsilon\right\} \le 13\exp\left(-\frac{n\varepsilon^2\alpha}{128\cdot 2304K_1^2}\right)$$

The proof is left to the reader.

STEP 4. Introduction of additional randomness by random signs.
Let $U_1,\dots,U_n$ be independent uniformly distributed over $\{-1,1\}$ and independent of $Z_1,\dots,Z_n$, $Z_1',\dots,Z_n'$. Because of the independence and identical distribution of $Z_1,\dots,Z_n$, $Z_1',\dots,Z_n'$, the joint distribution of $Z_1^n, Z_1'^n$ doesn't change if one randomly interchanges corresponding components of Z_1^n and $Z_1'^n$. Hence

$$\mathbb{P}\left\{\exists f\in\mathcal{F}: \left|\frac{1}{n}\sum_{i=1}^n f(Z_i)-\frac{1}{n}\sum_{i=1}^n f(Z_i')\right| >\right.$$

$$\left.\frac{\varepsilon\alpha}{4}+\frac{(1-\varepsilon)\varepsilon}{4(1+\varepsilon)K_2}\left(\frac{1}{n}\sum_{i=1}^n f(Z_i)^2+\frac{1}{n}\sum_{i=1}^n f(Z_i')^2\right)\right\}$$

$$=\mathbb{P}\left\{\exists f\in\mathcal{F}: \left|\frac{1}{n}\sum_{i=1}^n U_i\left(f(Z_i)-f(Z_i')\right)\right| >\right.$$

$$\left.\frac{\varepsilon\alpha}{4}+\frac{(1-\varepsilon)\varepsilon}{4(1+\varepsilon)K_2}\left(\frac{1}{n}\sum_{i=1}^n f(Z_i)^2+\frac{1}{n}\sum_{i=1}^n f(Z_i')^2\right)\right\}$$

$$\le 2\mathbb{P}\left\{\exists f\in\mathcal{F}: \left|\frac{1}{n}\sum_{i=1}^n U_i f(Z_i)\right| > \frac{\varepsilon\alpha}{8}+\frac{(1-\varepsilon)\varepsilon}{4(1+\varepsilon)K_2}\frac{1}{n}\sum_{i=1}^n f(Z_i)^2\right\}.$$

STEP 5. Peeling.
We have

$$\mathbb{P}\left\{\exists f \in \mathcal{F} : \left|\frac{1}{n}\sum_{i=1}^{n} U_i f(Z_i)\right| > \frac{\varepsilon\alpha}{8} + \frac{(1-\varepsilon)\varepsilon}{4(1+\varepsilon)K_2}\frac{1}{n}\sum_{i=1}^{n} f(Z_i)^2\right\}$$

$$\le \sum_{k=1}^{\infty} \mathbb{P}\left\{\exists f \in \mathcal{F} : I_{\{k\neq 1\}} 2^{k-1}\frac{K_2(1+\varepsilon)\alpha}{2(1-\varepsilon)} \le \frac{1}{n}\sum_{i=1}^{n} f(Z_i)^2 \le 2^k \frac{K_2(1+\varepsilon)\alpha}{2(1-\varepsilon)}, \right.$$
$$\left. \left|\frac{1}{n}\sum_{i=1}^{n} U_i f(Z_i)\right| > \frac{\varepsilon\alpha}{8} + \frac{(1-\varepsilon)\varepsilon}{4(1+\varepsilon)K_2}\frac{1}{n}\sum_{i=1}^{n} f(Z_i)^2\right\}$$

$$\le \sum_{k=1}^{\infty} \mathbb{P}\left\{\exists f \in \mathcal{F} : \frac{1}{n}\sum_{i=1}^{n} f(Z_i)^2 \le 2^k \frac{K_2(1+\varepsilon)\alpha}{2(1-\varepsilon)}, \left|\frac{1}{n}\sum_{i=1}^{n} U_i f(Z_i)\right| > \frac{\varepsilon\alpha}{8}2^{k-1}\right\}$$

STEP 6. Application of Theorem 2.13.
Next we condition inside the above probabilities on $Z_1,\ldots,Z_n$, which is equivalent to considering for $z_1,\ldots,z_n \in \mathcal{X}$ and $k \in \mathcal{N}$

$$\mathbb{P}\left\{\exists f \in \mathcal{F} : \frac{1}{n}\sum_{i=1}^{n} f(z_i)^2 \le 2^k \frac{K_2(1+\varepsilon)\alpha}{2(1-\varepsilon)}, \left|\frac{1}{n}\sum_{i=1}^{n} U_i f(z_i)\right| > \frac{\varepsilon\alpha}{8}2^{k-1}\right\}. \tag{2.57}$$

The assumptions of Theorem 2.14 imply

$$\sqrt{n}\frac{\varepsilon\alpha}{8}2^{k-1} \ge$$

$$36\int_0^{\sqrt{\frac{2^k K_2(1+\varepsilon)\alpha}{2(1-\varepsilon)}}} \left(\log \mathcal{N}_2\left(u, \left\{f \in \mathcal{F} : \frac{1}{n}\sum_{i=1}^{n} f(z_i)^2 \le \frac{2^k K_2(1+\varepsilon)\alpha}{2(1-\varepsilon)}\right\}, z_1^n\right)\right)^{\frac{1}{2}} du$$

and

$$\sqrt{n}\frac{\varepsilon\alpha}{8}2^{k-1} \ge 36\sqrt{\frac{2^k K_2(1+\varepsilon)\alpha}{2(1-\varepsilon)}}.$$

Hence we can conclude by Theorem 2.13 that (2.57) is bounded from above by

$$4\exp\left(-\frac{n\left(\frac{\varepsilon\alpha}{8}2^{k-1}\right)^2}{2304\frac{2^k K_2(1+\varepsilon)\alpha}{2(1-\varepsilon)}}\right) = 4\exp\left(-\frac{n\varepsilon^2(1-\varepsilon)\alpha}{64\cdot 2304(1+\varepsilon)K_2}2^{k-1}\right).$$

It follows

$$\sum_{k=1}^{\infty} \mathbb{P}\left\{ \exists f \in \mathcal{F} : \frac{1}{n}\sum_{i=1}^{n} f(Z_i)^2 \leq 2^k \frac{K_2(1+\varepsilon)\alpha}{2(1-\varepsilon)}, \left| \frac{1}{n}\sum_{i=1}^{n} U_i f(Z_i) \right| > \frac{\varepsilon\alpha}{8} 2^{k-1} \right\}$$

$$\leq \sum_{k=1}^{\infty} 4\exp\left(-\frac{n\varepsilon^2(1-\varepsilon)\alpha}{64\cdot 2304(1+\varepsilon)K_2} 2^{k-1} \right)$$

$$\leq \sum_{k=1}^{\infty} 4\exp\left(-\frac{n\varepsilon^2(1-\varepsilon)\alpha}{64\cdot 2304(1+\varepsilon)K_2} \cdot k \right)$$

$$= \frac{4}{1-\exp\left(-\frac{n\varepsilon^2(1-\varepsilon)\alpha}{64\cdot 2304(1+\varepsilon)K_2}\right)} \exp\left(-\frac{n\varepsilon^2(1-\varepsilon)\alpha}{64\cdot 2304(1+\varepsilon)K_2} \right).$$

Steps 1 to 6 imply the assertion. □

To bound the covering number we will use

Lemma 2.7. (VAN DE GEER (2000)) *Let $\mathcal{F}$ be a set of functions $f : \mathcal{R}^d \to \mathcal{R}$. Assume that $\mathcal{F}$ is a linear vector space of dimension D. Then one has for arbitrary $R > 0$, $\varepsilon > 0$ and $z_1, \ldots, z_n \in \mathcal{R}^d$:*

$$\mathcal{N}_2\left(\varepsilon, \left\{ f \in \mathcal{F} \; : \; \frac{1}{n}\sum_{i=1}^{n} |f(z_i)|^2 \leq R^2 \right\}, z_1^n \right) \leq \left(\frac{4R+\varepsilon}{\varepsilon} \right)^D.$$

PROOF. For $f, g : \mathcal{R}^d \to \mathcal{R}$ set

$$< f, g >_n := \frac{1}{n}\sum_{i=1}^{n} f(z_i) g(z_i) \text{ and } \|f\|_n^2 := < f, f >_n .$$

Let $f_1, \ldots, f_N \in \{ f \in \mathcal{F} \; : \; \|f\|_n \leq R \}$ be such that

$$\|f_i - f_j\|_n > \varepsilon \quad \text{for all } 1 \leq i < j \leq N.$$

It suffices to show

$$N \leq \left(\frac{4R+\varepsilon}{\varepsilon} \right)^D. \tag{2.58}$$

In order to show (2.58) let $B_1, \ldots, B_D$ be a basis of the linear vector space $\mathcal{F}$. Then for any $a_1, b_1, \ldots, a_D, b_D \in \mathcal{R}$

$$\| \sum_{j=1}^{D} a_j B_j - \sum_{j=1}^{D} b_j B_j \|_n^2 = < \sum_{j=1}^{D} (a_j - b_j) B_j, \sum_{j=1}^{D} (a_j - b_j) B_j >_n = (a-b)^T B (a-b),$$

where

$$B = (<B_j, B_k>_n)_{1 \le j,k \le D} \text{ and } (a-b) = (a_1 - b_1, \dots, a_D - b_D)^T.$$

Because of

$$a^T B a = \| \sum_{j=1}^{D} a_j B_j \|_n^2 \ge 0 \quad (a \in \mathcal{R}^D),$$

the symmetric matrix B is positive semidefinite, hence there exists a symmetric matrix $B^{1/2}$ such that

$$B = B^{1/2} \cdot B^{1/2}.$$

We have

$$\| \sum_{j=1}^{D} a_j B_j - \sum_{j=1}^{D} b_j B_j \|_n^2 = (a-B)^T B^{1/2} B^{1/2} (a-b) = \|B^{1/2}(a-b)\|^2,$$

where $\| \cdot \|$ is the euclidean norm in $\mathcal{R}^D$.

Because $f_i \in \mathcal{F}$ we get

$$f_i = \sum_{j=1}^{D} a_j^{(i)} B_j$$

for some

$$a^{(i)} = (a_1^{(i)}, \dots, a_D^{(i)})^T \in \mathcal{R}^D$$

$(i = 1, \dots, N)$. It follows

$$\|B^{1/2} a^{(i)}\| = \|f_i\|_n \le R$$

and

$$\|B^{1/2} a^{(i)} - B^{1/2} a^{(j)}\| = \|f_i - f_j\|_n > \varepsilon$$

for all $1 \le i < j \le N$. Hence the N balls in $\mathcal{R}^d$ with centers $B^{1/2}a^{(1)}, \dots, B^{1/2}a^{(N)}$ and radius $\varepsilon/4$ are disjoint subsets of the ball with center zero and radius $R + \varepsilon/4$. By comparing the volumes of the balls one gets

$$N \cdot c_D \cdot \left(\frac{\varepsilon}{4}\right)^D \le c_D \cdot \left(R + \frac{\varepsilon}{4}\right)^D,$$

where c_D is the volume of a ball with radius one in $\mathcal{R}^D$. This implies the assertion. □

We consider next univariate least squares spline estimates. We start with the definition of univariate polynomial spline spaces. Let $K \in \mathcal{N}$ and $M \in \mathcal{N}_0$. Denote by $S^1_{K,M}$ the set of all functions $f : \mathcal{R} \to \mathcal{R}$ which satisfy

$$f(x) = 0 \quad \text{for } x \notin [0,1], \tag{2.59}$$

$$f_{|[0,\frac{1}{K})}, f_{|[\frac{1}{K},\frac{2}{K})}, \dots, f_{|[\frac{K-1}{K},1]} \quad \text{are polynomials of degree } M \text{ (or less)}, \tag{2.60}$$

and

$$\text{if } M > 0, \text{ then } f \text{ is } M-1 \text{ times continuously differentiable on } [0,1]. \tag{2.61}$$

EXAMPLE 2.1. *For $M = 0$ the functions in $S^1_{K,0}$ are constant on each set $[\frac{i}{K}, \frac{i+1}{K})$. For $M = 1$ the functions in $S^1_{K,1}$ are linear on each set $[\frac{i}{K}, \frac{i+1}{K})$ and continuous on $[0,1]$.*

Let $\{B_{j,M,K} \,:\, j = 1,...,K+M\}$ be the usual B–spline basis of $S^1_{K,M}$ (see de Boor (1978)).

EXAMPLE 2.2. *$B_{j,0,K} = I_{[\frac{j}{K}, \frac{j+1}{K})}$ is the characteristic function of the interval $[\frac{j}{K}, \frac{j+1}{K})$. $B_{j,1,K}$ is equal to one at $\frac{j-1}{K}$, zero at $\frac{i}{K}$ for all $i \neq j-1$ and piecewise linear between (so-called hat function).*

Multivariate spline functions are defined as tensor products of univariate spline functions. More precisely, for $j = (j_1, \ldots, j_d) \in \mathcal{Z}^d$ define the multivariate B–spline $B_{j,M,K} : \mathcal{R}^d \to \mathcal{R}$ by

$$B_{j,M,K}(x_1, \ldots, x_d) = B_{j_1,M,K}(x_1) \cdot \ldots \cdot B_{j_d,M,K}(x_d) \quad (x_1, \ldots, x_d \in \mathcal{R}).$$

The spline spaces which we will use for our estimates are defined by

$$S_{K,M} = \left\{ \sum_{j \in \{1,\ldots,K+M\}^d} a_j B_{j,M,K} \;:\; |a_j| \leq L_{M,d} \quad (j \in \{1,\ldots,K+M\}^d) \right\}.$$

Here $L_{M,d} > 0$ is a constant which is specified below and which depends only on M and d.

It follows from the definition of $S^1_{K,M}$ that the functions in $S_{K,M}$ are equal to a multivariate polynomial of degree M (or less, in each coordinate) on every d–dimensional product of the intervals $[0, \frac{1}{K}), \ldots, [\frac{K-1}{K}, 1]$, that they are zero outside of $[0,1]^d$ and that they are $(M-1)$–times continuously differentiable on $[0,1]^d$. Furthermore,

$$B_{j,M,K}(x) \geq 0 \quad (j = 1, \ldots, K+M) \quad \text{and} \quad \sum_{j=1}^{K+M} B_{j,M,K}(x) = 1 \quad \text{for } x \in [0,1]$$

(see de Boor (1978)) imply that the functions in $S_{K,M}$ are bounded in absolute value by $L_{M,d}$.

We are now in the position to define our estimates: Depending on the data $(X_1,Y_1), \ldots, (X_n,Y_n)$ choose a function

$$m_{n,(K,M)}(\cdot) = m_{n,(K,M)}(\cdot, (X_1,Y_1), \ldots, (X_n,Y_n)) \in S_{K,M}$$

such that

$$\frac{1}{n}\sum_{i=1}^{n} |m_{n,(K,M)}(X_i) - Y_i|^2 \leq \inf_{f \in S_{K,M}} \frac{1}{n}\sum_{i=1}^{n} |f(X_i) - Y_i|^2 + \frac{1}{n}.$$

Our next result gives a bound on the expected L_2 error of $m_{n,(K,M)}$:

Theorem 2.15. (KOHLER (2000)) *Let $L > 0$ be arbitrary. Then*

$$\mathbf{E}\int |m_{n,(K,M)}(x) - m(x)|^2 \mu(dx) \leq c_1 \cdot \frac{(M+1)^d K^d}{n} + 2 \inf_{f \in S_{K,M}} \int |f(x) - m(x)|^2 \mu(dx)$$

for every distribution of (X,Y) with $|Y| \leq L$ a.s. Here c_1 is a constant which depends only on d, L and $L_{M,d}$.

COROLLARY 2.1. *Let $C,L>0$ and $p=k+\beta\geq 1$ with $k\in\mathcal{N}_0$ and $\beta\in(0,1]$. Set $M=k$, $L_{M,d}=(2M+3)^d 9^{M\cdot d}L$ and $K_n=\lceil C^{2/(2p+d)}n^{1/(2p+d)}\rceil$. Then*

$$\mathbb{E}\int|m_{n,(K_n,M)}(x)-m(x)|^2\mu(dx)=O\left(C^{\frac{2d}{2p+d}}n^{-\frac{2p}{2p+d}}\right)$$

for every distribution of (X,Y) with $X\in[0,1]^d$ a.s., $|Y|\leq L$ a.s. and m (p,C)–smooth.

PROOF. Theorems 12.5–12.7 of Schumaker (1981) imply that for m (p,C)–smooth

$$\inf_{f\in S_{K,M}}\int|f(x)-m(x)|^2\mu(dx)\leq\sup_{x\in[0,1]^d}|f(x)-m(x)|^2\leq c\cdot C^2\left(\frac{1}{K}\right)^{2p} \tag{2.62}$$

Here c is a constant which depends only on k and d. This together with Theorem 2.15 implies the assertion. □

In the proof of Theorem 2.15 we will apply the following lemma:

Lemma 2.8. *Let (X,Y), (X_1,Y_1), …, (X_n,Y_n) be independent identically distributed $\mathcal{R}^d\times\mathcal{R}$–valued random variables. Assume $|Y|\leq L$ a.s. for some $L\geq 1$. Let $L_{M,d}\geq L$, $K\in\mathcal{N}$, $M\in\mathcal{N}_0$ and define the spline space*

$$S_{K,M}=\left\{\sum_{j\in\{1,\dots,K+M\}^d}a_jB_{j,M,K}\quad:\quad|a_j|\leq L_{M,d}\quad(j\in\{1,\dots,K+M\}^d)\right\}$$

as above. Then one has for any $\alpha\geq c_{M,d}\frac{(M+1)^dK^d}{n}$

$$\mathbb{P}\left\{\sup_{f\in S_{K,M}}\frac{\mathbb{E}\{|f(X)-Y|^2-|m(X)-Y|^2\}-\frac{1}{n}\sum_{i=1}^n\{|f(X_i)-Y_i|^2-|m(X_i)-Y_i|^2\}}{\alpha+\mathbb{E}\{|f(X)-Y|^2-|m(X)-Y|^2\}}>\frac{1}{2}\right\}$$

$$\leq 50\exp\left(-\frac{n\alpha}{128^2\cdot 2304\cdot L_{M,d}^4}\right).$$

Here $c_{M,d}=8\cdot(288\cdot 512\cdot L_{M,d}^2)^2(1+2^d)$ is a constant which depends only on d and $L_{M,d}$.

PROOF. Set $Z=(X,Y)$ and $Z_i=(X_i,Y_i)$ $(i=1,\dots,n)$. For $f:\mathcal{R}^d\to\mathcal{R}$ define $g_f:\mathcal{R}^d\times\mathcal{R}\to\mathcal{R}$ by

$$g_f(x,y)=(|f(x)-y|^2-|m(x)-y|^2)\cdot I_{[-L,L]}(y).$$

Set

$$\mathcal{G}=\{g_f\quad:\quad f\in S_{K,M}\}.$$

Then

$$\mathbb{P}\left\{\sup_{f\in S_{K,M}}\frac{\cdots}{\cdots}>\frac{1}{2}\right\}=\mathbb{P}\left\{\exists g\in\mathcal{G}:\frac{\mathbb{E}\{g(Z)\}-\frac{1}{n}\sum_{i=1}^{n}g(Z_i)}{\alpha+\mathbb{E}\{g(Z)\}}>\frac{1}{2}\right\}.$$

As we have seen above, the functions in $S_{K,M}$ are bounded in absolute value by $L_{M,d}$, hence

$$|g(z)|\le 2L_{M,d}^2+2L^2\le 4L_{M,d}^2 \quad (z\in\mathcal{R}^d\times\mathcal{R})$$

for all $g\in\mathcal{G}$. Furthermore,

$$\begin{aligned}\mathbb{E}\{g_f(Z)^2\}&=\mathbb{E}\left\{\left(|f(X)-Y|^2-|m(X)-Y|^2\right)^2\right\}\\&=\mathbb{E}\left\{|(f(X)+m(X)-2Y)\cdot(f(X)-m(X))|^2\right\}\\&\le(L_{M,d}+3L)^2\mathbb{E}\{|f(X)-m(X)|^2\}\\&=(L_{M,d}+3L)^2\mathbb{E}\{g_f(Z)\}\quad(f\in S_{K,M}).\end{aligned}$$

The functions in $S_{K,M}$ are multivariate piecewise polynomials of degree M (or less, in each coordinate) with respect to some partition of $[0,1]^d$ consisting of K^d rectangles. Hence, $f\in S_{K,M}$ implies that f^2 is a multivariate piecewise polynomial of degree $2M$ (or less, in each coordinate) with respect to the same partition. This together with

$$g_f(x,y)=\left(f^2(x)-2f(x)y+y^2-|m(x)-y|^2\right)\cdot I_{[-L,L]}(y)$$

yields that $\mathcal{G}$ is a subset of a linear vector space of dimension

$$D=K^d((2M+1)^d+(M+1)^d+1),$$

thus by Lemma 2.7

$$\log\mathcal{N}_2\left(u,\left\{g\in\mathcal{G}:\frac{1}{n}\sum_{i=1}^{n}g(z_i)^2\le 4\delta\right\},z_1^n\right)\le D\cdot\log\frac{8\sqrt{\delta}+u}{u}.$$

Because of

$$\begin{aligned}\int_0^{\sqrt{\delta}}\left(D\cdot\log\frac{8\sqrt{\delta}+u}{u}\right)^{\frac{1}{2}}du &= \sqrt{D}\sqrt{\delta}\int_1^{\infty}\frac{\sqrt{\log(1+8v)}}{v^2}dv\\&\le \sqrt{D}\sqrt{\delta}\int_1^{\infty}\frac{\sqrt{8v}}{v^2}dv\\&\le 4\sqrt{2}\sqrt{D}\sqrt{\delta}\end{aligned}$$

the assumptions of Theorem 2.14 are satisfied whenever

$$\sqrt{n}\frac{1}{4}\delta \geq 288 \cdot 32 \cdot L_{M,d}^2 \cdot 4\sqrt{2}\sqrt{D}\sqrt{\delta}$$

for all $\delta \geq \alpha/4$, which is implied by

$$\alpha \geq c_{M,d}\frac{(M+1)^d K^d}{n}.$$

Thus we can conclude by Theorem 2.14

$$\mathbb{P}\left\{\exists g \in \mathcal{G} : \frac{\mathbb{E}\{g(Z)\} - \frac{1}{n}\sum_{i=1}^n g(Z_i)}{\alpha + \mathbb{E}\{g(Z)\}} > \frac{1}{2}\right\} \leq 50\exp\left(-\frac{n\alpha\frac{1}{4}(1-\frac{1}{2})}{128 \cdot 2304 \cdot 16 L_{M,d}^4}\right)$$

□

PROOF OF THEOREM 2.15.

$$\begin{aligned}\int |m_{n,(K,M)}(x) - m(x)|^2\mu(dx) &= \mathbb{E}\{|m_{n,(K,M)}(X) - Y|^2|\mathcal{D}_n\} - \mathbb{E}\{|m(X) - Y|^2\} \\ &= T_{1,n} + T_{2,n}\end{aligned}$$

where

$$T_{1,n} = 2\frac{1}{n}\sum_{i=1}^n \{|m_{n,(K,M)}(X_i) - Y_i|^2 - |m(X_i) - Y_i|^2\}$$

and

$$T_{2,n} = \mathbb{E}\{|m_{n,(K,M)}(X) - Y|^2|\mathcal{D}_n\} - \mathbb{E}\{|m(X) - Y|^2\} - T_{1,n}.$$

By definition of the estimate,

$$T_{1,n} \leq 2\inf_{f \in S_{K,M}} \frac{1}{n}\sum_{i=1}^n \{|f(X_i) - Y_i|^2 - |m(X_i) - Y_i|^2\} + \frac{2}{n}$$

which implies

$$\mathbb{E}T_{1,n} \leq 2\inf_{f \in S_{K,M}} \int |f(x) - m(x)|^2\mu(dx) + \frac{2}{n}. \tag{2.63}$$

Furthermore, Lemma 2.8 implies that for $t \geq c_{M,d}\frac{(M+1)^d K^d}{n}$

$$\begin{aligned}&\mathbb{P}\{T_{2,n} > t\} \\ &\leq \mathbb{P}\Big\{\exists f \in S_{K,M} : 2\mathbb{E}\{|f(X) - Y|^2 - |m(X) - Y|^2\} \\ &\qquad - \frac{2}{n}\sum_{i=1}^n \{|f(X_i) - Y_i|^2 - |m(X_i) - Y_i|^2\} \\ &\qquad > t + \mathbb{E}\{|f(X) - Y|^2 - |m(X) - Y|^2\}\Big\} \\ &\leq 50\exp\left(-\frac{nt}{c_3}\right).\end{aligned}$$

Hence,

$$\begin{aligned}\mathbf{E}\{T_{2,n}\} &\leq \int_0^\infty \mathbf{P}\{T_{2,n} > t\}\,dt \leq c_{M,d}\frac{(M+1)^d K^d}{n} + \int_{c_{M,d}\frac{(M+1)^d K^d}{n}}^\infty \mathbf{P}\{T_{2,n} > t\}\,dt \\ &\leq c_{M,d}\frac{(M+1)^d K^d}{n} + \frac{50c_3}{n}\exp\left(-\frac{c_{M,d}}{c_3}(M+1)^d K^d\right).\end{aligned}$$

This together with (2.63) implies the assertion. □

The definition of the estimate in Corollary 2.1 depends on the smoothness (p,C) of the regression function, which is usually unknown in an application. This can be avoided by using the method of complexity regularization (for a motivation see, e.g., Barron, Birgé and Massart (1999)).

Let $M_0 \in \mathcal{N}_0$ and set

$$\mathcal{P}_n = \left\{(K,M) \in \mathcal{N} \times \mathcal{N}_0 \quad : \quad 1 \leq K \leq n \text{ and } 0 \leq M \leq M_0\right\}.$$

For $(K,M) \in \mathcal{P}_n$ define $m_{n,(K,M)}$ as above. Depending on the data $(X_1,Y_1),\ldots,(X_n,Y_n)$ choose

$$(K^*,M^*) \in \mathcal{P}_n$$

such that

$$\begin{aligned}&\frac{1}{n}\sum_{i=1}^n |m_{n,(K^*,M^*)}(X_i) - Y_i|^2 + pen_n(K^*,M^*) \\ &\qquad = \min_{(K,M)\in\mathcal{P}_n}\left\{\frac{1}{n}\sum_{i=1}^n |m_{n,(K,M)}(X_i) - Y_i|^2 + pen_n(K,M)\right\},\end{aligned}$$

where

$$pen_n(K,M) = c_{M,d}\cdot\frac{(M+1)^d K^d}{n}$$

is a penalty term penalizing the complexity of $S_{K,M}$ and $c_{M,d}$ is a constant which is given in Lemma 2.8 above and which depends only on d and $L_{M,d}$. Set

$$m_n(x,(X_1,Y_1),\ldots,(X_n,Y_n)) = m_{n,(K^*,M^*)}(x,(X_1,Y_1),\ldots,(X_n,Y_n)).$$

Theorem 2.16. (KOHLER (2000)) *Let $L > 0$ be arbitrary. Then*

$$\begin{aligned}&\mathbf{E}\int |m_n(x) - m(x)|^2 \mu(dx) \leq \\ &\qquad \min_{(K,M)\in\mathcal{P}_n}\left\{2c_{M,d}\cdot\frac{(M+1)^d K^d}{n} + 2\inf_{f\in S_{K,M}}\int |f(x) - m(x)|^2\mu(dx)\right\} + \frac{c_2}{n}\end{aligned}$$

for every distribution of (X,Y) with $|Y| \leq L$ a.s. Here c_2 is a constant which depends only on d, L, M_0 and $\{L_{M,d} \,:\, 0 \leq M \leq M_0\}$.

COROLLARY 2.2. *Let $L > 0$, $M \le M_0$ and $K \in \mathcal{N}$. Then*

$$\mathbb{E}\int |m_n(x) - m(x)|^2 \mu(dx) = O\left(\frac{1}{n}\right)$$

for every distribution of (X,Y) with $|Y| \le L$ a.s. and $m \in S_{K,M}$.

PROOF. The assertion follows directly from Theorem 2.16. □

COROLLARY 2.3. *Let $L > 0$ be arbitrary. Set $L_{M,d} = (2M+3)^d 9^{M \cdot d} L$ $(0 \le M \le M_0)$. Then for any $1 \le p < M_0 + 1$ and any $C > 0$*

$$\mathbb{E}\int |m_n(x) - m(x)|^2 \mu(dx) = O\left(C^{\frac{2d}{2p+d}} n^{-\frac{2p}{2p+d}}\right)$$

for every distribution of (X,Y) with $X \in [0,1]^d$ a.s., $|Y| \le L$ a.s. and m (p,C)–smooth.

PROOF. The assertion follows directly from Theorem 2.16 and (2.62). □

PROOF OF THEOREM 2.16.

$$\begin{aligned}\int |m_n(x) - m(x)|^2 \mu(dx) &= \mathbb{E}\{|m_{n,(K^*,M^*)}(X) - Y|^2 | \mathcal{D}_n\} - \mathbb{E}\{|m(X) - Y|^2\} \\ &= T_{1,n} + T_{2,n}\end{aligned}$$

where

$$T_{1,n} = 2\frac{1}{n}\sum_{i=1}^n \{|m_{n,(K^*,M^*)}(X_i) - Y_i|^2 - |m(X_i) - Y_i|^2\} + 2pen_n(K^*,M^*)$$

and

$$T_{2,n} = \mathbb{E}\{|m_{n,(K^*,M^*)}(X) - Y|^2 | \mathcal{D}_n\} - \mathbb{E}\{|m(X) - Y|^2\} - T_{1,n}.$$

By definition of the estimate,

$$T_{1,n} \le 2 \min_{(K,M)\in\mathcal{P}_n}\left\{\inf_{f\in S_{K,M}} \frac{1}{n}\sum_{i=1}^n |f(X_i) - Y_i|^2 + pen_n(K,M)\right\} + \frac{2}{n} - \frac{2}{n}\sum_{i=1}^n |m(X_i) - Y_i|^2,$$

which implies

$$\mathbb{E}T_{1,n} \le 2 \min_{(K,M)\in\mathcal{P}_n}\left\{pen_n(K,M) + \inf_{f\in S_{K,M}} \int |f(x) - m(x)|^2 \mu(dx)\right\} + \frac{2}{n}. \tag{2.64}$$

Hence it suffices to show

$$\mathbb{E}T_{2,n} \le \frac{c_2 - 2}{n}. \tag{2.65}$$

To this end let $t > 0$ be arbitrary. Then Lemma 2.8 implies

$$\begin{aligned}
&\mathbf{P}\{T_{2,n} > t\} \\
&\leq \sum_{K=1}^{n} \sum_{M=0}^{M_0} \mathbf{P}\Bigg\{ \exists f \in S_{K,M} : 2\mathbf{E}\{|f(X)-Y|^2 - |m(X)-Y|^2\} \\
&\qquad\qquad -\frac{2}{n}\sum_{i=1}^{n}\{|f(X_i)-Y_i|^2 - |m(X_i)-Y_i|^2\} \\
&\qquad\qquad > t + 2pen_n(M,K) + \mathbf{E}\{|f(X)-Y|^2 - |m(X)-Y|^2\} \Bigg\} \\
&\leq \sum_{K=1}^{n} \sum_{M=0}^{M_0} 50 \exp\left(-\frac{n(t+2pen_n(K,M))}{c_3}\right) \\
&= \exp\left(-\frac{nt}{c_3}\right) \cdot \sum_{K=1}^{n} \sum_{M=0}^{M_0} 50 \exp\left(-\frac{2c_{M,d}}{c_3} \cdot (M+1)^d K^d\right) \\
&\leq (M_0+1) \cdot c_4 \cdot \exp\left(-\frac{nt}{c_3}\right).
\end{aligned}$$

Hence

$$\mathbf{E}T_{2,n} \leq \int_0^\infty \mathbf{P}\{T_{2,n} > t\}\, dt \leq \frac{c_3 c_4 (M_0+1)}{n}.$$

□

2.8 References

Antos, A., Györfi, L. and Kohler, M. (2000). Lower bounds on the rate of convergence of nonparametric regression estimates. *J. Statistical Planning and Inference*, 83, pp. 91-100.

Barron, A. R., Birgé, L. and Massart, P. (1999). Risk bounds for model selection via penalization. *Probability Theory and Related Fields*, 113, pp. 301-413.

Beirlant, J. and Györfi, L. (1998). On the asymptotic L_2-error in partitioning regression estimation. *J. Statistical Planning and Inference*, 71, pp. 93-107.

Birgé, L. (1986). On estimating a density using Hellinger distance and some other strange facts. *Probability Theory and Related Fields*, 71, pp. 271-291.

Devroye, L. (1995). Another proof of a slow convergence result of Birgé. *Statistics and Probability Letters*, 23, pp. 63-67.

de Boor, C. (1978). *A Practical Guide to Splines.* Springer, New York.

Devroye, L. and Györfi, L. (1983). Distribution-free exponential upper bound on the L_1 error of partitioning estimates of a regression function. In *Proceedings of the Fourth Pannonian Symposium on Mathematical Statistics*, Konecny F., Mogyoródi, J. and Wetz, W. Eds., pp. 67-76, Budapest, Akadémiai Kiadó.

Devroye, L., Györfi, L. and Lugosi, G. (1996). *A Probabilistic Theory of Pattern Recognition.* Springer Verlag.

Devroye, L., Györfi, L., Krzyżak, A. and Lugosi, G. (1994). On the strong universal consistency of nearest neighbor regression function estimates. *Annals of Statistics*, 22, pp. 1371–1385.

Devroye, L. and Krzyżak, A.(1989). An equivalence theorem for L1 convergence of the kernel regression estimate. *Journal of Statistical Planning and Inference*, 23, pp. 71–82.

Devroye, L. and Wagner, T. J. (1980). Distribution-free consistency results in nonparametric discrimination and regression function estimation. *Annals of Statistics*, 8, pp. 231–239.

Dudley, R. (1978). Central limit theorems for empirical measures. *Annals of Probability* **6** 899–929.

Györfi, L. (1991). Universal consistencies of regression estimate for unbounded regression functions. in *Nonparametric functional estimation and related topics, ed. G. Roussas*, pp. 329–338. Dordrecht: Kluwer Academic Publishers.

Györfi, L., Kohler, M., Krzyzak, A. and Walk, H.(2002). *A Distribution-Free Theory of Nonparametric Regression.* Springer, (book under preparation)

Györfi, L., Kohler, M. and Walk, H.(1998). Weak and strong universal consistency of semi-recursive partitioning and kernel regression estimates. *Statistics and Decisions*, 16, pp. 1-18.

Györfi, L. and Walk, H. (1996). On the strong universal consistency of a series type regression estimate. *Mathematical Methods in Statistics*, 5, pp. 332-342.

Györfi, L. and Walk, H. (1997). Consistency of a recursive regression estimate by Pál Révész. *Statistics and Probability Letters*, 31, pp. 177-183.

Ibragimov, I. A. and Khasminskii, R. Z. (1980). On nonparametric estimation of regression. *Doklady Acad. Nauk SSSR* 252, 780–784.

Ibragimov, I. A. and Khasminskii, R. Z. (1981). *Statistical Estimation: Asymptotic Theory.* Springer Verlag, New York.

Ibragimov, I. A. and Khasminskii, R. Z. (1982). On the bounds for quality of nonparametric regression function estimation. *Theory of Probability and its Applications* 27, 81–94.

Kohler, M. (2000). Inequalities for uniform deviations of averages from expectations with applications to nonparamtric regression. *Journal of Statistical Planning and Inference* 89, 1–23.

Kohler, M. and Krzyżak, A. (2001). Nonparametric regression estimation using penalized least squares. To appear in *IEEE Trans. Inf. Theory.*

Korostelev, A. P. and Tsybakov, A. B. (1993). *Minimax Theory of Image Reconstruction.* Springer Verlag, Berlin.

Lee, W. S., Bartlett, P. L., Williamson, R. C. (1996). Efficient agnostic learning of neural networks with bounded fan–in. *IEEE Trans. Inform. Theory* **42**, 2118–2132.

Nadaraya, E. A. (1964). On estimating regression. *Theory of Probability and its Applications*, 9, pp. 141–142.

Nadaraya, E. A. (1970). Remarks on nonparametric estimates for density functions and regression curves. *Theory of Probability and its Applications*, 15, pp. 134–137.

Pollard, D. (1984). *Convergence of Stochastic Processes.* Springer, New York.

Schumaker, L. (1981). *Spline functions: Basic Theory.* Wiley, New York.

Spiegelman, C. and Sacks, J. (1980). Consistent window estimation in nonparametric regression. *Annals of Statistics*, 8, pp. 240–246.

Steele, J. (1975) *Combinatorial entropy and uniform limit laws.* PhD Thesis, Stanford University, Stanford, CA.

Stone, C. J. (1977). Consistent nonparametric regression. *Annals of Statistics*, 5, pp. 595–645.

Stone, C. J. (1982). Optimal global rates of convergence for nonparametric regression. *Annals of Statistics* 10, 1040–1053.

Tukey, J. W. (1947). Nonparametric estimation II. Statistically equivalent blocks and tolerance regions. *Annals of Mathematical Statistics*, 18, pp. 529-539.

Tukey, J. W. (1961). Curves as parameters and touch estimation. *Proceedings of the Fourth Berkeley Symposium*, pp. 681-694.

van de Vaart, A. W. and Wellner, J. A. (1996). *Weak convergence and empirical processes.* Springer, New York.

Vapnik, V. N. and Chervonenkis, A. Y. (1974). *Theory of Pattern Recognition.* Nauka, Moscow. (in Russian); German translation: *Theorie der Zeichenerkennung*, Akademie Verlag, Berlin, 1979.

Vajda, I. (1989). *Theory of Statistical Inference and Information.* Kluwer Academic Publishers.

van de Geer, S. (1987). A new approach to least squares estimation, with applications. *Ann. Statist.* **15**, 587–602.

van de Geer, S. (1988). *Regression Analysis and Empirical Processes.* CWI-tract 45, Centre for Mathematics and Computer Science, Amsterdam.

van de Geer, S. (1990). Estimating a regression function. *Ann. Statist.* **18**, 907–924.

van de Geer, S. (2000). *Empirical Processes in M-Estimation.* Cambridge University Press, Cambridge.

Walk, H. (2000). Strong universal consistency of kernel and partitioning regression estimates. (submitted for publications)

Watson, G. S. (1964). Smooth regression analysis. *Sankhya Series A*, 26, pp. 359-372.

3

UNIVERSAL PREDICTION

N. Cesa-Bianchi
Universita di Milano, Milano, Italy

3.1 Introduction

Consider the problem of forecasting the elements of a data sequence $y_1, y_2 \ldots$ where the prediction $\widehat{p}_t$ for the t-th element y_t can only be based on the past data y_s, $s < t$, but not on the future ones. The goal of the forecaster is to minimize the cumulative loss $\ell(\widehat{p}_1, y_1) + \ell(\widehat{p}_2, y_2) + \ldots$, where ℓ is a given loss function measuring the discrepancy between predictions $\widehat{p}_t$ and true outcomes y_t. The traditional statistical approach to such problems first assumes a probabilistic model for the generating mechanism of the data sequence and then investigates the possibilities and limitations of the prediction of such random sequences. For example, in many applications the data sequence is assumed to be a realization of some stationary stochastic process. This approach works in many cases, when a tractable statistical model reasonably describes the underlying process. However, some processes may be inherently hard to model probabilistically and, in such cases, more robust prediction methods are required.

In this chapter we look at the sequential prediction problem without making any statistical assumption on the way the data sequence is generated. In this scenario, the basic question faced by the forecaster is how to predict well the individual sequence it is going to observe. Of course, without knowing the future, a forecaster cannot expect to perform uniformly well on any possible sequence. We call *bias* the tendency of the forecaster to predict certain sequences better than others. An elegant way to design a forecaster S with a certain bias is to use a set of *reference forecasters*, which compete against S on the same prediction task. For example, if our task is to predict every day tomorrow's weather in a certain area, then our pool of reference forecasters might include a collection of software packages for weather forecasting and, possibly, some of the local weather men. To implement a forecaster S with a certain bias, we require that S exhibit, on every possible data sequence, a performance close to that of the reference forecaster that turned out to be best on that data sequence. As the reference forecasters are possibly chosen based on a concrete prediction problem, we will call them *experts*.

3.1.1 Prediction with expert advice

We now introduce our general framework for sequential prediction problems. Given a *decision space* $\mathcal{X}$, an *outcome space* $\mathcal{Y}$, and a nonnegative *loss function* $\ell : \mathcal{X} \times \mathcal{Y} \to \mathcal{R}$, a forecaster is used to predict an arbitrary hidden sequence $y_1, y_2, \ldots \in \mathcal{Y}$. At each prediction round $t = 1, 2, \ldots$, the forecaster chooses its guess $\widehat{p}_t \in \mathcal{X}$ for the next element y_t of the hidden sequence. Then y_t is revealed and the forecaster incurs a loss $\ell(\widehat{p}_t, y_t)$. We are interested in forecasting strategies that are *nonanticipative*. That is, at each round t, the guess $\widehat{p}_t$ can only depend on the past elements $y^{t-1} = (y_1, \ldots, y_{t-1})$ but not on the future ones.

In the model of *prediction with expert advice*, the forecaster can base its guess $\widehat{p}_t$ also on the prediction $x_{E,t} \in \mathcal{X}$ of each expert E in a given set $\mathcal{E}$. We call $\{x_{E,t} : E \in \mathcal{E}\}$ the *advice* of the experts in $\mathcal{E}$ for outcome y_t and use $\mathcal{X}^{\mathcal{E}}$ to denote the set to which this advice belongs. As the forecaster's prediction $\widehat{p}_t$ depends both on the previously observed outcomes $y_1, \ldots, y_{t-1}$ and on the past expert advice $\{x_{E,s} : E \in \mathcal{E}\}$, $s < t$, we may define a *forecasting strategy* S by a sequence $S_1, S_2, \ldots$ of functions $S_t : \left(\mathcal{Y} \times \mathcal{X}^{\mathcal{E}}\right)^{t-1} \to \mathcal{X}$. Thus we write $\widehat{p}_t = S_t(z_1, \ldots, z_{t-1})$, where each $z_t = (y_{t-1}, \{x_{E,t} : E \in \mathcal{E}\})$ denotes the information received by the forecaster between round $t-1$ and round t, i.e. the previous outcome y_{t-1} and the expert advice for the next outcome y_t.

Given a sequence $z_1, z_2, \ldots \in \mathcal{Y} \times \mathcal{X}^{\mathcal{E}}$ of outcomes and expert advice, the *regret* of a forecasting strategy S with respect to expert $E \in \mathcal{E}$ at round t is defined by $r_{E,t} = \ell\big(S_t(z^{t-1}), y_t\big) - \ell(x_{E,t}, y_t)$ and the regret after n rounds, or cumulative regret, is defined by

$$R_{E,n} = \sum_{t=1}^{n} r_{E,t} = \sum_{t=1}^{n} \ell\big(S_t(z^{t-1}), y_t\big) - \sum_{t=1}^{n} \ell(x_{E,t}, y_t) \ . \tag{3.1}$$

We may view $R_{E,n}$ as the additional loss incurred by the forecaster for not having consistently followed the advice of expert E for the first n rounds of prediction. Note that $R_{E,n}$ may be negative for some (or all) experts $E \in \mathcal{E}$.

The goal of the forecaster is to keep the (cumulative) regret with respect to each expert as small as possible over the sequence of rounds.

3.2 Potential-based forecasters

Assume there is a finite number of experts indexed by $1, 2, \ldots, N$. We now describe a general method to derive forecasting strategies that minimize the cumulative regret (3.1) in a strong sense. More precisely, we will prove bounds on the *worst-case regret* that, for a given forecasting strategy S, is defined by

$$V_n(S,N) \stackrel{\text{def}}{=} \sup_{z^n} \left(\sum_{t=1}^{n} \ell(S_t(z^{t-1}), y_t) - \min_{1 \le i \le N} \sum_{t=1}^{n} \ell(x_{i,t}, y_t) \right) . \tag{3.2}$$

Here the supremum is taken over all $z^n \in (\mathcal{Y} \times \mathcal{X}^N)^n$; i.e. over the choice of both the hidden sequence y^n and the expert advice $(x_{1,t}, \ldots, x_{N,t})$, $1 \le t \le n$. We say that a forecasting strategy S is *consistent* if

$$\lim_{n \to \infty} \frac{V_n(S,N)}{n} = 0 .$$

To design regret-minimizing (or consistent) forecasting strategies, it is helpful to think of the cumulative regret as a point $R_t = (R_{1,t}, \ldots, R_{N,t})$ in $\mathcal{R}^N$. This point "drifts" with the number t of prediction rounds. That is, $R_t = R_{t-1} + r_t$, where $r_t = (r_{1,t}, \ldots, r_{N,t})$ is the regret vector representing the amount of drift at time t. The goal of the forecaster, as we defined it before, is to choose $\widehat{p}_t$ so that the components of the drifted point be do not increase too much, no matter which outcome y_t comes next. To design prediction strategies that achieve this, we define on the regret space $\mathcal{R}^N$ a suitable convex and positive *potential function* Φ, and then look for a prediction $\widehat{p}_t$ such that

$$\sup_{y_t \in \mathcal{Y}} \nabla\Phi(R_{t-1}) \cdot r_t \le 0 , \tag{3.3}$$

where $0 = (0, \ldots, 0)$ and $u \cdot v = u_1 v_1 + \ldots + u_N v_N$ denotes the internal product between two vectors. Condition (3.3) helps to keep R_t close to the minumum of the potential by ensuring that the drift vector point away from the gradient of the current potential for all possible outcomes y_t. To keep things simple, we will always use *additive* potentials of the form

$$\Phi(R_{t-1}) = \sum_{i=1}^{N} \phi(R_{i,t})$$

where $\phi : \mathcal{R} \to \mathcal{R}$ is twice differentiable, and such that $\phi \ge 0$, $\phi' > 0$, and $\phi'' \ge 0$.

The additive potential allows to prove a simple bound on the regret of any prediction strategy satisfying condition (3.3) in terms of the potential of the current regret vector. More precisely, by definition of additive potential we have, for any $1 \le i \le N$, $R_{i,n} \le \phi^{-1}(\Phi(R_n))$. That is, the components of the regret are uniformly bounded by a function of the current potential. In the next theorem, we bound $\Phi(R_n)$ in terms of $\Phi(R_0)$ by bounding the increments of $\Phi(R_t)$ for each $t = 1, \ldots, n$ via Taylor's theorem. Actually, bounding $f(\Phi(R_t))$ for a suitably chosen function $f : \mathcal{R} \to \mathcal{R}$ will lead to better bounds. Note that f has a role in the analysis but not in the definition of the predictor.

Theorem 3.1. *Choose an additive potential Φ and a sequence $r_1, r_2, \ldots \in \mathcal{R}^N$ of vectors such that condition (3.3) is satisfied for all $t \ge 1$. Let $f : \mathcal{R} \to \mathcal{R}$ be a twice differentiable auxiliary* *function such that* $f \geq 0$, $f' > 0$, $f'' \leq 0$, *and*

$$\sup_{u \in \mathcal{R}^N} f'(\Phi(u)) \sum_{i=1}^{N} \phi''(u_i) r_{i,t}^2 \leq C(r_t)$$

for each $t \geq 1$ *and for some nonnegative function* $C : \mathcal{R}^N \to \mathcal{R}$. *Then, for all* $n \geq 1$,

$$f(\Phi(R_n)) \leq f(\Phi(0)) + \frac{1}{2} \sum_{t=1}^{n} C(r_t) .$$

PROOF. Note that $\nabla f(\Phi(R_{t-1})) = f'(\Phi(R_{t-1})) \nabla \Phi(R_{t-1})$. Thus, by Taylor's theorem we obtain

$$\begin{aligned}
f(\Phi(R_t)) &= f(\Phi(R_{t-1} + r_t)) \\
&= f(\Phi(R_{t-1})) + f'(\Phi(R_{t-1}))\, \nabla \Phi(R_{t-1}) \cdot r_t \\
&\quad + \frac{1}{2} \sum_{i=1}^{N} \sum_{j=1}^{N} \left. \frac{\partial^2 f(\Phi(u))}{\partial u_i \partial u_j} \right|_{\xi} r_{i,t} r_{j,t} \quad \text{(where } \xi \text{ is some vector in } \mathcal{R}^N\text{)} \\
&\leq f(\Phi(R_{t-1})) + \frac{1}{2} \sum_{i=1}^{N} \sum_{j=1}^{N} \left. \frac{\partial^2 f(\Phi)}{\partial u_i \partial u_j} \right|_{\xi} r_{i,t} r_{j,t}
\end{aligned}$$

where the inequality follows by (3.3) and the fact that $f' > 0$. Now straightforward calculation shows that

$$\begin{aligned}
&\sum_{i=1}^{N} \sum_{j=1}^{N} \left. \frac{\partial^2 f(\Phi(u))}{\partial u_i \partial u_j} \right|_{\xi} r_{i,t} r_{j,t} \\
&= f''(\Phi(\xi)) \sum_{i=1}^{N} \sum_{j=1}^{N} \phi'(\xi_i) \phi'(\xi_j) r_{i,t} r_{j,t} + f'(\Phi(\xi)) \sum_{i=1}^{N} \phi''(\xi_i) r_{i,t}^2 \\
&= f''(\Phi(\xi)) \left(\sum_{i=1}^{N} \phi'(\xi_i) r_{i,t} \right)^2 + f'(\Phi(\xi)) \sum_{i=1}^{N} \phi''(\xi_i) r_{i,t}^2 \\
&\leq f'(\Phi(\xi)) \sum_{i=1}^{N} \phi''(\xi_i) r_{i,t}^2 \qquad \text{(since } f'' \leq 0\text{)} \\
&\leq C(r_t) .
\end{aligned}$$

Thus, we have obtained $f(\Phi(R_t)) \leq f(\Phi(R_{t-1})) + C(r_t)/2$. The proof is finished by iterating the argument. □

3.3 Convex loss functions

We show two applications of Theorem 3.1 in the case of convex loss functions defined over a convex decision space $\mathcal{X}$.

Assumption 2 (CONVEXITY) *The decision space $\mathcal{X}$ is a convex subset of a linear space and the loss function ℓ is convex in its first argument.*

A fact that makes convex loss functions appealing is that, for such loss functions, condition (3.3) is satisfied by simple weighted average forecasters. The convexity of the decision space $\mathcal{X}$ ensures that weighted average predictions still belong to $\mathcal{X}$.

FACT 3.1. *Under Assumption 2, inequality (3.3) is satisfied by the weighted average forecaster whose predictions are defined by*

$$\widehat{p}_t = \frac{\sum_{i=1}^N \phi'(R_{i,t-1})x_{i,t}}{\sum_{j=1}^N \phi'(R_{j,t-1})}. \tag{3.4}$$

PROOF. By convexity of the loss function, for all $y \in \mathcal{Y}$,

$$\ell(\widehat{p}_t, y) = \ell\left(\frac{\sum_{i=1}^N \phi'(R_{i,t-1})x_{i,t}}{\sum_{j=1}^N \phi'(R_{j,t-1})}, y\right) \le \frac{\sum_{i=1}^N \phi'(R_{i,t-1})\ell(x_{i,t}, y)}{\sum_{j=1}^N \phi'(R_{j,t-1})}.$$

Rearranging, we obtain (3.3). □

We first apply Theorem 3.1 to the forecaster defined by the polynomial potential based on $\phi(x) = x_+^p$, where $p \ge 2$ and a_+ denotes $\max\{0, a\}$. Plugging this choice of ϕ into (3.4) we obtain the *polynomially weighted average forecaster* (denoted by S^{poly}) whose predictions are defined by

$$\widehat{p}_t = \frac{\sum_{i=1}^N \left(\sum_{s=1}^{t-1} (\ell(\widehat{p}_s, y_s) - \ell(x_{i,s}, y_s))\right)_+^{p-1} x_{i,t}}{\sum_{j=1}^N \left(\sum_{s=1}^{t-1} (\ell(\widehat{p}_s, y_s) - \ell(x_{j,s}, y_s))\right)_+^{p-1}}.$$

We now show that S^{poly} is consistent with a worst-case regret that converges to zero at rate $O(1/\sqrt{n})$.

COROLLARY 3.1. (POLYNOMIAL POTENTIAL) *Suppose Assumption 2 holds and, moreover, suppose that the loss function takes values in $[0,1]$. Then, for any $p \ge 2$ and $N \ge 1$, the polynomially weighted average forecaster S^{poly} satisfies*

$$V_n(S^{\text{poly}}, N) \le \sqrt{n(p-1)N^{2/p}}$$

for all $n \ge 1$.

Note that the choice $p = 2$ yields a simple algorithm, but the dependence of the upper bound on the number N of experts is $\Theta(\sqrt{N})$. A better dependence is obtained choosing $p = 2\ln N$ (for $N > 2$), which leads to the bound

$$V_n(S^{\text{poly}}, N) \le \sqrt{ne(2\ln N - 1)}. \tag{3.5}$$

PROOF OF COROLLARY 3.1. Apply Theorem 3.1 with $f(x) = x^{2/p}$, $x \geq 0$. Clearly,

$$f'(x) = \frac{2}{p x^{(p-2)/p}} \quad \text{and} \quad \phi''(x) = p(p-1)x_+^{p-2}.$$

By Hölder's inequality, for $t \geq 1$,

$$\begin{aligned}
\sum_{i=1}^N \phi''(u_i) r_{i,t}^2 &= p(p-1)\sum_{i=1}^N (u_i)_+^{p-2} r_{i,t}^2 \\
&\leq p(p-1)\left(\sum_{i=1}^N \left((u_i)_+^{p-2}\right)^{p/(p-2)}\right)^{(p-2)/p} \left(\sum_{i=1}^N |r_{i,t}|^p\right)^{2/p}.
\end{aligned}$$

Thus,

$$f'\left(\sum_{i=1}^N \phi(u_i)\right)\sum_{i=1}^N \phi''(u_i) \leq (p-1)\left(\sum_{i=1}^N |r_{i,t}|^p\right)^{2/p}$$

and therefore the conditions of Theorem 3.1 are satisfied with $C(r_t) = (p-1)\|r_t\|_p^2$. Since $\Phi(0) = 0$, Theorem 3.1 and the boundedness of the loss function imply that

$$\left(\sum_{i=1}^N (R_{i,n})_+^p\right)^{2/p} = f(\Phi(R_n)) \leq (p-1)\sum_{t=1}^n \|r_t\|_p^2 \leq n(p-1)N^{2/p}.$$

Since

$$V_n(S^{\text{poly}}, N) = \max_{1 \leq i \leq N} R_{i,n} \leq \left(\sum_{i=1}^N (R_{i,n})_+^p\right)^{1/p}$$

the result follows. □

Our second example is the *exponentially weighted average forecaster* based on $\phi(x) = e^{\eta x}$, where η is a positive parameter, and denoted by $S^{\exp}$. For this potential, the prediction of the weighted average forecaster (3.4) simplifies to

$$\begin{aligned}
\widehat{p}_t &= \frac{\sum_{i=1}^N \exp\left(\eta \sum_{s=1}^{t-1} (\ell(\widehat{p}_s, y_s) - \ell(x_{i,s}, y_s))\right) x_{i,t}}{\sum_{j=1}^N \exp\left(\eta \sum_{s=1}^{t-1} (\ell(\widehat{p}_s, y_s) - \ell(x_{j,s}, y_s))\right)} \\
&= \frac{\sum_{i=1}^N \exp\left(-\eta \sum_{s=1}^{t-1} \ell(x_{i,s}, y_s)\right) x_{i,t}}{\sum_{j=1}^N \exp\left(-\eta \sum_{s=1}^{t-1} \ell(x_{j,s}, y_s)\right)}. \qquad (3.6)
\end{aligned}$$

The beauty of the exponentially weighted average forecaster is that it only depends on the past performance of the experts, whereas the predictions made using other general potentials depend on the past predictions $\widehat{p}_s$, $s < t$, as well. A simple application of Theorem 3.1 reveals the following performance bound for $S^{\exp}$.

COROLLARY 3.2. (EXPONENTIAL POTENTIAL) *Suppose Assumption 2 holds and, moreover, suppose that the loss function takes values in* $[0,1]$. *Then, for any* $N \geq 1$ *and any* $\eta > 0$, *the exponentially weighted average forecaster* $S^{\exp}$ *satisfies*

$$V_n(S^{\exp},N) \leq \frac{\ln N}{\eta} + \frac{n\eta}{2} \tag{3.7}$$

for all $n \geq 1$.

According to this analysis, for any constant value of η this forecaster is not consistent, as the right-hand side of (3.7) is $\Omega(n)$. On the other hand, setting $\eta = \sqrt{2(\ln N)/n}$ the upper bound becomes $\sqrt{2n\ln N}$, which is slightly better than the previous best bound (3.5), obtained using $\phi(x) = x_+^p$ with $p = 2\ln N$.

PROOF OF COROLLARY 3.2. We use Theorem 3.1 with $f(x) = (1/\eta)\ln x$. Then, using the boundedness of the loss,

$$f'(\Phi(u))\sum_{i=1}^{N}\phi''(u_i)r_{i,t}^2 \leq \eta \max_{1\leq i\leq N} r_{i,t}^2 \leq \eta\,.$$

Using $\Phi(0) = N$, Theorem 3.1 implies

$$V_n(S^{\text{poly}},N) = \max_{1\leq i\leq N} R_{i,n} \leq \frac{1}{\eta}\ln\Phi(R_n) \leq \frac{\ln N}{\eta} + \frac{n\eta}{2}$$

as desired.

3.3.1 *An optimal regret bound*

In the previous section we have seen that, with convex loss functions, the weigthed average forecasters S^{poly} and $S^{\exp}$ have both regret bounds of order $\sqrt{n\ln N}$ with the best leading constant obtained by $S^{\exp}$. In this section we show a refined analysis of $S^{\exp}$ that improves on this leading constant. Later, we will see that this improved bound is asymptotically exact.

Theorem 3.2. (TUNED EXPONENTIAL POTENTIAL) *Suppose Assumption 2 holds and, moreover, suppose that the loss function takes values in* $[0,1]$. *Then, for any* $N \geq 1$ *and* $\eta > 0$, *the exponentially weighted average forecaster* $S^{\exp}$ *satisfies*

$$V_n(S^{\exp},N) \leq \frac{\ln N}{\eta} + \frac{n\eta}{8}$$

for all $n \geq 1$. *In particular, with* $\eta = \sqrt{8(\ln N)/n}$, *the upper bound becomes* $\sqrt{(n/2)\ln N}$.

Let

$$L_i(y^n) = \sum_{t=1}^{n}\ell(x_{i,t},y_t) \qquad \text{and} \qquad L_S(y^n) = \sum_{t=1}^{n}\ell(\widehat{p}_t,y_t)$$

denote the cumulative loss of, respectively, expert i and forecasting strategy S on the hidden sequence y^n for a given choice of the expert advice $(x_{1,t},\ldots,x_{N,t})$, $1 \le t \le n$.

The proof of Theorem 3.2 is similar in spirit to that of Corollary 3.2, but now, instead of bounding the evolution of $\ln\left(\sum_i \exp(\eta R_{i,t})\right)$, we consider the sum of weights

$$W_t = \sum_{i=1}^{N} w_{i,t} = \sum_{i=1}^{N} \exp\left(-\eta L_i(y^t)\right)$$

for $t \ge 1$, where $W_0 = N$.

PROOF OF THEOREM 3.2. First observe that

$$\begin{aligned}
\ln \frac{W_n}{W_0} &= \ln\left(\sum_{i=1}^{N} \exp\left(-\eta L_i(y^n)\right)\right) - \ln N \\
&\ge \ln\left(\max_{1\le i\le N} \exp\left(-\eta L_i(y^n)\right)\right) - \ln N \\
&= -\eta \min_{1\le i\le N} L_i(y^n) - \ln N\,.
\end{aligned} \tag{3.8}$$

On the other hand, for each $t = 1,\ldots,n$

$$\begin{aligned}
\ln \frac{W_t}{W_{t-1}} &= \ln \frac{\sum_{i=1}^{N} \exp\left(-\eta \ell(x_{i,t},y_t)\right) \exp\left(-\eta L_i(y^{t-1})\right)}{\sum_{j=1}^{N} \exp\left(-\eta L_j(y^{t-1})\right)} \\
&= \ln \frac{\sum_{i=1}^{N} w_{i,t-1} \exp\left(-\eta \ell(x_{i,t},y_t)\right)}{\sum_{j=1}^{N} w_{j,t-1}}\,.
\end{aligned}$$

Using Chernoff-Hoeffding bounds, we observe that the quantity above may be upper bounded by

$$\begin{aligned}
-\eta \frac{\sum_{i=1}^{N} w_{i,t-1} \ell(x_{i,t},y_t)}{\sum_{j=1}^{N} w_{j,t-1}} + \frac{\eta^2}{8} &\le -\eta \ell\left(\frac{\sum_{i=1}^{N} w_{i,t-1} x_{i,t}}{\sum_{j=1}^{N} w_{j,t-1}}\right) + \frac{\eta^2}{8} \\
&= -\eta \ell(\widehat{p}_t, y) + \frac{\eta^2}{8}
\end{aligned}$$

where we used the convexity of the loss function and the definition (3.6) of $S^{\exp}$. Summing over $t = 1,\ldots,n$ we get

$$\ln \frac{W_n}{W_0} \le -\eta L_{S^{\exp}}(y^n) + \frac{\eta^2}{8} n\,.$$

Combining this with the lower bound (3.8) and solving for $L_{S^{\exp}}(y^n)$ we find that

$$L_{S^{\exp}}(y^n) \le \frac{\ln N}{\eta} + \frac{\eta}{8} n\,.$$

As the choice of the hidden sequence and of the expert advice was arbitrary, we get the statement of the theorem. □

3.3.2 A data-dependent regret bound

The exponentially weighted average predictor can be also used to obtain regret bounds that explicitely depend on the loss of the best expert on the observed data sequence. If the loss of the best expert is small compared to n, these bounds may be significantly better than the worst-case bound of Theorem 3.2.

Throughout this section, we suppose Assumption 2 holds and that the loss function takes values in $[0,1]$. For simplicity, we will write $L^*(y^n) = \min_{1\le i\le N} L_i(y^n)$. Our main tool is the following result.

Theorem 3.3. *For any $\eta \ge 0$ and for any $N \ge 1$, the exponentially weighted average predictor $S^{\exp}$ satisfies*

$$L_{S^{\exp}}(y^n) \le \frac{\ln N + \eta L^*(y^n)}{1-e^{-\eta}} \qquad \text{for all } n \ge 1.$$

PROOF. Recall from the proof of Theorem 3.2 that

$$-\eta L^*(y^n) - \ln N \le \ln \frac{W_n}{W_0} = \sum_{t=1}^{n} \ln \frac{W_t}{W_{t-1}} = \sum_{t=1}^{n} \ln \frac{\sum_{i=1}^{N} w_{i,t-1} e^{-\eta \ell(x_{i,t},y_t)}}{\sum_{j=1}^{N} w_{j,t-1}}\,.$$

Letting $e^{-\eta} = \beta$, and using the linear interpolation $\beta^x \le 1-(1-\beta)x$ which holds for all $0 \le x \le 1$ and for all $0 < \beta \le 1$, we get

$$\begin{aligned}
\sum_{t=1}^{n} \ln \frac{\sum_{i=1}^{N} w_{i,t-1}\beta^{\ell(x_{i,t},y_t)}}{\sum_{j=1}^{N} w_{j,t-1}} &\le \sum_{t=1}^{n} \ln \left(1 - \frac{(1-\beta)\sum_{i=1}^{N} w_{i,t-1}\ell(x_{i,t},y_t)}{\sum_{j=1}^{N} w_{j,t-1}}\right) \\
&\le -(1-\beta)\sum_{t=1}^{n} \frac{\sum_{i=1}^{N} w_{i,t-1}\ell(x_{i,t},y_t)}{\sum_{j=1}^{N} w_{j,t-1}} && (3.9) \\
&\le -(1-\beta)\sum_{t=1}^{n} \ell(\widehat{p}_t,y_t) && (3.10) \\
&= -(1-\beta)L_{S^{\exp}}(y^n)
\end{aligned}$$

where we used the bound $\ln(1-x) \le -x$ for all $x \ge 0$ in (3.9) and the convexity of ℓ in (3.10). Solving for $L_{S^{\exp}}(y^n)$ yields the result. □

The next result shows how to bound the regret in Theorem 3.2 by choosing η as a function of any konwn constant M bounding $L^*(y^n)$ from the above.

COROLLARY 3.3. *For any $N \ge 1$ and for $\eta = \ln\left(1+\sqrt{(2\ln N)/M}\right)$, the exponentially weighted average predictor $S^{\exp}$ satisfies*

$$L_{S^{\exp}}(y^n) - L^*(y^n) \le \sqrt{2M\ln N} + \ln N$$

for all $y^n \in \mathcal{Y}^n$ and for all choices of the expert advice such that $L^*(y^n) \le M$.

PROOF. Using Theorem 3.3, we just need to show that

$$\frac{\ln N + \eta L^*}{1-e^{-\eta}} \le L^* + \ln N + \sqrt{2M \ln N} \tag{3.11}$$

whenever η is chosen as in the statement of the Corollary. We start from the elementary inequality $(e^{\eta} - e^{-\eta})/2 = \sinh(\eta) \ge \eta$ which holds for all $\eta \ge 0$. Rearranging, we get

$$\eta \le \frac{1-e^{-2\eta}}{2e^{-\eta}} \; .$$

Replacing the η at the numerator in the left-hand side of (3.11) with the above upper bound, and rearranging a bit, we get that the Corollary is implied by

$$\frac{\ln N}{1-e^{-\eta}} + \frac{1-e^{-\eta}}{2e^{-\eta}} M \le \ln N + \sqrt{2M \ln N} \; .$$

Replacing η with the choice given in the statement of the Corollary yields, after a little algebra, the desired result. □

The next result, whose lengthy proof is omitted, shows that a proper time-varying potential makes $S^{\exp}$ a consistent forecaster.

Theorem 3.4. (DYNAMIC EXPONENTIAL POTENTIAL) *For any $N \ge 1$, the exponentially weighted average predictor $S^{\exp}$ using, at each round t, the parameter*

$$\eta_t = -\ln\left(1 - \min\left\{\frac{1}{4}, \sqrt{\frac{2\ln N}{L^*(y^{t-1})}}\right\}\right)$$

satisfies for all $n \ge 1$, for all $y^n \in \mathcal{Y}^n$, and for all choices of the expert advice,

$$L_{S^{\exp}}(y^n) - L^*(y^n) \le (2.83 + o(1))\sqrt{L^*(y^n)\ln N}$$

where $o(1) \to 0$ for $L^*(y^n) \to \infty$.

3.4 Exp-concave loss functions

Corollaries 3.1 and 3.2 show that, under very general assumptions—only convexity and boundedness of the loss function is required—it is possible to achieve that the worst-case regret with respect to the best expert be of order $\sqrt{n \ln N}$ for all possible outcome sequences. However, there is an important class of loss functions for which even better bounds can be achieved. This is summarized in the next result.

Theorem 3.5. *Assume that the loss function ℓ is such that the function $F(z) = e^{-\eta\ell(z,y)}$ is concave for all $y \in \mathcal{Y}$. Then the exponentially weighted average forecaster $S^{\exp}$, for all $n \geq 1$, achieves*

$$V_n(S^{\exp}, N) \leq \frac{\ln N}{\eta} .$$

For the loss functions satisfying the assumption of the theorem, the worst-case regret of $S^{\exp}$ is bounded by a quantity which is independent of the length n of the sequence of outcomes. Note that, apart from the the assumption of "exp-concavity", we do not assume anything else about the loss function. In particular, we do not explicitely assume that the loss function is convex and that it is bounded.

PROOF. It suffices to show that $\Phi(R_t) \leq \Phi(0)$ where Φ is the exponential potential function $\Phi(u) = \sum_{i=1}^{N} e^{\eta u_i}$. The result then follows immediately from the fact that $\Phi(0) = N$. Thus, it suffices to show that the value of the potential function can never increase, that is, $\Phi(R_t) \leq \Phi(R_{t-1})$ or, equivalently, that

$$\sum_{i=1}^{N} \exp\left(-\eta L_i(y^{t-1})\right) \exp\left(\eta(\ell(\widehat{p}_t, y_t) - \ell(x_{i,t}, y_t))\right) \leq \sum_{i=1}^{N} \exp\left(-\eta L_i(y^{t-1})\right) .$$

Denoting $w_{i,t-1} = \exp\left(-\eta L_i(y^{t-1})\right)$, this may be re-written as

$$\exp\left(-\eta\ell(\widehat{p}_t, y_t)\right) \geq \frac{\sum_{i=1}^{N} w_{i,t-1} \exp\left(-\eta\ell(x_{i,t}, y_t)\right)}{\sum_{j=1}^{N} w_{j,t-1}} .$$

But since $\widehat{p}_t = \sum_{i=1}^{N} w_{i,t-1} x_{i,t} \Big/ \sum_{i=1}^{N} w_{i,t-1}$, this follows by the concavity of $F(z)$ and Jensen's inequality. □

Some simple and important loss functions satisfy the exp-concavity assumption of Theorem 3.5.

Logarithmic loss.

Let $\mathcal{Y} = \{0,1\}$ and $\mathcal{X} = [0,1]$ and consider the logarithmic loss

$$\ell(z,y) = \mathbb{I}_{\{y=1\}} \ln \frac{1}{x} + \mathbb{I}_{\{y=0\}} \ln \frac{1}{1-x} .$$

If we take $\eta = 1$, then

$$F(z) = e^{-\eta\ell(z,y)} = \mathbb{I}_{\{y=1\}} z + \mathbb{I}_{\{y=0\}} (1-z)$$

which is a linear function of z for both values of and therefore it is concave. Note that this is an unbounded loss function.

Quadratic loss.

Let $\mathcal{X} = \mathcal{Y} = [0,1]$ and consider the quadratic loss function $\ell(z,y) = (z-y)^2$ A straightforward calculation shows that $F(z)$ is concave if and only if, for all z, $(z-y)^2 \leq 1/(2\eta)$. This is clearly guaranteed if $\eta \leq 1/2$.

Bibliographic remarks

Our model of sequential prediction without statistical assumptions on the data-generating mechanism finds its roots in the theory of repeated games. Hannan (1957) first showed the existence of consistent (randomized) predictors, and also derived the optimal rate of convergence $O(1/\sqrt{n})$. The problem of predicting an individual data sequence as good as the best element in a set of strategies has been studied in various contexts: by Foster (1991) and Foster and Vohra (1993) in game theory, by Dawid (1984) in statistics, and by several others in computer science — see the book by Borodin and El-Yaniv (1998). An important connection, which is not explored here, is with Bayesian decision theory. In fact, many of the forecasters which we described have a natural Bayesian interpretation — see e.g. Cesa-Bianchi, Helmbold and Panizza (1998). The framework of prediction with experts was introduced by Littlestone and Warmuth (1991, 1994) and Vovk (1990, 1992, 1998). However, the earliest result of this kind is probably the analysis of De Santis, Markowsky, and Wegman (1988). The notion of worst-case regret for the experts framework was first investigated in Cesa-Bianchi, Helmbold, Haussler, Freund, Schapire, and Warmuth (1997) and Chung (1994). The use of potential functions in sequential prediction is due to Hart and Mas-Colell (2001), who introduced condition (3.3) in a game-theoretic context, and to Grove, Littlestone, and Schuurmans (1997), who used exactly the same condition for the analysis of certain variants of the Perceptron algorithm (see Section 3.7). Our Theorem 3.1 is inspired by, and partially builds on, Hart and Mas-Colell's analysis of Λ-strategies for playing iterated games (Hart and Mas-Colell (2001)) and on the analysis of the quasi-additive algorithm of Grove, Littlestone and Schuurmans (2001). The unified framework for sequential prediction based on potential framework which we describe here was first introduced in Cesa-Bianchi and Lugosi (2001). A prominent example of a prediction strategy based on a potential is due to Blackwell (1956), who used it to prove his celebrated approachability theorem — generalizing to vector-valued payoffs von Neumann's minimax theorem — and also obtained Hannan's result as a corollary. Blackwell's strategy is equivalent to the polynomially weighted average forecaster with p set to 2. The polynomial potential was used by Hart and Mas-Colell (1999), by Grove et al., and by others since then. The forecaster based on the exponential potential is known in game theory as smooth fictitious player (Fudenberg and Levine (1998)). In learning theory exponential forecasters were introduced and analysed by Littlestone and Warmuth (1991, 1994) — the Weighted Majority algorithm — and by Vovk (1990, 1992, 1998) — the Aggregating Algorithm

— who gave a general analysis for arbitrary loss functions. The trick of setting the parameter p of the polynomial potential to $2\ln N$ is due to Gentile (2001). The analysis in Section 3.3.1 is from Cesa-Bianchi (1999). The data-dependent bounds of Section 3.3.2 are from a variety of sources: Theorem 3.3 is from Littlestone and Warmuth (1994), Corollary 3.3 is due to Freund and Schapire (1997), and Theorem 3.4 is from Auer, Cesa-Bianchi, and Gentile (2001). A simpler but less effective technique for tuning the exponential potential is the "doubling trick" of Cesa-Bianchi, Helmbold, Haussler, Freund, Schapire, and Warmuth (1997). Log-concave loss functions are related to the "perfectly mixable" prediction games of Vovk (1998). and to the class of losses studied by Haussler, Kivinen and Warmuth (1998) for the case $\mathcal{Y} = \{0,1\}$. Theorem 3.5 is due to Kivinen and Warmuth (1999).

3.5 Absolute loss

In this section we take a closer look at the sequential prediction problem when the outcome space is $\mathcal{Y} = \{0,1\}$, the decision space is $\mathcal{X} = [0,1]$, and the loss function is the "absolute loss" $\ell(\widehat{p},y) = |\widehat{p} - y|$. This model is equivalent to the case when $\mathcal{Y} = \mathcal{X} = \{0,1\}$, $\ell(x,y) = \mathbb{I}_{\{x \neq y\}}$ and the predictor makes randomized predictions $\widehat{Y}$ such that $\mathbb{P}(\widehat{Y} = 1) = \widehat{p}$. Hence, the cumulative absolute loss can be viewed as the expected number of mistakes made by a randomized forecaster in a binary prediction problem.

The absolute loss is convex. So, by Theorem 3.2 from Section 3.3.1, we know that the best bound on the worst-case regret is $\sqrt{(n/2)\ln N}$, achieved by the tuned exponentially weighted average forecaster. Can this bound be improved if the forecaster knows all the future expert advice in advance? For instance, suppose that $N = 2$ (there are only two experts) and the forecaster knows that $x_{1,t} = x_{2,t}$ for all but K rounds t. Then, by predicting as the experts if $x_{1,t} = x_{2,t}$ and predicting $1/2$ otherwise, a forecaster can achieve a regret bound of $K/2$, a quantity independent of n and thus significantly better than $\Theta\left(\sqrt{n}\right)$ which we got from 3.2.

In the rest of this section, we will study the predictive performance of forecasting strategies assuming that all future expert advice, for all possible choices of the hidden sequence y^n, is made preliminarily available to the forecasting strategy. In this respect, we define an *expert* F as a sequence $F = F_1, F_2, \ldots$ of functions $F_t : \mathcal{Y}^{t-1} \to \mathcal{X}$ mapping the observed outcomes y^{t-1} to a prediction $x_{F,t} = F_t(y^{t-1})$. As $\mathcal{F}$ is known to the forecasting strategy S, we may also assume that each forecaster prediction $\widehat{p}_t$ depend only on y^{t-1}, dropping the dependence on the expert advice. Hence, S will be defined here by the sequence $S_1, S_2, \ldots$ of functions $S_t : \mathcal{Y}^{t-1} \to \mathcal{X}$. Note that, unlike the definition given in the Introduction, here a forecasting strategy is formally equivalent to an expert.

Our main quantity of interest in this section is the *worst-case regret with respect to a set $\mathcal{F}$ of* *experts*. For a given forecasting strategy S, this is defined by

$$V_n^{\text{abs}}(S,\mathcal{F}) \stackrel{\text{def}}{=} \max_{\mathcal{Y}^n} \sup_{\mathcal{F}} \left(L_S(y^n) - L_F(y^n)\right) .$$

3.5.1 *Convex hulls of experts*

As we saw in the simple example with two experts, the forecaster can exploit the knowledge that the predictions of the experts are "similar". In general, this knowledge could be used to reduce the "effective size" of the set of experts, as we explain in the next example. Suppose that the predictions of all experts lie in the convex hull of the predictions of some (small) set of experts. Formally, if $F^{(1)},\dots,F^{(N)}$ are experts then their convex hull $\text{co}(F^{(1)},\dots,F^{(N)})$ contains any expert F for which there exist coefficients $q_1,\dots,q_N \geq 0$ with $q_1+\dots+q_N = 1$ such that, for any hidden sequence y and for all $t \geq 1$,

$$F_t(y^{t-1}) = \sum_{j=1}^{N} F_t^{(j)}(y^{t-1})\, q_j .$$

The next result shows that the hardness of minimizing the worst-case regret with respect to any subset $\mathcal{F} \subseteq \text{co}(F^{(1)},\dots,F^{(N)})$ depends only on N and not on the number of experts in $\mathcal{F}$.

Theorem 3.6. *If $\mathcal{F}$ is a subset of $\text{co}(F^{(1)},\dots,F^{(N)})$, then the exponentially weighted average forecaster S^{exp}, run on $\{F^{(1)},\dots,F^{(N)}\}$ with parameter $\eta = \sqrt{8\ln N/n}$, satisfies*

$$V_n^{\text{abs}}(S^{\text{exp}},\mathcal{F}) \leq \sqrt{\frac{n}{2}\ln N} .$$

PROOF. The key property is that for any bit sequence $y^n \in \{0,1\}^n$ and expert $F = \sum_{j=1}^{N} F^{(j)} q_j \in \mathcal{F}$ there exists an expert among $F^{(1)},\dots,F^{(N)}$ whose loss on y^n is not larger than that of F. To see this note that

$$\begin{aligned}
L_F(y^n) &= \sum_{t=1}^{n} |F_t(y^{t-1}) - y_t| \\
&= \sum_{t=1}^{n} \left| \sum_{j=1}^{N} F_t^{(j)}(y^{t-1}) - y_t \right| q_j \\
&= \sum_{t=1}^{n} \sum_{j=1}^{N} \left| F_t^{(j)}(y^{t-1}) - y_t \right| q_j \\
&= \sum_{j=1}^{N} \left(\sum_{t=1}^{n} \left| F_t^{(j)}(y^{t-1}) - y_t \right| \right) q_j \\
&= \sum_{j=1}^{N} L_j(y^n) q_j \\
&\geq \min_{1 \leq i \leq N} L_i(y^n) .
\end{aligned}$$

Thus, for all $y^n \in \{0,1\}^n$,

$$\inf_{\mathcal{F}} L_F(y^n) \geq \min_{1 \leq i \leq N} L_i(y^n) .$$

An application of Theorem 3.2 then concludes the proof. □

3.5.2 Geometry of expert sets

A notion that quantifies the extent to which the experts in a set $\mathcal{F}$ make similar predictions is r-covering. An *r-covering* of $\mathcal{F}$ is any set $\mathcal{F}_r$ of experts (not necessarily all belonging to $\mathcal{F}$) such that, for all $F \in \mathcal{F}$,

$$\max_{y^n} \min_{G \in \mathcal{F}_r} \sum_{t=1}^{n} \left| G_t(y^{t-1}) - F_t(y^{t-1}) \right| \leq r .$$

The *r-covering number* of $\mathcal{F}$, denoted by $N_1(\mathcal{F}, r)$, is the size of the smallest r-covering of $\mathcal{F}$.

If the initial set $\mathcal{F}$ of experts is very large, so large that the bound of Theorem 3.2 is not useful, then a forecaster might listen only to the advice of the experts in the smallest r-covering. By doing that, he can trade-off the radius of the covering number with the additional loss incurred by ignoring the advice of experts in $\mathcal{F}$ which are not in the r-covering.

COROLLARY 3.4. *For any set $\mathcal{F}$ of experts and for any $r \geq 0$, if the exponentially weighted average forecaster $S^{\exp}$ is run on a r-covering of $\mathcal{F}$ of size $N_1(\mathcal{F}, r)$ with $\eta = \sqrt{8 \ln N_1(\mathcal{F}, r)/n}$, then*

$$V_n^{\text{abs}}(S^{\exp}, \mathcal{F}) \leq r + \sqrt{\frac{n}{2} \ln N_1(\mathcal{F}, r)} .$$

PROOF. It is enough to note that, for all $y^n \in \{0,1\}^n$ and for all $F \in \mathcal{F}$ there exists $G \in \mathcal{F}_r$ such that

$$\begin{aligned}
&L_S(y^n) - L_F(y^n) \\
&= L_S(y^n) - \sum_{t=1}^{n} \left|F_t(y^{t-1}) - y_t\right| \\
&\le L_S(y^n) - \sum_{t=1}^{n} \left(\left|F_t(y^{t-1}) - G_t(y^{t-1})\right| - \left|G_t(y^{t-1}) - y_t\right|\right) \\
&\le L_S(y^n) - \sum_{t=1}^{n} \left|G_t(y^{t-1}) - y_t\right| + r
\end{aligned}$$

Then, by Theorem 3.2 and using the fact that $G \in \mathcal{F}_r$ with $|\mathcal{F}_r| = N_1(\mathcal{F}, r)$,

$$L_S(y^n) - \sum_{t=1}^{n} \left|F_t(y^{t-1}) - y_t\right| \le \sqrt{\frac{n}{2} \ln N_1(\mathcal{F}, r)}$$

concluding the proof. □

Monotone experts.

We may apply Corollary 3.4 to the set $\mathcal{F}$ of all monotone experts. An expert F is *monotone* if either $F_t \le F_{t+1}$ for all $t \ge 1$ or $F_t \ge F_{t+1}$ for all $t \ge 1$. So $\mathcal{F}$ is a large "nonparametric" set of experts. We now upper bound the log of $N_1(\mathcal{F}, r)$ for any $0 < r < n$. Let $\mathcal{F}_r$ be the r-covering of $\mathcal{F}$ including all monotone experts taking values in

$$\{(2k+1)r/n : k = 0, 1, \ldots, m\}$$

where m is the largest integer such that $(2m+1)r/n \le 1$. Then $m \le \lfloor n/(2r) \rfloor$. Let $d = m+1$ be the cardinality of the range of the functions in $\mathcal{F}_r$. Clearly, $N_1(\mathcal{F}, r) \le |\mathcal{F}_r| \le 2\binom{n+d}{d}$. Using

$$\ln \binom{n+d}{d} \le d(\ln(1 + n/d) + 1)$$

and $n/d = O(r)$ we get $\ln N_1(\mathcal{F}, r) = O((n/r)\ln r)$. Therefore, applying Corollary 3.4 with $r = n^{2/3}(\ln n)^{1/3}$ yields $V_n^{\text{abs}}(\mathcal{F}) = O(n^{2/3}(\ln n)^{1/3})$.

3.5.3 A minimax optimal forecaster for the absolute loss

So far we have seen how the exponentially weighted forecaster, but the same could be argued for any other potential-based forecaster, can be modified so to exploit knowledge about correlations between the experts' predictions. The purpose of this section is to design a general forecasting strategy that exploits this knowledge in the best possible way.

Let $\mathcal{F}$ be a set of experts and fix some $n \geq 1$. We say that a forecasting strategy S^* is *minimax optimal* if

$$S^* = \operatorname*{arginf}_S V_n^{\text{abs}}(S, \mathcal{F})$$

where the infimum is taken with respect to all nonanticipative strategies S. We call the quantity

$$V_n^{\text{abs}}(\mathcal{F}) \stackrel{\text{def}}{=} \inf_S V_n^{\text{abs}}(S, \mathcal{F})$$

minimax regret for the expert set $\mathcal{F}$.

We will determine the optimal forecasting strategy S^* "backwards", starting with the prediction at time n. Assume that the first $n-1$ outcomes y^{n-1} have been revealed and we want to determine, optimally, the prediction $\widehat{p}_n = S_n^*(y^{n-1})$. Recalling the definition of $V_n(\mathcal{F})$, we need to choose $\widehat{p}_n$ in order to minimize

$$L_{S^*}(y^{n-1}) + \max\left\{\ell(\widehat{p}_n, 0) - \inf_{\mathcal{F}} L_F(y^{n-1}0),\ \ell(\widehat{p}_n, 1) - \inf_{\mathcal{F}} L_F(y^{n-1}1)\right\}$$

where $y^{n-1}0$ denotes the string of n bits whose first $n-1$ bits are y^{n-1} and the last bit is 0. Minimizing this quantity is equivalent to minimizing

$$\max\left\{\widehat{p}_n - \inf_{\mathcal{F}} L_F(y^{n-1}0),\ 1 - \widehat{p}_n - \inf_{\mathcal{F}} L_F(y^{n-1}1)\right\}\ .$$

If we write $A_n(y^n) = -\inf_{\mathcal{F}} L_F(y^n)$ then this minimization is achieved by

$$\widehat{p}_n = \begin{cases} 0 & \text{if } A_n(y^{n-1}0) > A_n(y^{n-1}1) + 1 \\ 1 & \text{if } A_n(y^{n-1}0) + 1 < A_n(y^{n-1}1) \\ \dfrac{1}{2} + \dfrac{A_n(y^{n-1}1) - A_n(y^{n-1}0)}{2} & \text{otherwise.} \end{cases}$$

A crucial observation is that $\widehat{p}_n$ does not depend on the previous predictions $\widehat{p}_1, \ldots, \widehat{p}_{n-1}$. We now determine optimally $\widehat{p}_{n-1}$, assuming that at time n the optimal prediction is used. Define

$$\begin{aligned} A_{n-1}(y^{n-1}) &\stackrel{\text{def}}{=} \min_{\widehat{p}_n \in [0,1]} \max\left\{\widehat{p}_n - \inf_{\mathcal{F}} L_F(y^{n-1}0),\ 1 - \widehat{p}_n - \inf_{\mathcal{F}} L_F(y^{n-1}1)\right\} \\ &= \min_{\widehat{p}_n \in [0,1]} \max\left\{\widehat{p}_n + A_n(y^{n-1}0),\ 1 - \widehat{p}_n + A_n(y^{n-1}1)\right\}\ . \end{aligned}$$

Determining $\widehat{p}_{n-1}$ is then equivalent to minimizing

$$L_{S^*}(y^{n-2}) + \max\left\{\ell(\widehat{p}_{n-1}, 0) + A_{n-1}(y^{n-2}0),\ \ell(\widehat{p}_{n-1}, 1) + A_{n-1}(y^{n-2}1)\right\}\ ,$$

or, equivalently, to minimizing

$$\max\left\{\widehat{p}_{n-1} + A_{n-1}(y^{n-2}0),\ 1 - \widehat{p}_{n-1} + A_{n-1}(y^{n-2}1)\right\}\ .$$

Algorithm: Minimax Optimal Forecaster.
Initialization: $A_n(y^n) = -\inf_{\mathcal{F}} L_F(y^n)$

Recurrence: For $t = n, n-1, \ldots, 1$

$$A_{t-1}(y^{t-1}) = \begin{cases} A_t(y^{t-1}0) & \text{if } A_t(y^{t-1}0) > A_t(y^{t-1}1) + 1 \\ A_t(y^{t-1}1) & \text{if } A_t(y^{t-1}0) + 1 < A_t(y^{t-1}1) \\ \dfrac{A_t(y^{t-1}1) + A_t(y^{t-1}0) + 1}{2} & \text{otherwise} \end{cases}$$

and

$$\widehat{p}_t = \begin{cases} 0 & \text{if } A_t(y^{t-1}0) > A_t(y^{t-1}1) + 1 \\ 1 & \text{if } A_t(y^{t-1}0) + 1 < A_t(y^{t-1}1) \\ \dfrac{A_t(y^{t-1}1) - A_t(y^{t-1}0) + 1}{2} & \text{otherwise.} \end{cases}$$

FIGURE 3.1: A minimax optimal forecaster for a given set of experts.

Thus, the optimal prediction $\widehat{p}_{n-1}$ has the same formula as the formula for $\widehat{p}_n$ where n is replaced by $n-1$. Clearly, the procedure may be continued in the same way until we determine $\widehat{p}_1$. Formally, given a set $\mathcal{F}$ of experts and a positive integer n, the minimax predictor is described in Figure 3.1.

The algorithm for calculating the minimax optimal predictions has an important by-product: the value A_0 of the quantity $A_{t-1}(y^{t-1})$ at the last step $t = 1$ of the recurrence gives the minimax regret

$$A_0 = \max_{\mathcal{Y}^n} \left(\sum_{t=1}^{n} \ell(\widehat{p}_t, y_t) + A_n(y^n) \right) = V_n^{\text{abs}}(\mathcal{F}) .$$

3.5.4 *Static experts*

In this section, we focus our attention to a special type of experts which we call static. An expert F is *static* if, for all $t \le n$ and all $y^{t-1} \in \mathcal{Y}^{t-1}$, $F_t(y^{t-1}) = f_t$, where $f_t \in \mathcal{X}$ is a constant depending on F and t only. In other words, the predictions of a static expert do not depend on the past outcomes. For example, the expert which predicts always 0 regardless of the past outcomes is static, but the expert whose prediction is the average of all previously seen outcomes is not static. We will use f_t to denote the prediction of a static expert F at round t.

The next simple technical result has some surprising consequences. Its simple inductive proof is left as an exercise.

Lemma 3.1. *Let $\mathcal{F}$ be an arbitrary set of static experts. Then, for all $t \le n$ and all $y^{t-1} \in \mathcal{Y}^{t-1}$, $|A_t(y^{t-1}1) - A_t(y^{t-1}0)| \le 1$.*

This lemma implies the following interesting result.

Theorem 3.7. *If $\mathcal{F}$ is a set of static experts, then*

$$V_n^{\mathrm{abs}}(\mathcal{F}) = \frac{n}{2} - \mathbb{E}\left[\inf_{\mathcal{F}} L_F(Y^n)\right]$$

where $Y_1, \ldots, Y_n$ are independent symmetric Bernoulli random variables, i.e. $\mathbb{P}(Y_t = 0) = \mathbb{P}(Y_t = 1) = 1/2$.

PROOF. Lemma 3.1 and the definition of A_t given in Figure 3.1 imply that, for all $t = 1, \ldots, n$,

$$A_{t-1}(y^{t-1}) = \frac{A_t(y^{t-1}1) + A_t(y^{t-1}0) + 1}{2} .$$

Applying this equation recursively, we obtain

$$A_0 = \frac{1}{2^n} \sum_{\mathcal{Y}^n} A_n(y^n) + \frac{n}{2} .$$

Recalling that $V_n^{\mathrm{abs}}(\mathcal{F}) = A_0$ and $A_n(y^n) = -\inf_{\mathcal{F}} L_F(y^n)$ concludes the proof. □

Note that the proof of Theorem 3.7 also reveals that, in case of static experts, the minimax optimal forecaster of Figure 3.1 takes the simple form

$$\widehat{p}_t = \frac{1}{2} + \mathbb{E}\left[\frac{\inf_{\mathcal{F}} L_F(y^{t-1}1Y^{n-t}) - \inf_{\mathcal{F}} L_F(y^{t-1}0Y^{n-t})}{2}\right]$$

where $Y_1, \ldots, Y_n$ are independent symmetric Bernoulli random variables.

To understand better the behavior of the value $V_n^{\mathrm{abs}}(\mathcal{F})$ it will be advantageous to reformulate the obtained expression. Recall that, since experts are static, each expert F may be represented by a vector $(f_1, \ldots, f_n) \in [0,1]^n$ where f_t is the prediction of F at time t. Since $\ell(f_t, y) = |f_t - y|$, $L_F(y^n) = \sum_{t=1}^n |f_t - y_t|$. Then, by Theorem 3.7,

$$\begin{aligned} V_n^{\mathrm{abs}}(\mathcal{F}) &= \frac{n}{2} - \mathbb{E}\left[\inf_{\mathcal{F}} \sum_{t=1}^n |f_t - Y_t|\right] \\ &= \mathbb{E}\left[\sup_{\mathcal{F}} \sum_{t=1}^n \left(\frac{1}{2} - |f_t - Y_t|\right)\right] \\ &= \mathbb{E}\left[\sup_{\mathcal{F}} \sum_{t=1}^n \left(\frac{1}{2} - f_t\right) Z_t\right] \qquad (3.12) \end{aligned}$$

where $Z_t = 1 - 2Y_t$ are i.i.d. random variables. In statistics, such random variables Z_t are called *Rademacher*.

Putting $V_n^{\mathrm{abs}}(\mathcal{F})$ in this form opens the toolbox of probability theory in which expected values of this form have been very well understood. For illustration purposes we first consider a simple example that already offers us important messages. In this section and in the next one we will obtain general upper and lower bounds on $V_n^{\mathrm{abs}}(\mathcal{F})$ based on its characterization in terms of Rademacher averages.

Two maximally distant experts.

Assume that $\mathcal{F}$ only contains the two "extreme" static experts $F^{(1)} = (0,\dots,0)$ and $F^{(2)} = (1,\dots,1)$. Recall that, by Theorem 3.2,

$$V_n^{\mathrm{abs}}(\mathcal{F}) \le \sqrt{\frac{n \ln 2}{2}} \approx 0.5887\sqrt{n}\,.$$

Next we contrast this bound with bounds obtained directly using (3.12). Since $f_t^{(1)} = 0$ and $f_t^{(2)} = 1$ for all t,

$$V_n^{\mathrm{abs}}(\mathcal{F}) = \frac{1}{2}\mathbb{E}\left[\max\left\{\sum_{t=1}^n Z_t, \sum_{t=1}^n -Z_t\right\}\right] = \frac{1}{2}\mathbb{E}\left|\sum_{t=1}^n Z_i\right|\,.$$

We may easily bound this quantity, from above, using the Cauchy-Schwarz inequality,

$$V_n^{\mathrm{abs}}(\mathcal{F}) = \frac{1}{2}\mathbb{E}\left|\sum_{t=1}^n Z_t\right| \le \frac{1}{2}\sqrt{\mathbb{E}\left[\left(\sum_{t=1}^n Z_t\right)^2\right]} = \frac{1}{2}\sqrt{\mathbf{Var}\left[\sum_{t=1}^n Z_t\right]} = \frac{\sqrt{n}}{2}\,.$$

Observe that this bound has the same order of magnitude as the bound obtained by Theorem 3.2 but it has a slightly better constant.

We may obtain a similar lower bound as an easy consequence of Khinchine's inequality, which we recall here.

Theorem 3.8. (KHINCHINE'S INEQUALITY) *Let $a_1,\dots,a_n$ be real numbers, and let $Z_1,\dots,Z_n$ be i.i.d. Rademacher random variables. Then*

$$\mathbb{E}\left|\sum_{i=1}^n a_i Z_i\right| \ge \frac{1}{\sqrt{2}}\sqrt{\sum_{i=1}^n a_i^2}\,.$$

Applying Theorem 3.8 to our problem we obtain the lower bound

$$V_n^{\mathrm{abs}}(\mathcal{F}) \ge \sqrt{\frac{n}{8}}\,.$$

Summarizing the upper and lower bounds, for every n, we have

$$0.3535 \le \frac{V_n^{\text{abs}}(\mathcal{F})}{\sqrt{n}} \le 0.5 \,.$$

For example, for $n = 100$, there exists a prediction strategy such that for any sequence $y_1, \ldots, y_{100}$ the expected number of mistakes is not more than that of the best expert plus 5, but for any prediction strategy there exists a sequence $y_1, \ldots, y_{100}$ such that the regret is at least 3. The proof of the exact asymptotic value

$$\lim_{n\to\infty} V_n^{\text{abs}}(\mathcal{F})/\sqrt{n} = 1/\sqrt{2\pi} \approx 0.3989$$

is left as an exercise.

We now use Theorem 3.7 and the rich theory of empirical processes to obtain an upper bound on $V_n^{\text{abs}}(\mathcal{F})$ for general classes $\mathcal{F}$ of static experts. This upper bound is expressed in terms of the covering numbers of $\mathcal{F}$, a notion that we introduced in Section 3.5.2 for sets of general (nonstatic) experts. The covering numbers suitable for the analysis of the minimax regret for static experts are defined as follows. For any class $\mathcal{F}$ of static experts, let $N_2(\mathcal{F}, r)$ be the minimum cardinality of a set $\mathcal{F}_r$ of static experts (possibly not all belonging to $\mathcal{F}$) such that, for all $F = (f_1, \ldots, f_n) \in \mathcal{F}$ there exists some $G = (g_1, \ldots, g_n) \in \mathcal{F}_r$ such that

$$\sqrt{\sum_{t=1}^{n} (f_t - g_t)^2} \le r \,.$$

Note that this metric is simply the Euclidean distance between the real vectors representing the experts. The following bound shows how $V_n^{\text{abs}}(\mathcal{F})$ may be bounded from the above in terms of these covering numbers.

Theorem 3.9. *For any class $\mathcal{F}$ of static experts,*

$$V_n^{\text{abs}}(\mathcal{F}) \le 12 \int_0^{\sqrt{n}/2} \sqrt{\ln(N_2(\mathcal{F}, r) + 1)}\, dr \,.$$

The result is a straightforward corollary of Theorem 3.7, Chernoff-Hoeffding bounds, and the following classical result of empirical process theory due to Dudley — see e.g. Dudley (2001). To state the result in a general form, consider a family $\mathcal{T} = \{T_F : F \in \mathcal{F}\}$ of zero mean random variables indexed by a metric space $(\mathcal{F}, \rho)$. Let $N_\rho(\mathcal{F}, r)$ denote the covering number of the metric space $\mathcal{F}$ with respect to the metric ρ. The family $\mathcal{T}$ is called *subgaussian* in the metric ρ whenever

$$\mathbb{E}\left[e^{\lambda(T_F - T_G)}\right] \le e^{\lambda^2 \rho(F,G)^2/2}$$

holds for any $F, G \in \mathcal{F}$ and $\lambda > 0$. The family $\mathcal{T}$ is called *sample continuous* if for any sequence $F^{(1)}, F^{(2)}, \ldots \in \mathcal{F}$ converging in the metric ρ to some $F \in \mathcal{F}$, we have $T_{F^{(n)}} - T_F \to 0$ almost surely.

Theorem 3.10. *If $\{T_F : F \in \mathcal{F}\}$ is subgaussian and sample continuous in the metric ρ, then*

$$\mathbb{E}\left[\sup_{F\in\mathcal{F}} T_F\right] \le 12\int_0^{D/2} \sqrt{\ln N_\rho(\mathcal{F},\varepsilon)}\, d\varepsilon$$

where D is the diameter of $\mathcal{F}$.

Monotone experts revisited.

The class $\mathcal{F}$ of monotone experts introduced in Section 3.5.2 is static. Therefore, we may apply Theorem 3.9 to get an upper bound on $V_n^{\mathrm{abs}}(\mathcal{F})$. We first derive an upper bound on $N_2(\mathcal{F},r)$ Proceeding similarly to what we did for calculating an upper bound on $N_1(\mathcal{F},r)$, we get

$$\ln N_2(\mathcal{F},r) = O\left(\frac{\sqrt{n}}{r}\ln(rn)\right) .$$

An application of Theorem 3.9 then yields

$$V_n^{\mathrm{abs}}(\mathcal{F}) = O\left(\sqrt{n}\ln n\right)$$

a much better bound than what we got using the exponentially weighted average forecaster of Corollary 3.4.

Sets of experts with bounded VC-dimension.

In the special case when all experts $F \in \mathcal{F}$ are "resolute", i.e. $f_t \in \{0,1\}$ for all $t \ge 1$, then a quantity which may be used to obtain good bounds on the covering numbers is the Vapnik-Chervonenkis (VC) dimension. If a set $\mathcal{F}$ has VC-dimension bounded by a positive constant d, then

$$N_2(\mathcal{F},r) = O\left((n/r^2)^d\right) .$$

For such sets, Theorem 3.9 gives the bound

$$V_n^{\mathrm{abs}}(\mathcal{F}) = O\left(\sqrt{dn}\right) .$$

This bound is best possible in the sense that, for each $d \ge 1$, there exists sets $\mathcal{F}$ of VC-dimension d for which $V_n^{\mathrm{abs}}(\mathcal{F}) = \Omega\left(\sqrt{dn}\right)$.

3.5.5 A general lower bound

Equipped with the results obtained for static experts, we now return to the scenario, introduced in Section 3.2, where the expert advice is not available to the forecaster. Our main quantity of interest is now the worst-case regret for a strategy S defined by

$$V_n^{\mathrm{abs}}(S,N) \stackrel{\mathrm{def}}{=} \sup_{z^n}\left(\sum_{t=1}^{n}\left|S_t(z^{t-1})-y_t\right| - \min_{1\le i\le N}\sum_{t=1}^{n}|x_{i,t}-y_t|\right)$$

which in (3.2) we defined for general losses. Recall that $z_t = (y_t, (x_{1,t}, \ldots, x_{N,t}))$, where $x_{i,t} \in \mathcal{X}$ is the advice of expert i at round t.

For the absolute loss, the best upper bound we derived was

$$V_n^{\text{abs}}(S^{\text{exp}}, N) \le \sqrt{(n/2)\ln N} \tag{3.13}$$

achieved by the exponentially weighted average forecaster (see Theorem 3.2). In this section, we build on the results of previous sections and prove a lower bound on the quantity

$$V_n^{\text{abs}}(N) \stackrel{\text{def}}{=} \inf_S V_n^{\text{abs}}(S, N) \tag{3.14}$$

asymptotically matching the upper bound (3.13) for S^{exp}. This shows that the exponentially weighted average forecaster is, in a sense, optimal when the set of experts is unknown.

Theorem 3.11.

$$\sup_{n,N \ge 1} \frac{V_n^{\text{abs}}(N)}{\sqrt{(n/2)\ln N}} \ge 1 .$$

PROOF. Using (3.14), we lower bound $V_n^{\text{abs}}(N)$ by computing a lower bound on $V_n^{\text{abs}}(\mathcal{F})$ for a randomly chosen set $\mathcal{F}$ of N experts. More precisely, let $\mathcal{F}$ contain experts of the form $F^{(i)} = (F_1^{(i)}, \ldots, F_n^{(i)})$, $1 \le i \le N$, where each $F_t^{(i)}$ is a symmetric Bernoulli random variable. Then, by (3.12),

$$\begin{aligned} V_n^{\text{abs}}(\mathcal{F}) &= \mathbb{E}\left[\sup_{F \in \mathcal{F}} \sum_{t=1}^n \left(\frac{1}{2} - F_t\right) Z_t\right] \\ &\ge \frac{1}{2}\mathbb{E}\left[\max_{1 \le i \le N} \sum_{t=1}^n Z'_{i,t} Z_t\right] \\ &= \frac{1}{2}\mathbb{E}\left[\max_{1 \le i \le N} \sum_{t=1}^n Z'_{i,t}\right] \end{aligned}$$

where $\{Z'_{i,t}\}$ is a set of $N \times n$ i.i.d. Rademacher random variables. By the Central Limit Theorem, for each $i = 1, \ldots, N$, the random variable $n^{-1/2}\sum_{t=1}^n Z_{i,t}$ converges to a standard normal random variable. In fact, it is not difficult to show (proof omitted) that

$$\lim_{n \to \infty} \mathbb{E}\left[\max_{1 \le i \le N} \frac{1}{\sqrt{n}} \sum_{t=1}^n Z_{i,t}\right] = \mathbb{E}\left[\max_{1 \le i \le N} G_i\right]$$

where $G_1, \ldots, G_N$ are independent standard normal random variables. it is then well known (see e.g. Galambos (1987)) that

$$\lim_{N \to \infty} \frac{\mathbb{E}[\max_{1 \le i \le N} G_i]}{\sqrt{2 \ln N}} = 1$$

which concludes the proof. □

Bibliographic remarks

The earliest results on absolute loss, viewed as probability of making a mistake in a binary decision problem, are special cases of theorems by Hannan (1957) and Blackwell (1956). The analysis of absolute loss in the prediction with experts framework was carried out by Littlestone and Warmuth (1991, 1994) with their Continuous Weighted Majority algorithm and by Vovk (1990, 1998). A pioneering paper is the one by Feder, Merhav, and Gutman (1992), where the problem of finding a consistentl forecaster for the expert set of all finite state automata is solved by covering this set with the subclass of Markov automata, and then deriving a consistent forecaster for this latter expert set. The minimax worst case regret for the absolute loss was investigated in Cesa-Bianchi, Helmbold, Haussler, Freund, Schapire, and Warmuth (1997) for finite set of experts and in Cesa-Bianchi and Lugosi (1999) for infinite sets of experts. The minimax optimal forecaster was defined in Cesa-Bianchi, Helmbold, Haussler, Freund, Schapire, and Warmuth (1997) and further studied, also for general losses, in Chung (1994). Theorem 3.7 for static experts is from Cesa-Bianchi, Helmbold, Haussler, Freund, Schapire, and Warmuth (1997), whereas the result for two maximally distant experts is due to Cover (1965). The connection between empirical process theory, as well as all the remaining results of Section 3.5, are from Cesa-Bianchi and Lugosi (1999). For excellent surveys on empirical process theory see Pollard (1990) and Giné (1996).

3.6 Logarithmic loss

This section is entirely devoted to the investigation of a special loss function, the *logarithmic* loss, sometimes also called *self information* loss. The reason for this special attention is that, as we will see, this loss function has a meaningful interpretation in various sequential decision problems, including gambling, data compression, and portfolio management.

We describe the setup for the entire section. Let $m > 1$ be a fixed positive integer, and let the outcome space be $\mathcal{Y} = \{1, 2, \ldots, m\}$. The decision space $\mathcal{X}$ is the simplex

$$\left\{ (Q(1), \ldots, Q(m)) : \sum_{j=1}^{m} Q(j) = 1,\ Q(j) \geq 0, j = 1, \ldots, m \right\}. \tag{3.15}$$

Viewing an element Q of this simplex as a probability distribution over the set $\mathcal{Y}$, we may naturally apply this setup to those cases where the forecaster is required to assign a probability, which may be viewed as the forecaster's belief, to each one of the possible outcomes. For example, wheather forecasters predictions usually have the form "the possibility of rain is 40%".

As in Section 3.5, we will study the predictive performance of forecasting strategies assuming that the expert advice is completely accessible to the strategy. Hence, similarly to Section 3.5, we define a forecaster/expert F (forecasters and experts are again defined the same way) as a sequence $F_1, F_2, \ldots$ of functions $F_t : \mathcal{Y}^{t-1} \to \mathcal{X}$ mapping the observed outcomes y^{t-1} to a probability distribution $F_t(\cdot|y^{t-1}) \in \mathcal{X}$, where

$$F_t(\cdot|y^{t-1}) = \left(F_t(1|y^{t-1}), \ldots, F_t(m|y^{t-1})\right) .$$

Note that, as suggested by the notation, $F_t(j|y^{t-1})$ may be interpreted as the *conditional* probability F assigns to $j \in \mathcal{Y}$ given the past y^{t-1}.

Appealing to the analogy with probability distributions, we may introduce, for any forecaster/expert F and for all $y^n \in \mathcal{Y}^n$, the notation

$$F(y^n) = \prod_{s=1}^{n} F_s(y_s|y^{s-1}) \qquad \text{and} \qquad F(y^t) = \sum_{y_{t+1},\ldots,y_n \in \mathcal{Y}} F(y_1,\ldots,y_n) \quad \text{for all } t = 1,\ldots,n.$$

Observe that

$$\sum_{\mathcal{Y}^n} F(y^n) = 1$$

and, therefore, the forecaster/expert F defines a probability distribution F over the set $\mathcal{Y}^n$ of all sequences of length n. Conversely, any probability distribution Q over the set $\mathcal{Y}^n$ induces a forecaster/expert defined by the conditional distributions

$$Q_t(y_t|y^{t-1}) = \frac{Q(y^t)}{Q(y^{t-1})} .$$

The loss function we consider is defined by

$$\ell(Q,y) = \sum_{j=1}^{m} \mathbb{I}_{[y=j]} \ln \frac{1}{Q(j)} = \ln \frac{1}{Q(y)} .$$

For an outcome sequence y^n, the cumulative loss of expert F and forecaster S are, respectively,

$$L_F(y^n) = \sum_{t=1}^{n} \ln \frac{1}{F_t(y_t|y^{t-1})} \qquad L_S(y^n) = \sum_{t=1}^{n} \ln \frac{1}{S_t(y_t|y^{t-1})} .$$

Given a set $\mathcal{F}$ of experts, the regret may now be written as

$$\begin{aligned} L_S(y^n) - \inf_{\mathcal{F}} L_F(y^n) &= \sum_{t=1}^{n} \ln \frac{1}{S_t(y_t|y^{t-1})} - \inf_{\mathcal{F}} \sum_{t=1}^{n} \ln \frac{1}{F_t(y_t|y_{t-1})} \\ &= \sup_{\mathcal{F}} \ln \frac{F(y^n)}{S(y^n)} . \end{aligned}$$

Thus, the regret may be interpreted as the logarithm of the ratio of the total probabilities that are sequentially assigned to the outcome sequence by the forecaster and the experts.

In these notes we restrict our discussion to the case when the outcome space is a finite set. Note, however, that most results may be extended easily to the more general case when $\mathcal{Y}$ is a measurable space. In such cases the decision space becomes the set of all densities over $\mathcal{Y}$ with respect to some fixed common dominating measure, and the loss is the negative logarithm of the density evaluated at the outcome.

3.6.1 *Mixture forecasters*

Recall from Section 3.4 that the logarithmic loss function is exp-concave and so Theorem 3.5 applies. Thus, if the set $\mathcal{F}$ of experts is finite and $|\mathcal{F}| = N$, then the exponentially weighted average forecaster $S^{\exp}$ with parameter $\eta = 1$ satisfies

$$L_{S^{\exp}}(y^n) - \inf_{\mathcal{F}} L_F(y^n) \le \ln N$$

or equivalently,

$$S^{\exp}(y^n) \ge \frac{1}{N} \sup_{\mathcal{F}} F(y^n) \, .$$

It is worth noting that $S^{\exp}$ has an interesting interpretation in this special case. Observe that

$$S_t^{\exp}(y_t|y^{t-1}) = \frac{\sum_{\mathcal{F}} F_t(y_t|y^{t-1}) e^{-\eta L_F(y^{t-1})}}{\sum_{\mathcal{F}} e^{-\eta L_F(y^{t-1})}}$$

with $\eta = 1$ reduces to

$$S_t^{\exp}(y_t|y^{t-1}) = \frac{\sum_{\mathcal{F}} F_t(y_t|y^{t-1}) F(y^{t-1})}{\sum_{\mathcal{F}} F(y^{t-1})} = \frac{\sum_{\mathcal{F}} F(y^t)}{\sum_{\mathcal{F}} F(y^{t-1})} \, .$$

Thus, the total probability $S^{\exp}$ assigns to a sequence y^n is just

$$S^{\exp}(y^n) = \prod_{t=1}^{n} \frac{\sum_{\mathcal{F}} F(y^t)}{\sum_{\mathcal{F}} F(y^{t-1})} = \frac{\sum_{\mathcal{F}} F(y^n)}{N}$$

as $F(y^0)$ is defined to be equal to one. In other words, the probability distribution $S^{\exp}$ defines over the set $\mathcal{Y}^n$ of all sequences of length n is just the uniform *mixture* of the distributions defined by the experts. Thus, we will sometimes call $S^{\exp}$ the *mixture forecaster*.

Countably many experts.

This latter observation suggests a straightforward extension to the case when $\mathcal{F}$ may contain countably many experts. Denote these experts by $F^{(i)}$, $i \ge 1$. To define the mixture forecaster, we assign a nonnegative number $q_i \ge 0$ to each expert $F^{(i)} \in \mathcal{F}$ such that $\sum_{i=1}^{\infty} q_i = 1$. The numbers

q_i may be called *prior probabilities.* Now we introduce the forecasting strategy S "backwards", by defining the joint probability it assigns to each sequence:

$$S(y^n) = \sum_{i=1}^{\infty} q_i F^{(i)}(y^n) .$$

Note that S indeed defines a valid probability distribution S over $\mathcal{Y}^n$. Mimicking the argument above, we see that this probability distribution defines the weighted average forecaster

$$\begin{aligned} S_t(y_t|y^{t-1}) &= \frac{\sum_{i=1}^{\infty} q_i F_t^{(i)}(y_t|y^{t-1}) F^{(i)}(y^{t-1})}{\sum_{j=1}^{\infty} q_j F^{(j)}(y^{t-1})} \\ &= \frac{\sum_{i=1}^{\infty} q_i F_t^{(i)}(y_t|y^{t-1}) e^{-\eta L_i(y^{t-1})}}{\sum_{j=1}^{\infty} q_j e^{-\eta L_j(y^{t-1})}} \end{aligned}$$

where we used L_i to denote $L_{F^{(i)}}$. Using the trivial bound

$$S(y^n) = \sum_{j=1}^{\infty} q_j F^{(j)}(y^n) \geq q_i F^{(i)}(y^n)$$

which holds for all $i \geq 1$, we obtain the following.

Theorem 3.12. *Let $\mathcal{F} = \left\{F^{(i)} : i \geq 1\right\}$ be an arbitrary set of experts and let $\{q_i : i \geq 1\}$ be a probability distribution over the set of positive integers. Then S defined in (3.16) satisfies, for all $y^n \in \mathcal{Y}$,*

$$L_S(y^n) \leq \inf_{i \geq 1} \left(L_i(y^n) + \ln \frac{1}{q_i} \right) .$$

Thus, the cumulative loss of the forecaster exceeds the loss of each expert by at most a constant, but the constant depends on the expert. If we write the forecaster in the form

$$S_t(y_t|y^{t-1}) = \frac{\sum_{i=1}^{\infty} F_t^{(i)}(y_t|y^{t-1}) \exp\left(-\eta \left(L_i(y^{t-1}) + \frac{1}{\eta} \ln \frac{1}{q_i}\right)\right)}{\sum_{j=1}^{\infty} \exp\left(-\eta \left(L_j(y^{t-1}) + \frac{1}{\eta} \ln \frac{1}{q_j}\right)\right)}$$

then we see that the quantity $(1/\eta)\ln(1/q_i)$ may be regarded as a "penalty" we add to the cumulative loss of expert i at each time t. Theorem 3.12 is a so-called "oracle inequality" stating that the the mixture forecaster achieves a cumulative loss matching the best penalized cumulative loss of the experts.

3.6.2 *The minimax optimal forecaster for the log loss*

In this section we investigate, in the logarithmic loss case, the *minimax regret* which we defined for the absolute loss in Section 3.5.3. For a given set $\mathcal{F}$ of experts, for an integer $n > 0$, and a forecasting strategy S, the worst-case regret of S with respect to the logarithmic loss is defined by

$$V_n^{\log}(S,\mathcal{F}) \stackrel{\text{def}}{=} \sup_{\mathcal{Y}^n}\left(L_S(y^n) - \inf_{\mathcal{F}} L_F(y^n)\right)$$

The minimax regret is then defined by

$$V_n^{\log}(\mathcal{F}) \stackrel{\text{def}}{=} \inf_S V_n^{\log}(S,\mathcal{F})\,. \tag{3.16}$$

As for the absolute loss, we will identify explicitly a strategy achieving the minimax loss. Furthermore, we will show that this is the unique minimax strategy. The next theorem shows that the strategy S^{mle}, defined by the *normalized maximum likelihood* probability

$$S^{\text{mle}}(y^n) \stackrel{\text{def}}{=} \frac{\sup_{\mathcal{F}} F(y^n)}{\sum_{x^n\in\mathcal{Y}^n}\sup_{\mathcal{F}} F(x^n)}\,,$$

has this property. Note that S^{mle} is indeed a probability distribution over the set $\mathcal{Y}^n$ and recall also that any probability distribution defines a strategy by the corresponding conditional probabilities $S_t^{\text{mle}}(y_t|y_1^{t-1})$.

Theorem 3.13. *For any class $\mathcal{F}$ of experts and integer $n > 0$, the normalized maximum likelihood forecaster S^{mle} is the unique forecaster such that*

$$\sup_{\mathcal{Y}^n}\left(L_{S^{\text{mle}}}(y^n) - \inf_{\mathcal{F}} L_F(y^n)\right) = V_n^{\log}(\mathcal{F})\,.$$

Moreover, S^{mle} is an equalizer, that is, for all $y^n \in \mathcal{Y}^n$,

$$\ln\frac{\sup_{\mathcal{F}} F(y^n)}{S^{\text{mle}}(y^n)} = \ln\sum_{x^n\in\mathcal{Y}^n}\sup_{\mathcal{F}} F(x^n) = V_n^{\log}(\mathcal{F})\,.$$

PROOF. We show the second part of the statement first. Note that, by the definition of S^{mle}, its regret satisfies

$$\begin{aligned} L_{S^{\text{mle}}}(y^n) - \inf_{\mathcal{F}} L_F(y^n) &= \ln\frac{\sup_{\mathcal{F}} F(y^n)}{S^{\text{mle}}(y^n)} \\ &= \ln\sum_{x^n\in\mathcal{Y}^n}\sup_{\mathcal{F}} F(x^n)\,, \end{aligned}$$

which is independent of y^n and so S^{mle} is indeed an equalizer.

To show that S^{mle} is minimax optimal, let $S \neq S^{\text{mle}}$ be an arbitrary forecasting strategy. Then, since $\sum_{\mathcal{Y}^n} S(y^n) = \sum_{\mathcal{Y}^n} S^{\text{mle}}(y^n) = 1$, for some $x^n \in \mathcal{Y}^n$ we must have $S(x^n) < S^{\text{mle}}(x^n)$. But then, for this x^n,

$$\ln\frac{\sup_{\mathcal{F}} F(x^n)}{S(x^n)} > \ln\frac{\sup_{\mathcal{F}} F(x^n)}{S^{\text{mle}}(x^n)} = V_n^{\log}(S^{\text{mle}},\mathcal{F})$$

by the equalizer property. Hence,

$$V_n^{\log}(S,\mathcal{F}) = \sup_{y^n} \ln \frac{\sup_{\mathcal{F}} F(y^n)}{S(y^n)} > V_n^{\log}(S^{\text{mle}},\mathcal{F})$$

which proves the theorem. □

Even though we have been able to determine the minimax optimal forecasting strategy explicitely, note that the practical implementation of this strategy may be problematic. First of all, we have determined the strategy via the joint probabilities it assigns to all strings of length n, and calculation of the actual predictions $S_t^{\text{mle}}(y_t|y^{t-1})$ involves sums of exponentially many terms. Also, previous knowledge of the total length n of the sequence to be predicted is necessary to determine S^{mle}. On the other hand, the theorem gives a useful formula for the minimax cumulative regret $V_n^{\log}(\mathcal{F})$, which we study in detail in the subsequent sections.

3.6.3 Examples

Next we work out a few simple examples to understand better the behavior of the minimax regret $V_n^{\log}(\mathcal{F})$.

Finite sets.

To start with the simplest possible case, consider a finite set of experts with $|\mathcal{F}| = N$. Then clearly

$$\begin{aligned} V_n^{\log}(\mathcal{F}) &= \ln \sum_{y^n} \sup_{\mathcal{F}} F(y^n) \\ &\le \ln \sum_{y^n} \sum_{\mathcal{F}} F(y^n) \\ &= \ln \sum_{\mathcal{F}} \sum_{y^n} F(y^n) \\ &= \ln N\,. \end{aligned}$$

Of course, we already knew this. In fact, the mixture forecaster described in Section 3.6.1 achieves the same bound. Furthermore, it has obvious computational advantages over the normalized maximum likelihood forecaster and it does not suffer from the "horizon-dependence" of the latter mentioned above. These bounds suggest that one does not lose much by using the simple uniform mixture forecaster instead of the optimal, but horizon-dependent, forecaster. This fact remains true in more general settings, even for certain infinite sets of experts. However, even if $\mathcal{F}$ is finite, in some cases $V_n^{\log}(\mathcal{F})$ may be significantly smaller than the worst-case regret achievable by any mixture forecaster.

In the next example we derive an asymptotic expression for the upper bound on the minimax regret for a simple parametric set of uncountably many experts. Later, we will prove a general upper bound for any set of experts which is finitable coverable according to a certain metric.

Constant experts.

Consider the set $\mathcal{F}$ of all experts such that, for each $F \in \mathcal{F}$, $F_t(j|y^{t-1}) = f_j$ (with $f_j \geq 0$ and $\sum_{j=1}^{m} f_j = 1$) independently of t and y^{t-1}. In other words, $\mathcal{F}$ contains all static forecasters F such that the associated probability distribution over $\mathcal{Y}^n$ is a product distribution with identical components. If $m = 2$, (i.e., $\mathcal{Y} = \{1,2\}$ and $\mathcal{X} = \{(q, 1-q) \in \mathcal{R}^2 : q \in [0,1]\}$), each expert in $\mathcal{F}$ may be identified with a number $q \in [0,1]$ representing $q = f_1$. Thus, this expert predicts, at each time t, according to the vector $(q, 1-q) \in \mathcal{X}$, regardless of t and the past outcomes $y_1, \ldots, y_{t-1}$. We call this set $\mathcal{F}$ the set of *constant experts*. Next, we state the asymptotic value of the minimax regret $V_n^{\log}(\mathcal{F})$.

Theorem 3.14. *If $\mathcal{F}$ is the set of all constant experts over the alphabet $\mathcal{Y} = \{1,2\}$, then*

$$V_n^{\log}(\mathcal{F}) = \frac{1}{2}\ln n + \frac{1}{2}\ln\frac{\pi}{2} + o(1)$$

where $o(1) \to 0$ as $n \to \infty$.

PROOF SKETCH. Recall that, by Theorem 3.13,

$$V_n^{\log}(\mathcal{F}) = \ln \sum_{\mathcal{Y}^n} \sup_{\mathcal{F}} F(y^n) .$$

Now assume that the number of 1's in the sequence $y^n \in \{1,2\}^n$ is k and the number of 2's is $n-k$. Then, for the expert F which predicts according to $(q, 1-q)$,

$$F(y^n) = q^k(1-q)^{n-k} .$$

It is easy to see (for example by differentiation) that the above expression is maximized for $q = k/n$, and therefore

$$V_n^{\log}(\mathcal{F}) = \ln \sum_{k=0}^{n} \binom{n}{k} \left(\frac{k}{n}\right)^k \left(\frac{n-k}{n}\right)^{n-k} \tag{3.17}$$

since there are $\binom{n}{k}$ sequences containing exactly k ones. The proof is finished by taking a careful asymptotic expansion of the right-hand side of (3.17). □

In the general case, when $m \geq 2$ is a positive integer, Theorem 3.14 becomes

$$V_n^{\log}(\mathcal{F}) = \frac{m-1}{2}\ln n + \ln\frac{\Gamma(1/2)^m}{\Gamma(m/2)} + o(1)$$

where Γ denotes the Gamma function. The fact that the minimax regret grows as a constant times $\ln n$, the constant being the half of the number of "free parameters", is a general phenomenon.

3.6.4 A general upper bound

We now investigate the minimax regret $V_n^{\log}(\mathcal{F})$ for general sets of experts. Similarly to what we did in Section 3.5.2, we will derive a bound showing how the geometry of the set $\mathcal{F}$ affects the regret. To any set $\mathcal{F}$ of experts, we associate the metric ρ defined as

$$\rho(F,G) = \sqrt{\sum_{t=1}^{n} \rho_t(F,G)^2},$$

where

$$\rho_t(F,G) = \sup_{\mathcal{Y}^t} \left(\ln F_t(y_t|y^{t-1}) - \ln G_t(y_t|y^{t-1})\right) .$$

Denote by $N_{\log}(\mathcal{F},\varepsilon)$ the ε-covering number of $\mathcal{F}$ under the metric ρ. Recall that, for any $\varepsilon > 0$, the ε-covering number is the cardinality of the smallest subset $\mathcal{G}$ (not necessarily included in $\mathcal{F}$) such that for all $F \in \mathcal{F}$ there exists a $G \in \mathcal{G}$ for which$d(F,G) \leq \varepsilon$.

Theorem 3.15. *For any set $\mathcal{F}$ of experts,*

$$V_n^{\log}(\mathcal{F}) \leq \inf_{\varepsilon>0} \left(\ln N_{\log}(\mathcal{F},\varepsilon) + 24 \int_0^{\varepsilon} \sqrt{\ln N_{\log}(\mathcal{F},\delta)}\, d\delta \right) .$$

Note that if $\mathcal{F}$ is a finite set then the right-hand side converges to $\ln|\mathcal{F}|$ as $\varepsilon \to 0$, and therefore we recover our earlier general bound. However, even for finite classes, the right-hand side may be significantly smaller than $\ln|\mathcal{F}|$. Also, this result allows us to derive upper bounds for very general sets of experts.

As a first step in the proof of Theorem 3.15 we obtain a weak bound for $V_n^{\log}(\mathcal{F})$. This weak bound, which will be later refined to prove the stronger bound of Theorem 3.15, is similar to the bound for static experts with the absolute loss proven in Theorem 3.9, Section 3.7.

Lemma 3.2. *For any set $\mathcal{F}$ of experts,*

$$V_n^{\log}(\mathcal{F}) \leq 24 \int_0^{D/2} \sqrt{\ln N(\mathcal{F},\varepsilon)}\, d\varepsilon$$

where D is the diameter of $\mathcal{F}$.

PROOF. Recall that, by the equalizer property of the normalized maximum likelihood forecaster S^{mle} established in Theorem 3.13, for all $y^n \in \mathcal{Y}^n$, $V_n^{\log}(\mathcal{F}) = \sup_{\mathcal{F}} \ln\left(F(y^n)/S^{\text{mle}}(y^n)\right)$. Since the right-hand side is a constant function of y^n, we may take any weighted average of it without changing its value. The trick is to weight the average according to the probability distribution defined by S^{mle}. Thus, we may write

$$V_n^{\log}(\mathcal{F}) = \sum_{\mathcal{Y}^n} \left(\sup_{\mathcal{F}} \ln \frac{F(y^n)}{S^{\text{mle}}(y^n)} \right) S^{\text{mle}}(y^n) .$$

If we introduce a vector $Y^n = (Y_1, \dots, Y_n)$ of random variables distributed according to S^{mle}, we obtain

$$\begin{aligned} V_n^{\log}(\mathcal{F}) &= \mathbb{E}\left[\sup_{\mathcal{F}} \ln \frac{F(Y^n)}{S^{\mathrm{mle}}(Y^n)}\right] \\ &= \mathbb{E}\left[\sup_{\mathcal{F}} \sum_{t=1}^{n} \ln \frac{F_t(Y_t|Y^{t-1})}{S_t^{\mathrm{mle}}(Y_t|Y^{t-1})}\right] \\ &\le \mathbb{E}\left[\sup_{\mathcal{F}} \sum_{t=1}^{n} \left(\ln \frac{F_t(Y_t|Y^{t-1})}{S_t^{\mathrm{mle}}(Y_t|Y^{t-1})} - \mathbb{E}\left[\ln \frac{F_t(Y_t|Y^{t-1})}{S_t^{\mathrm{mle}}(Y_t|Y^{t-1})}\middle| Y^{t-1}\right]\right)\right] \end{aligned}$$

where the last step follows from the nonnegativity of the Kullback-Leibler divergence of the conditional densities. That is, from the fact that

$$\mathbb{E}\left[\ln \frac{S_t^{\mathrm{mle}}(Y_t|Y^{t-1} = y^{t-1})}{F_t(Y_t|Y^{t-1} = y^{t-1})}\right] \ge 0$$

(see, e.g., Cover and Thomas (1991)). Now, for each $F \in \mathcal{F}$ let

$$T_F(y^n) = \frac{1}{2} \sum_{t=1}^{n} \left(\ln \frac{F_t(y_t|y^{t-1})}{S_t^{\mathrm{mle}}(y_t|y^{t-1})} - \mathbb{E}\left[\ln \frac{F_t(Y_t|Y^{t-1})}{S_t^{\mathrm{mle}}(Y_t|Y^{t-1})}\middle| Y^{t-1}\right]\right)$$

so that we have $V_n^{\log}(\mathcal{F}) \le 2\mathbb{E}\left[\sup_{\mathcal{F}} T_F\right]$.

To obtain a suitable upper bound for this quantity, we apply Theorem 3.10. To do this, we need to show that the process $\{T_F : F \in \mathcal{F}\}$ is indeed a subgaussian family under the metric ρ (sample continuity of the process is obvious.) To this end, note that for any $F, G \in \mathcal{F}$,

$$T_F(y^n) - T_G(y^n) = \sum_{t=1}^{n} Z_t(y^t),$$

where

$$Z_t(y^t) = \frac{1}{2}\left(\ln \frac{F_t(y_t|y^{t-1})}{G_t(y_t|y^{t-1})} - \mathbb{E}\left[\ln \frac{F_t(Y_t|Y^{t-1} = y^{t-1})}{G_t(Y_t|Y^{t-1} = y^{t-1})}\right]\right).$$

Now it is easy to see that $T_F - T_G = T_F(Y^n) - T_G(Y^n)$ is a sum of bounded martingale differences with respect to the sequence $Y_1, Y_2, \dots, Y_n$, that is, each term Z_t has zero conditional mean and range bounded by $2\rho_t(F,G)$. Then the Chernoff-Azuma inequality (see Hoeffding (1963) and Azuma (1967)) implies that, for all $\lambda > 0$,

$$\mathbb{E}\left[e^{\lambda(T_F - T_G)}\right] \le \exp\left(\frac{\lambda^2}{2}\rho(F,G)^2\right).$$

Thus, the family $\{T_F : F \in \mathcal{F}\}$ is indeed subgaussian. Hence, using $V_n^{\log}(\mathcal{F}) \le 2\mathbb{E}\left[\sup_{F \in \mathcal{F}} T_F\right]$ and applying Theorem 3.10 we obtain the statement of the lemma. □

PROOF OF THEOREM 3.15. To prove the main inequality, we partition $\mathcal{F}$ into small subsets and calculate the minimax predictor for each subset. Lemma 3.2 is then applied in each subset. Finally, the optimal forecasters for these subsets are combined using the weighted average method.

Fix an arbitrary $\varepsilon > 0$ and let $\mathcal{G} = \{G^{(1)}, \ldots, G^{(N)}\}$ be an ε-covering of $\mathcal{F}$ of minimum size $N = N_{\log}(\mathcal{F}, \varepsilon)$. Determine the subsets $\mathcal{F}_1, \ldots, \mathcal{F}_N$ of $\mathcal{F}$ by

$$\mathcal{F}_i = \left\{ F \in \mathcal{F} : \rho(F, G^{(i)}) \leq \min_{1 \leq j \leq N} \rho(F, G^{(j)}) \right\},$$

that is, $\mathcal{F}_i$ contains all experts which are closest to $G^{(i)}$ in the covering (ties broken arbitrarily). Clearly, the union of $\mathcal{F}_1, \ldots, \mathcal{F}_N$ is $\mathcal{F}$. For each $i = 1, \ldots, N$, let $S^{(i)}$ denote the normalized maximum likelihood forecaster S^{mle} for $\mathcal{F}_i$,

$$S^{(i)}(y^n) = \frac{\sup_{F \in \mathcal{F}_i} F(y^n)}{\sum_{x^n \in \mathcal{Y}^n} \sup_{G \in \mathcal{F}_i} G(x^n)}.$$

Now let the forecaster $S^{(\varepsilon)}$ be the uniform mixture of the $S^{(1)}, \ldots, S^{(N)}$. Clearly, $V_n^{\log}(\mathcal{F}) \leq \inf_{\varepsilon > 0} V_n^{\log}(S^{(\varepsilon)}, \mathcal{F})$. So all we have to do is to bound the regret of $S^{(\varepsilon)}$. To this end, fix any $y^n \in \mathcal{Y}^n$ and let $k = k(y^n)$ be such that

$$\ln \sup_{\mathcal{F}} F(y^n) = \ln \sup_{\mathcal{F}_k} F(y^n).$$

Then

$$\ln \frac{\sup_{\mathcal{F}} F(y^n)}{S^{(\varepsilon)}(y^n)} = \ln \frac{G^{(k)}(y^n)}{S^{(\varepsilon)}(y^n)} + \ln \frac{\sup_{\mathcal{F}_k} F(y^n)}{G^{(k)}(y^n)}.$$

On the one hand, by the upper bound for the loss of the mixture predictor,

$$\sup_{\mathcal{Y}^n} \ln \frac{G^{(k)}(y^n)}{S^{(\varepsilon)}(y^n)} \leq \ln N.$$

On the other hand,

$$\sup_{\mathcal{Y}^n} \ln \frac{\sup_{\mathcal{F}_k} F(y^n)}{G^{(k)}(y^n)} \leq \max_{1 \leq i \leq N} \sup_{\mathcal{Y}^n} \ln \frac{\sup_{\mathcal{F}_i} F(y^n)}{G^{(i)}(y^n)} = \max_{1 \leq i \leq N} V_n^{\log}(\mathcal{F}_i).$$

Hence, we get

$$V_n^{\log}(S^{(\varepsilon)}, \mathcal{F}) \leq \ln N + \max_{1 \leq i \leq N} V_n^{\log}(\mathcal{F}_i).$$

Now note that the diameter of each element of the partition $\mathcal{F}_1, \ldots, \mathcal{F}_N$ is at most 2ε. Hence, applying Lemma 3.2 to each $\mathcal{F}_i$ we find that

$$\begin{aligned}
V_n^{\log}(S^{(\varepsilon)}, \mathcal{F}) &\leq \ln N_{\log}(\mathcal{F}, \varepsilon) + \max_{1 \leq i \leq N} 24 \int_0^{\varepsilon} \sqrt{\ln N_{\log}(\mathcal{F}_i, \delta)}\, d\delta \\
&\leq \ln N_{\log}(\mathcal{F}, \varepsilon) + 24 \int_0^{\varepsilon} \sqrt{\ln N_{\log}(\mathcal{F}, \delta)}\, d\delta
\end{aligned}$$

concluding the proof.

3.6.5 Equivalence with sequential data compression

An important motivation for the study of the logarithmic loss function has its roots in information theory and, more concretely, in sequential data compression. Let $\mathcal{Y}$ be a finite set and suppose we want to encode elements of $\mathcal{Y}^n$ using sequences of bits. A binary *code* for the source $\mathcal{Y}^n$ is a one-to-one mapping assigning to each $y^n \in \mathcal{Y}^n$ a *codeword*; that is, a sequence of bits (possibly of different lengths for different y^n) that represents y^n. Shannon-Fano coding is a technique to construct a code based on an arbitrary probability distribution Q over $\mathcal{Y}^n$ so that the codeword for each y^n has length $\lceil -\log_2 Q(y^n) \rceil$. This codeword length is almost optimal in the following sense. Suppose that Y^n is a random variable whose distribution is indeed Q. Then, the expected value of the codeword length $\lceil -\log_2 Q(Y^n) \rceil$ is at most one bit longer than the expected codeword length for the best possible code for Y^n. By Shannon-Fano coding, we can thus view any set $\mathcal{F}$ of experts for the logarithmic loss framework as a set of codes for $\mathcal{Y}^n$. Given such a set of codes, and given a sequence $y^n \in \mathcal{Y}^n$ to encode, we may look for the code $F \in \mathcal{F}$ that minimizes the Shannon-Fano length $\lceil -\log_2 F(y^n) \rceil$ or, equivalently, that minimizes the cumulative logarithmic loss $L_F(y^n)$.

In *sequential data compression*, it is assumed that the elements of the string y^n are revealed one by one and the goal is to compress it almost as well as the best code in $\mathcal{F}$. Using a method called arithmetic coding, is possible to sequentially construct a codeword for y^n of length $\lceil -\log_2 S(y^n) \rceil + 1$, where S is any forecaster/expert. The quantity

$$\max_{\mathcal{Y}^n} [-\log_2 S(y^n) - (-\log_2 F(y^n))] = \frac{1}{\ln 2} \max_{\mathcal{Y}^n} \left(L_S(y^n) - \inf_{\mathcal{F}} L_F(y^n) \right) .$$

is called the *redundancy* of the code S with respect to the class of codes $\mathcal{F}$. The redundancy of the best possible S for the set $\mathcal{F}$ of codes,

$$\frac{1}{\ln 2} \inf_S \max_{\mathcal{Y}^n} \left(L_S(y^n) - \inf_{\mathcal{F}} L_F(y^n) \right) ,$$

is known as *minimax redundancy.* As minimax redundancy is formally equivalent to the minimax regret $V_n^{\log}(\mathcal{F})$ defined in (3.16), the problem of sequential data compression is equivalent to sequential prediction under the logarithmic loss.

3.6.6 Equivalence with gambling

Imagine that we want to bet in a horse race in which m horses run many times. In the t-th race we bet our entire fortune on the m horses according to proportions $\widehat{p}_t(1), \ldots, \widehat{p}_t(m)$. If horse j

wins the t-th race, we multiply our money bet on this horse by a factor of $o_t(j)$, i.e. the odds for horse j in the t-th race, and we lose it otherwise. In other words, if y_t denotes the index of the winner of the t-th race, we multiply our wealth by a factor of

$$\sum_{j=1}^{m} \mathbb{I}_{[y_t=j]}\, \widehat{p}_t(j)\, o_t(j) = \widehat{p}_t(y_t)\, o_t(y_t)\,.$$

To make it explicit that both the odds $o_t(j)$ and the proportions $\widehat{p}_t(j)$ according which we bet may depend on the results of previous races, we write $o_t(j) = o_t(j|y^{t-1})$ and $\widehat{p}_t(j) = S_t(j|y^{t-1})$, where S is formally equivalent to a forecaster in the prediction with logarithmic loss framework. If our initial capital is C euros, after n races our wealth is

$$C\prod_{t=1}^{n} S_t(y_t|y^{t-1})o_t(y_t|y^{t-1})\,.$$

Now assume that, before each race, we ask the advice of a set $\mathcal{F}$ of experts and our goal is to win as much as the best one of these experts. If expert F divides his wealth in the t-th race according to proportions $F_t(j|y_1^{t-1})$, $1 \le j \le m$, and starts with the same initial wealth C, then his wealth after n races is

$$C\prod_{t=1}^{n} F_t(y_t|y^{t-1})\, o_t(y_t|y^{t-1})\,.$$

The ratio between the wealth of the best expert and ours is thus

$$\frac{\sup_{\mathcal{F}} C\prod_{t=1}^{n} F_t(y_t|y^{t-1})o_t(y_t|y^{t-1})}{C\prod_{t=1}^{n} S_t(y_t|y^{t-1})o_t(y_t|y^{t-1})} = \sup_{\mathcal{F}} \frac{F(y^n)}{S(y^n)}$$

independently of the odds. The logarithm of this quantity is just the difference of the cumulative logarithmic loss of the forecaster S and that of the best expert. Therefore, we have a perfect equivalence between gambling and sequential prediction with logarithmic loss.

3.6.7 Equivalence with portfolio investment

In this section we show that the sequential prediction problem with logarithmic loss is also equivalent, under certain conditions, to the problem of managing a portfolio of assets with the goal of achieving the performance of the best investment strategy computed in hindsight.

A *market vector* $x = (x_1, \ldots, x_m)$ for m assets is a vector of nonnegative real numbers representing price relatives for a given trading period: the quantity x_i denotes the ratio of closing to opening price of the i-th asset for that period. Hence, an initial wealth invested on the m assets according to fractions $\widehat{p}(1)\ldots,\widehat{p}(m)$ changes by a factor of $\sum_{i=1}^{m} x_i\,\widehat{p}(i)$ at the end of period. The market behavior during n trading periods is represented by a sequence $x^n = (x_1, \ldots, x_n)$. A (nonanticipating) *investment strategy* S^{inv} for n trading periods is a sequence $S_1^{\mathrm{inv}}, \ldots, S_n^{\mathrm{inv}}$ of functions taking values in the simplex (3.15). Each function S_t^{inv} is such that $S_t^{\text{inv}}(i|x^{t-1})$ denotes the fraction of the current wealth invested on the i-th asset at the beginning of the t-th period given the past market behavior x^{t-1}. We use

$$G(S^{\text{inv}},x^n) = \prod_{t=1}^{n}\left(\sum_{i=1}^{m} x_{t,i}\, S_t^{\text{inv}}(i|x^{t-1})\right)$$

to denote the wealth factor of strategy S^{inv} after n trading periods. Let $\mathcal{Q}$ be an arbitrary set of investment strategies. The *minimax wealth ratio* of S^{inv} with respect to $\mathcal{Q}$ is

$$W_n(\mathcal{Q}) = \inf_{S^{\text{inv}}} \sup_{x^n} \sup_{Q} \ln \frac{G(Q,x^n)}{G(S^{\text{inv}},x^n)} .$$

This is the the worst-case ratio, over all market sequences, between the wealth factor of the best strategy chosen in hindsight from $\mathcal{Q}$ and the wealth factor of the best possible nonanticipating investment strategy.

A *static* investment strategy, denoted by B, satisfies $B_t(i|x^{t-1}) = B_t(i)$ for each $1 \le t \le n$, where B_t is a probability distribution on $\{1,\dots,m\}$. In other words, static strategies allocate wealth for each trading period without using information on the the past market behavior. Note that static experts for sequential prediction with logarithmic loss and outcome space $\mathcal{Y} = \{1,\dots,m\}$ are equivalent to static investment strategies. Hence, given a set $\mathcal{B}$ of static experts (or, equivalently, static investment strategies) one may ask whether prediction is easier or harder than investment with respect to the same static $\mathcal{B}$. We now prove that the minimax values of sequential prediction and sequential investment are always equal if the underlying set of experts is static.

Theorem 3.16. *For any set $\mathcal{B}$ of static investment strategies,*

$$W_n(\mathcal{B}) = V_n^{\log}(\mathcal{B}) .$$

In what follows, for any fixed m positive integer, we let $\mathcal{X}$ be the simplex (3.15) and we set $\mathcal{Y} = \{1,\dots,m\}$. Let us consider first *Kelly market vectors*. These are (quite unusual) market vectors such that the price relatives for each trading period are all zero except for one asset whose price relative is 1. We will denote with y^n a sequence of n Kelly market vectors, where $y_s \in \mathcal{Y}$ denotes the index of the asset with price relative equal to 1 for trading period s. For such a y^n, and for an arbitrary investment strategy S^{inv}, we have that the allocation $S_t^{\text{inv}}(\cdot|y^{t-1})$ at each trading period t can be viewed as the conditional of a distribution S^{inv} over $\mathcal{Y}^n$ defined by

$$S^{\text{inv}}(y^n) = \prod_{t=1}^{n} S_t^{\text{inv}}(y_t|y^{t-1}) .$$

As probability distributions over $\mathcal{Y}^n$ are forecasting strategies, any investment strategy can be used as a forecaster. Furthermore, note that the wealth factor of S^{inv} over a sequence y^n of Kelly market vectors equals the probability assigned to y^n by S^{inv},

$$G(S^{\text{inv}}, y^n) = S^{\text{inv}}(y^n)\,. \tag{3.18}$$

Based on these observations, we can show that minimax investment is not easier than minimax prediction.

Lemma 3.3. *For each investment strategy S^{inv} and for all sets $\mathcal{Q}$ of investment strategies, $W_n(S^{\text{inv}}, \mathcal{Q}) \geq V_n^{\log}(S^{\text{inv}}, \mathcal{Q})$.*

PROOF. Let S^{inv} be any investment strategy. Then

$$\begin{aligned}
\sup_{x^n}\sup_{Q}\ln\frac{G(Q,x^n)}{G(S^{\text{inv}},x^n)} &\geq \max_{\mathcal{Y}^n}\sup_{Q}\ln\frac{G(Q,y^n)}{G(S^{\text{inv}},y^n)} \\
&= \max_{\mathcal{Y}^n}\sup_{Q}\ln\frac{Q(y^n)}{S^{\text{inv}}(y^n)} \\
&\geq V_n^{\log}(S^{\text{inv}}, \mathcal{Q})
\end{aligned}$$

where we used (3.18) to prove the equality. □

We now define investment strategies based on forecasters. Given a forecasting strategy S we define the investment strategy S^{inv} based on S as follows

$$S_t^{\text{inv}}(y|x^{t-1}) = \frac{\sum_{\mathcal{Y}^{t-1}} S_t(y|y^{t-1})S(y^{t-1})\left(\prod_{s=1}^{t-1} x_{s,y_s}\right)}{\sum_{\mathcal{Y}^{t-1}} S(y^{t-1})\left(\prod_{s=1}^{t-1} x_{s,y_s}\right)}\,.$$

Note that $\prod_{s=1}^{t-1} x_{s,y_s}$ may be viewed as the return, in the first $t-1$ trading periods, of the "extremal" investment strategy (denoted by y^n) allocating, on each trading period t, everything on the y_t-th asset.

The next simple result shows that the return of the investment strategy based on a forecaster S equals to return of the strategy that assigns a fraction $S(y^n)$ of the initial wealth to each extremal strategy $y^n \in \mathcal{Y}^n$, and then it never rebalances again.

Lemma 3.4. *The total return of the investment strategy S^{inv} based on a forecaster S is*

$$G(S^{\text{inv}}, x^n) = \sum_{\mathcal{Y}^n}\left(\prod_{t=1}^{n} x_{t,y_t}\right)S(y^n)\,. \tag{3.19}$$

PROOF.

$$
\begin{aligned}
G(S^{\text{inv}},x^n) &= \prod_{t=1}^{n}\left(\sum_{\mathcal{Y}} x_{t,y} S_t^{\text{inv}}(y|x^{t-1})\right) \\
&= \prod_{t=1}^{n} \frac{\sum_{y\in\mathcal{Y}}\sum_{y^{t-1}\in\mathcal{Y}^{t-1}} S(y^{t-1},y)\,(\prod_{s=1}^{t} x_{s,y_s})}{\sum_{\mathcal{Y}^{t-1}} S(y^{t-1})\,(\prod_{s=1}^{t-1} x_{s,y_s})} \\
&= \prod_{t=1}^{n} \frac{\sum_{\mathcal{Y}^t} (\prod_{s=1}^{t} x_{s,y_s})\, S(y^t)}{\sum_{\mathcal{Y}^{t-1}} (\prod_{s=1}^{t-1} x_{s,y_s})\, S(y^{t-1})} \\
&= \sum_{\mathcal{Y}^n}\left(\prod_{t=1}^{n} x_{t,y_t}\right) S(y^n)\,.\square
\end{aligned}
$$

It will be convenient to express in a form similar to (3.19) also the return of a generic static investment strategy.

Lemma 3.5. *For any static investment strategy B,*

$$
G(B,x^n) = \sum_{\mathcal{Y}^n}\left(\prod_{t=1}^{n} x_{t,y_t}\right) B(y^n)\,.
$$

PROOF.

$$
\begin{aligned}
G(B,x^n) &= \prod_{t=1}^{n}\left(\sum_{\mathcal{Y}} x_{t,i} B_t(y)\right) \\
&= \sum_{\mathcal{Y}^n}\left(\prod_{t=1}^{n} x_{t,y_t} B_t(y_t)\right) \\
&= \sum_{\mathcal{Y}^n}\left(\prod_{t=1}^{n} x_{t,y_t}\right)\left(\prod_{t=1}^{n} B_t(y_t)\right) \\
&= \sum_{\mathcal{Y}^n}\left(\prod_{t=1}^{n} x_{t,y_t}\right) B(y^n)\,.\square
\end{aligned}
$$

We now restrict our attention to sets $\mathcal{B}$ of static investment strategies. First, we need a simple technical lemma whose proof is left as an exercise.

Lemma 3.6. *If* $\alpha_1,\ldots,\alpha_N,\beta_1,\ldots,\beta_N \geq 0$, *then*

$$
\frac{\sum_{i=1}^{N}\alpha_i}{\sum_{j=1}^{N}\beta_j} \leq \max_{1\leq k\leq N}\frac{\alpha_k}{\beta_k}\,.
$$

The next key lemma bounds from the above the wealth ratio of S^{inv} on an arbitrary market sequence with the worst-case regret of the forecasting strategy S on which S^{inv} is based.

Lemma 3.7. *For any investment strategy S^{inv} based on a forecaster S, for any class $\mathcal{B}$ of static investment strategies, and for any market sequence x^n*

$$\sup_{\mathcal{B}} \ln \frac{G(B,x^n)}{G(S^{\text{inv}},x^n)} \leq V_n^{\log}(S,\mathcal{B}) .$$

We are now ready to prove the equivalence between investment and prediction stated at the beginning of this section. PROOF. Fix any market sequence x^n and choose B' such that $G(B',x^n) = \sup_{\mathcal{B}} G(B,x^n)$. For notational convenience, let $G(y^n,x^n) = \prod_{t=1}^n x_{t,y_t}$. Then

$$\begin{aligned}
\frac{G(B',x^n)}{G(S^{\text{inv}},x^n)} &= \frac{\sum_{\mathcal{Y}^n} G(y^n,x^n)B'(y^n)}{\sum_{\mathcal{Y}^n} G(y^n,x^n)S(y^n)} \quad \text{(by Lemmas 3.4 and 3.5)} \\
&\leq \max_{\mathcal{Y}^n} \frac{B'(y^n)}{S(y^n)} \quad \text{(by Lemma 3.6)} \\
&\leq \max_{\mathcal{Y}^n} \sup_{\mathcal{B}} \frac{B(y^n)}{S(y^n)} . \square
\end{aligned}$$

PROOF OF THEOREM 3.16. Fix a class of static strategies $\mathcal{B}$. Let S^{inv} be the investment strategy based on the minimax forecaster S^{mle} for $\mathcal{B}$. By Lemma 3.7 we get

$$W_n(\mathcal{B}) \leq \sup_{x^n} \sup_{\mathcal{B}} \ln \frac{G(B,x^n)}{G(S^{\text{inv}},x^n)} \leq \max_{\mathcal{Y}^n} \sup_{\mathcal{B}} \ln \frac{B(y^n)}{S^{\text{mle}}(y^n)} = V_n^{\log}(\mathcal{B}) .$$

An application of Lemma 3.3 completes the proof. □

Bibliographic remarks

A recent survey summarizing most of the literature of prediction under the logarithmic loss is the one by Merhav and Feder (1998). The mixure forecaster for finite and countable sets of experts was analyzed by De Santis, Markowski, and Wegman (1988). Theorem 3.13 is due to Shtarkov (1987) and so is Theorem 3.14 (see also Freund (1996) and Xie and Barron (2000)). Xie and Barron also investigate the relatioships to data compression, gambling, and portfolio investment. The equivalence between prediction and portfolio investment for static investment strategies was apparently not known before. An excellent account of the basics of lossless data compression, including arithmetic coding, may be found in Cover and Thomas (1991). Theorem 3.15 is due to Cesa-Bianchi and Lugosi (2001). A weaker version, for classes of static experts was proven by Opper and Haussler (1987). Thigher bounds on the minimax regret, but only applicable to parametric sets of experts, have been shown by Rissanen (1996) and Yamanishi (1998).

3.7 Sequential pattern classification

In this section, we extend the framework of prediction with expert advice to the problem of sequential binary pattern classification with linear-threshold classifiers. In *sequential pattern classification* the forecaster observes a sequence of *attribute vectors* $x_t \in \mathcal{R}^N, t \geq 1$, representing a stream of data to be classified. After observing each x_t, the forecaster outputs its guess $\widehat{p}_t$ and then receives the correct label $y_t \in \{-1,1\}$ of x_t. If $\widehat{p}_t \neq y_t$, then a *mistake* occurs at round t. A *linear-threshold classifier* is a binary classifier $c : \mathcal{R}^N \to \{-1,1\}$ of the form

$$c(x) = \text{sign}(w \cdot x)$$

where sign is the signum function

$$\text{sign}(x) = \begin{cases} 1 & \text{if } x \geq 0, \\ -1 & \text{otherwise} \end{cases}$$

and $w \in \mathcal{R}^N$ is a *weight vector*. We will study forecasters that learn the best linear-threshold classifiers for the sequence of data being observed. These forecasters keep a weight vector w_t which they use to compute their prediction $\widehat{p}_t = \text{sign}(w_t \cdot x_t)$. After observing the true label y_t, the forecaster may update its weight vector via the assignment $w_{t+1} = F(w_t, x_t, y_t)$ where $F : \mathcal{R}^N \times \mathcal{R}^N \times \{-1,1\} \to \mathcal{R}^N$ is some update function.

The problem of sequential pattern classification can be easily fit into our framework of Section 3.1.1. The decision space $\mathcal{X}$ and the outcome space $\mathcal{Y}$ are both set equal to $\{-1,1\}$. The loss function is simply $\ell(\widehat{p}, y) = \mathbb{I}_{[y \neq \widehat{p}]}$ and the regret vector at time t is defined by

$$r_t = \mathbb{I}_{[y_t \neq \widehat{p}_t]} \, y_t x_t \; . \tag{3.20}$$

As we will see, the minimization of the number of mistakes in the classification problem will be achieved as a side-effect of the minimization of the regret (3.20). To minimize regret, we simply apply the potential-based forecasters of Section 3.2. Given an additive potential Φ, the forecaster based on Φ uses at round t the gradient of the current potential, $\nabla\Phi(R_{t-1})$, as weight vector w_t (compare this to the forecaster for convex losses of Section 3.3). Hence, the weight update function F is defined by

$$w_{t+i} = \nabla\Phi\left((\nabla\Phi)^{-1}(w_t) + r_{i,t}\right) \tag{3.21}$$

where $(\nabla\Phi)^{-1}$, when it exists, is the functional inverse of $\nabla\Phi$. We now check that condition (3.4) is satisfied. If $\widehat{p}_t = y_t$, then $r_t = (0, \ldots, 0)$ and the condition is trivially satisfied. Otherwise, $r_t \cdot \nabla\Phi(R_{t-1}) = \mathbb{I}_{[y_t \neq \widehat{p}_t]} y_t x_t \cdot w_t \leq 0$, and the condition is satisfied in this case as well.

Unlike the previous scenarios, here the loss of the experts (linear-threshold classifiers) will be measured differently from the loss of the forecaster. More precisely, we measure the loss of an expert $v \in \mathcal{R}^N$ (where, without loss of generality, we assume that v is a unit norm vector) with the cumulative deviation at margin γ. For an arbitrary sequence $\sigma^n = ((x_1, y_1), \ldots, (x_n, y_n))$ of labelled attribute vectors, the cumulative *deviation* of v with respect to margin $\gamma > 0$ is defined by

$$D_v(\sigma^n) \stackrel{\text{def}}{=} \sum_{t=1}^{n} \max\{0, \gamma - y_t x_t \cdot v\} . \tag{3.22}$$

Each term in the sum (3.22) tells whether, and by how much, the linear threshold classifier based on weight vector v missed to classify, to within a certain margin γ, the corresponding example. Thus the deviation measures a notion of loss, different from the number of misclassifications, associated to the weight vector v. If we let $\widehat{p} = \text{sign}(v \cdot x)$ then it is easy to see that

$$\mathbb{I}_{[y \neq \widehat{p}]} = \mathbb{I}_{[yv \cdot x < 0]} \leq \frac{\max\{0, \gamma - yx \cdot v\}}{\gamma}$$

implying that the deviation divided by the margin at which it is measured bounds the loss $\mathbb{I}_{[y \neq \widehat{p}]}$ from the above. In this respect, we may say that in the sequential pattern recognition problem the regret of the forecaster with respect to the linear-threshold classifier v is

$$L(\sigma^n) - \frac{D_v(\sigma^n)}{\gamma} = \sum_{t=1}^{n} \mathbb{I}_{[y_t \neq \widehat{p}_t]} - \sum_{t=1}^{n} \max\left\{0, 1 - \frac{y_t x_t \cdot v}{\gamma}\right\} . \tag{3.23}$$

This notion of regret is somewhat weaker than the notion of regret for the prediction with expert advice problem. In fact, we are measuring the loss of the forecaster using the number of mistakes and the loss of v using the deviation and, as we saw, the deviation upper bounds the number of mistakes.

3.7.1 *Polynomial potential classifier*

We now apply the polynomially weighted average forecaster to the sequential pattern classification problem and derive bounds on the number of mistakes made. However, instead of defining Φ based on $\phi(x) = (x)_+^p$, we will use the slight variant $\phi(x) = |x|^p$. Hence,

$$(\nabla\Phi(u))_i = \text{sign}(u_i)|u_i|^{p-1} \qquad \text{and} \qquad (\nabla\Phi(u))_i^{-1} = \text{sign}(u_i)|u_i|^{q-1}$$

where q is the dual norm of p, that is q is such that $1/p + 1/q = 1$. The weight update (3.21) can then be written as

$$\begin{aligned} w'_{i,t} &= \text{sign}(w_{i,t})|w_{i,t}|^{q-1} + r_{i,t} \\ w_{i,t+1} &= \text{sign}(w'_{i,t})|w'_{i,t}|^{p-1} . \end{aligned} \tag{3.24}$$

COROLLARY 3.5. *The number of mistakes made by the linear threshold forecaster based on the polynomial potential on an arbitrary sequence* $\sigma^n = ((x_1,y_1),\ldots,(x_n,y_n))$ *of labelled attribute vectors is at most*

$$L(\sigma^n) \;\le\; \frac{D_v(\sigma^n)}{\gamma} + (p-1)\left(\frac{X_p}{\gamma}\right)^2 + \sqrt{(p-1)\frac{D_v(\sigma^n)}{\gamma}\left(\frac{X_p}{\gamma}\right)^2}$$

where $X_p = \max_{1\le t\le n}\|x_t\|_p$ *and* $D_v(\sigma^n)$ *is the deviation with respect to margin* γ *for any* v *of unit q-norm (q being the dual norm of p) and any* $\gamma > 0$.

PROOF. An obvious adaptation of the proof of Corollary 3.1 yields

$$\|R_t\|_p^2 \le (p-1)X_p^2 L_S(\sigma^n)\,.$$

On the other hand, let $v \in \mathcal{R}^N$ be any vector such that $\|v\|_q = 1$. Then

$$\begin{aligned}
\|R_t\|_p &\ge R_t\cdot v \quad \text{(by Hölder's inequality)}\\
&= R_{t-1}\cdot v + \mathbb{I}_{[y_t\ne\widehat{p}_t]}y_t x_t\cdot v\\
&\ge R_{t-1}\cdot v + \mathbb{I}_{[y_t\ne\widehat{p}_t]}(\gamma - \max\{0,\gamma - y_t x_t\cdot v\})\\
&\ge \cdots \;\ge \gamma L_S(\sigma^n) - D_v(\sigma^n)\,.
\end{aligned}$$

Piecing together the two inequalities, and solving the resulting equation for $L_S(\sigma^n)$, yields the desired result. □

3.7.2 Exponential potential classifier

In this section, we derive an upper bound on the number of mistakes for the linear-threshold forecaster based on the exponential potential. As for the polynomial potential bound of Corollary 3.5, this result is expressed in terms of the cumulative normalized deviation $D_v(\sigma^n)/\gamma$ with respect to some margin $\gamma > 0$ and some unit norm vector v. However, as in previous cases, to get results comparable to those of Corollary 3.5 the parameter η of the potential needs to be tuned with respect to the horizon length n. As we did for the polynomial potential, we derive explicitely the weight updates. We have

$$(\nabla\Phi(u))_i = \eta e^{\eta u_i} \qquad \text{and} \qquad (\nabla\Phi(u))_i^{-1} = \frac{1}{\eta}\ln\frac{u_i}{\eta}\,.$$

In this case, the weight update (3.21) simplifies to

$$\begin{aligned}
w_{i,t+1} &= \eta\exp\left(\eta\left(\frac{1}{\eta}\ln\frac{w_{i,t}}{\eta} + r_{i,t}\right)\right)\\
&= \eta\exp(\ln w_{i,t} - \ln\eta + \eta r_{i,t}) = w_{i,t}\,e^{\eta r_{i,t}}\,.
\end{aligned}$$

COROLLARY 3.6. *The number of mistakes made by the linear threshold forecaster based on the exponential potential on an arbitrary sequence* $\sigma^n = ((x_1,y_1),\ldots,(x_n,y_n))$ *of labelled attribute vectors, and tuned with*

$$\eta = \begin{cases} \gamma/(X_\infty^2) & \text{if } n < 2(X_\infty/\gamma)\ln N \\ \sqrt{\dfrac{2\gamma \ln N}{n X_\infty^3}} & \text{otherwise} \end{cases}$$

$$\text{where } X_\infty = \max_{1\le t\le n} \|x_t\|_\infty,$$

is at most

$$L(\sigma^n) \le 6\,(X_\infty/\gamma)^2 \ln N \qquad \text{if } n < 2(X_\infty/\gamma)\ln N,$$

and at most

$$L(\sigma^n) \le \frac{D_v(\sigma^n)}{\gamma} + 2\left(\frac{X_\infty}{\gamma}\right)^2 \ln N + \sqrt{2n\left(\frac{X_\infty}{\gamma}\right)^2 \ln N}$$

otherwise.

PROOF. Let $L = L(\sigma^n)$ and $D = D_v(\sigma^n)$. Corollary 3.2 implies $\ln\Phi(R_t) \le \ln N + (\eta^2/2)X_\infty^2 L$. To obtain a lower bound on $\ln\Phi(R_t)$, consider any vector v of convex coefficients. Then we use the well-known "log-sum inequality" (see Cover and Thomas (1991)) which implies that, for any vectors $u, v \in \mathcal{R}^N$ of nonnegative numbers with $\sum_{i=1}^N v_i = 1$,

$$\ln \sum_{i=1}^N u_i \ge \sum_{i=1}^N v_i \ln u_i + H(v)\,,$$

where $H(v) = -\sum_{i=1}^N v_i \ln v_i$ is the *entropy* of v. Therefore, for any vector v of convex coefficients,

$$\ln\Phi(R_t) = \ln \sum_{i=1}^N e^{\eta R_{i,t}} \ge \eta R_t \cdot v + H(v) \ge \eta\,(\gamma L - D) + H(v)$$

where in the last step we proceeded just like in the proof of Corollary 3.5. Putting the upper and lower bounds for $\ln\Phi(R_t)$ together we obtain

$$\eta\,(\gamma L - D) + H(v) \le \ln N + (\eta^2/2) X_\infty^2 L$$

which, dropping the positive term $H(v)$, implies

$$L\left(1 - \frac{\eta}{2}\frac{X_\infty^2}{\gamma}\right) \le \frac{D}{\gamma} + \frac{\ln N}{\eta\gamma}\,. \tag{3.25}$$

The rest of the proof is left as an exercise. □

3.7.3 Kernels and maximum margin hyperplanes

For $p = 2$, the weight update (3.24) reduces to $w_{t+1} = w_t + \mathbb{I}_{[y_t \neq \widehat{p}_t]} y_t x_t$ so that we have

$$w_{t+1} = \sum_{s=1}^{t} \mathbb{I}_{[y_t \neq \widehat{p}_t]} y_t x_t \qquad \text{for all } t \geq 1. \tag{3.26}$$

This is the just the classical Perceptron algorithm of Rosenblatt (1958). Using (3.26), we can write the predictions of Perceptron in the equivalent "dual" form

$$\widehat{p}_t = \text{sign}(w_t \cdot x_t) = \text{sign}\left(\sum_{s=1}^{t-1} \mathbb{I}_{[y_s \neq \widehat{p}_s]} y_s x_s \cdot x_t\right). \tag{3.27}$$

An examination of (3.27) reveals that, by replacing each dot product $x_s \cdot x_t$ with a kernel operation $K(x_s, x_t)$, we could easily run the Perceptron algorithm in the feature space $\mathcal{H}$ induced by the kernel K (see, e.g., Cristianini and Shawe-Taylor (2000) for a introduction on kernel-based learning algorithms). To do that, we proceed as follows. At any prediction round t, we represent the current weight w_t with the sequence $\alpha^t = (\alpha_1, \ldots, \alpha_{t-1})$, where $\alpha_s = \mathbb{I}_{[y_s \neq \widehat{p}_s]} y_s$. The initial weight α^1 is thus the empty sequence. Predictions in $\mathcal{H}$ are then computed by simply replacing in (3.27) each dot product $x_s \cdot x_t$ performed in $\mathcal{R}^N$ with the dot product $K(x_s, x_t)$ performed in $\mathcal{H}$. We thus obtain

$$\widehat{p}_t = \text{sign}\left(\sum_{s=1}^{t-1} \alpha_s K(x_s, x)\right).$$

Finally, the weight update is carried out simply by adding the element $\alpha_t = \mathbb{I}_{[y_t \neq \widehat{p}_t]} y_t$ to the sequence α_t.

One of the currently most successful algorithms in pattern classification is Vapnik's Support Vector Machine (Vapnik, 1998). Given an abitrary set of binary labelled attribute vectors $\{(x_1, y_1), \ldots, (x_n, y_n)\}$ which is linearly separable in some kernel-induced feature space $\mathcal{H}$, say that a unit norm vector v has margin γ in $\mathcal{H}$ whenever

$$\max_{1 \leq t \leq n} y_t \frac{K(v, x_t)}{\sqrt{K(x_t, x_t)}} \geq \gamma \tag{3.30}$$

where $\sqrt{K(x_t, x_t)}$ is Euclidean norm of the image in feature space of the instance x_t. The maximum margin γ^* of $\{(x_1, y_1), \ldots, (x_n, y_n)\}$ in $\mathcal{H}$ is the largest margin γ achievable by any unit norm vector v. Vapnik showed that, among all separating hyperplanes, v^* has some desirable statistical properties and also works very well in practice. The Support Vector Machine is a general technique for finding the maximum margin separating hyperplane v^* in a generic kernel-induced feature space. We now show that a variant of the kernel-Perceptron (3.27) can be used to approximately find the maximum margin separating hyperplane.

Algorithm: **Maximum Margin Perceptron.**
Input parameters: Approximation factor $0 < \beta \leq 1$.
Initialization: Set the initial weight sequence α to the empty sequence. Set the update counter m to zero.

For $t \geq 1$

1. Read next example $(x_t, y_t) \in \mathcal{R}^N \times \{-1, 1\}$.
2. Set $\gamma_t = \frac{2}{\beta}\sqrt{\frac{2}{m+1}}$
3. If $y_t \sum_{s=1}^{t-1} \alpha_s K(x_s, x_t) \leq (1-\beta)\gamma_t \sqrt{K(x_t, x_t)}$ then

$$\eta_t = \sqrt{\frac{2/(m+1)}{K(x_t, x_t)}}, \qquad \alpha_t = \eta_t y_t, \qquad m \leftarrow m+1$$

$$W = \sum_{r=1}^{t} \sum_{s=1}^{t} \alpha_r \alpha_s K(x_r, x_s) \tag{3.28}$$

$$\text{For } s = 1, \ldots, t: \quad \alpha_s \leftarrow \frac{\alpha_s}{\sqrt{W}} \tag{3.29}$$

else $\alpha_t = 0$.

FIGURE 3.2: A variant of the kernel-Perceptron algorithm that finds an approximation of the maximum margin hyperplane.

By an easy modification of the proof of Corollary 3.5, we can prove that if we run the kernel-Perceptron cycling through a sequence σ with maximum margin γ^*, the algorithm will converge to a separating hyperplane after at most $1/(\gamma^*)^2$ mistakes. However, we have no guarantee on the margin of the hyperplane found. It turns out that the variant of the Perceptron, described in Figure 3.2, converges to a separating hyperplane w whose margin is at least $(1-\beta)\gamma^*$ after at most order of $1/(\beta\gamma^*)^2$ mistakes. In other words, we can approximate the maximum margin hyperplane to within a factor of $1-\beta$ with an algorithm whose rate of convergence is at most $1/\beta^2$ times slower than that of the standard Perceptron.

Theorem 3.17. *If the sequence* $\sigma^n = ((x_1,y_1),\dots,(x_n,y_n))$ *is linearly separable with margin* γ *in the feature space induced by a kernel K, then the algorithm in Figure 3.2, with input parameter* $0 < \beta \le 1$, *converges to an hyperplane* w *in* $\mathcal{H}$ *with margin at least* $(1-\beta)\gamma$ *after at most*

$$\frac{2}{\gamma^2}\left(\frac{2}{\beta}-1\right)^2+\frac{8}{\beta}-4$$

updates.

The maximum margin Perceptron algorithm of Figure 3.2 works like the kernel Perceptron (3.27) with a few important differences. Unlike the Perceptron, where the weight update occurs whenever a mistake occurs, here the update (3.29) takes place whenever the current example is classified with a margin smaller than $(1-\beta)\gamma_t$ — see condition at step 3. The quantity γ_t, set in step 2, is the current approximation to the unknown margin γ, see definition (3.30), and goes to zero at rate roughly $1/\sqrt{m}$, where m is the number of updates performed so far. Unlike the Perceptron weight update rule, here the algorithm uses a learning rate η_t that scales both with m and the norm $\sqrt{K(x_t,x_t)}$ (in feature space) of the attribute vector x_t. Moreover, in (3.28) the new weight vector is normalized to a unit norm vector. This normalization is needed to control the number of updates, as these updates are also functions of the length of the current weight vector.

Bibliographic remarks

The Perceptron algorithm was originally introduced by Rosenblatt (1958) for modeling human perception. The Perceptron convergence theorem (a special case of Corollary 3.5) was independently proven by Novikoff (1962) and Block (1962). The Winnow algorithm is due to Littlestone (1998, 1989), and the family of p-norm Perceptrons, which approximates Winnow for $p \to \infty$, was introduced by Grove, Littlestone and Schuurmans (2001). Gentile (2001a, 2001b) extended to the robust case the analysis of p-norm Perceptrons and developed the maximum margin Perceptron. Kernel Perceptrons were investigated, both theoretically and experimentally, by Freund and Schapire (1999).

3.8 References

P. Auer, N. Cesa-Bianchi, and C. Gentile. Adaptive and self-confident on-line learning algorithms. *Machine Learning*, 2001. To appear.

K. Azuma. Weighted sums of certain dependent random variables. *Tohoku Mathematical Journal*, 68:357–367, 1967.

D. Blackwell. An analog of the minimax theorem for vector payoffs. *Pacific Journal of Mathematics*, 6:1–8, 1956.

H.D. Block. The perceptron: A model for brain functioning. *Review of Modern Physics*, 34:123–135, 1962.

A. Borodin and R. El-Yaniv. *Online Computation and Competitive Analysis*. Cambridge University Press, 1998.

N. Cesa-Bianchi. Analysis of two gradient-based algorithms for on-line regression. *Journal of Computer and System Sciences*, 59(3):392–411, 1999.

N. Cesa-Bianchi, Y. Freund, D.P. Helmbold, D. Haussler, R. Schapire, and M.K. Warmuth. How to use expert advice. *Journal of the ACM*, 44(3):427–485, 1997.

N. Cesa-Bianchi, D.P. Helmbold, and S. Panizza. On Bayes methods for on-line Boolean prediction. *Algorithmica*, 22(1/2):112–137, 1998.

N. Cesa-Bianchi and G. Lugosi. On prediction of individual sequences. *Annals of Statistics*, 27(6):1865–1895, 1999.

N. Cesa-Bianchi and G. Lugosi. Minimax regret under log loss for general classes of experts. *Machine Learning*, 3(43):247–264, 2001.

T.H. Chung. Approximate methods for sequential decision making using expert advice. In *Proceedings of the 7th Annual ACM Workshop on Computational Learning Theory*, pages 183–189. ACM Press, 1994.

T.H. Chung. *Minimax Learning in Iterated Games via Distributional Majorization*. PhD thesis, Stanford University, 1994.

T. Cover. Behaviour of sequential predictors of binary sequences. In *Proceedings of the 4th Prague Conference on Information Theory, Statistical Decision Functions and Random Processes*, pages 263–272. Publishing house of the Czechoslovak Academy of Sciences, 1965.

T.M. Cover and J.A. Thomas. *Elements of Information Theory.* John Wiley and Sons, 1991.

N. Cristianini and J. Shawe-Taylor. *An Introduction to Support Vector Machines.* Cambridge University Press, 2001.

A. Dawid. Statistical theory: the prequential approach. *J. Roy. Statist. Soc. A*, pages 278–292, 1984.

R.M. Dudley. *Uniform Central Limit Theorems.* Cambridge University Press, 2001.

M. Feder, N. Merhav, and M. Gutman. Universal prediction of individual sequences. *IEEE Trans. on Information Theory*, 38:1258–1270, 1992.

D. Foster. Prediction in the worst-case. *Annals of Statistics*, 19:1084–1090, 1991.

D. Foster and R. Vohra. A randomized rule for selecting forecasts. *Operations Research*, 41:704–707, 1993.

Y. Freund. Predicting a binary sequence almost as well as the optimal biased coin. In *Proceedings of the 9th Annual Conference on Computational Learning Theory*, pages 89–98. ACM Press, 1996.

Y. Freund and R. Schapire. Large margin classification using the Perceptron algorithm. *Machine Learning*, pages 277–296, 1999.

Y. Freund and R.E. Schapire. A decision-theoretic generalization of on-line learning and an application to boosting. *Journal of Computer and System Sciences*, 55(1):119–139, 1997.

D. Fudenberg and D.K. Levine. *The Theory of Learning in Games.* MIT Press, 1998.

J. Galambos. *The Asymptotic Theory of Extreme Order Statistics.* R.E. Kreiger, 1987. Second edition.

C. Gentile. A new approximate maximal margin classification algorithm. Technical Report NC-TR-01-096, NeuroCOLT Working Group, 2001. An extended abstract appeared in the Proceedings of NIPS*2000.

C. Gentile. The robustness of the *p*-norm algorithms. An extended abstract (co-authored with N. Littlestone) appeared in the *Proceedings of the 12th Annual Conference on Computational Learning Theory*, pages 1–11. ACM Press, 1999., 2001.

E Giné. Empirical processes and applications: an overiew. *Bernoulli*, 2:1–28, 1996.

A.J. Grove, N. Littlestone, and D. Schuurmans. General convergence results for linear discriminant updates. In *Proceedings of the 10th Annual Conference on Computational Learning Theory*, pages 171–183. ACM Press, 1997.

A.J. Grove, N. Littlestone, and D. Schuurmans. General convergence results for linear discriminant updates. *Machine Learning*, 43(3):173–210, 2001.

J. Hannan. Approximation to Bayes risk in repeated play. *Contributions to the theory of games*, 3:97–139, 1957.

S. Hart and A. Mas-Colell. A simple adaptive procedure leading to correlated equilibrium. *Econometrica*, 68:1127–1150, 2000.

S. Hart and A. Mas-Colell. A general class of adaptive strategies. *Journal of Economic Theory*, 98(1):26–54, 2001.

D. Haussler, J. Kivinen, and M.K. Warmuth. Sequential prediction of individual sequences under general loss functions. *IEEE Transactions on Information Theory*, 44:1906–1925, 1998.

W. Hoeffding. Probability inequalities for sums of bounded random variables. *Journal of the American Statistical Association*, 58:13–30, 1963.

J. Kivinen and M.K. Warmuth. Averaging expert predictions. In *Proceedings of the Fourth European Conference on Computational Learning Theory*, pages 153–167. Lecture Notes in Artificial Intelligence, Vol. 1572. Springer, 1999.

N. Littlestone. Learning quickly when irrelevant attributes abound: a new linear-threshold algorithm. *Machine Learning*, 2(4):285–318, 1988.

N. Littlestone. *Mistake Bounds and Logarithmic Linear-threshold Learning Algorithms*. PhD thesis, University of California at Santa Cruz, 1989.

N. Littlestone and M.K. Warmuth. The weighted majority algorithm. Technical Report UCSC-CRL-91-28, University of California at Santa Cruz, 1991. An extended abstract appeared in the *Proceedings of the 30th Annual Symposium on the Foundations of Computer Science*.

N. Littlestone and M.K. Warmuth. The weighted majority algorithm. *Information and Computation*, 108:212–261, 1994.

N. Merhav and M. Feder. Universal prediction. *IEEE Transactions on Information Theory*, 44(6):2124–2147, 1998.

A.B.J. Novikoff. On convergence proofs of perceptrons. In *Proceedings of the Symposium on the Mathematical Theory of Automata*, volume XII, pages 615–622. 1962.

M. Opper and D. Haussler. Worst case prediction over sequences under log loss. In *The Mathematics of Information Coding, Extraction, and Distribution*. Springer Verlag, 1997.

D. Pollard. *Empirical Processes : Theory and Applications*, volume 2 of *NSF-CBMS Regional Conference Series in Probability and Statistics*. Institute of Math. Stat. and Am. Stat. Assoc., 1990.

J. Rissanen. Fischer information and stochastic complexity. *IEEE Transactions on Information Theory*, 42:40–47, 1996.

F. Rosenblatt. The perceptron: A probabilistic model for information storage and organization in the brain. *Psychological Review*, 65:386–408, 1958.

A. De Santis, G. Markowski, and M.N. Wegman. Learning probabilistic prediction functions. In *Proceedings of the 1st Annual Workshop on Computational Learning Theory*, pages 312–328. Morgan Kaufmann, 1988.

Y.M. Shtarkov. Universal sequential coding of single messages. Translated from: *Problems in Information Transmission*, 23(3):3–17, 1987.

V.N. Vapnik. *Statistical Learning Theory*. Wiley, 1998.

V.G. Vovk. Aggregating strategies. In *Proceedings of the 3rd Annual Workshop on Computational Learning Theory*, pages 372–383, 1990.

V.G. Vovk. Universal forecasting algorithms. *Information and Computation*, 96(2):245–277, 1992.

V.G. Vovk. A game of prediction with expert advice. *Journal of Computer and System Sciences*, 56(2):153–173, 1998.

Q. Xie and A.R. Barron. Asymptotic minimax regret for data compression, gambling, and prediction. *IEEE Transactions on Information Theory*, 46:431–445, 2000.

K. Yamanishi. A decision-theoretic extension of stochastic complexity and its application to learning. *IEEE Transactions on Information Theory*, 44:1424–1440, 1998.

4

LEARNING-THEORETIC METHODS IN VECTOR QUANTIZATION

T. Linder
Queens' University at Kingston, Kingston, Canada

4.1 Introduction

The principal goal of data compression (also known as source coding) is to replace data by a compact representation in such a manner that from this representation the original data can be reconstructed either perfectly, or with high enough accuracy. Generally, the representation is given in the form of a sequence of binary digits (bits) that can be used for efficient digital transmission or storage.

Certain types of data, such as general purpose data files on a computer, require perfect reconstruction. In this case the compression procedure is called *lossless*, and the goal is to find a representation that allows perfect reconstruction using the fewest possible bits. Other types of data, such as speech, audio, images, and video signals, do not require (or do not even admit) perfect reconstruction. In this case, the goal is to find an efficient digital representation from which the original data can be reconstructed with a prescribed level of accuracy, as measured by a fidelity criterion between the original and reconstructed data. Such a compression procedure is called *lossy*. In these notes, we will focus on lossy data compression.

In our model, the data to be compressed is a sequence of d-dimensional random vectors $Z_1, Z_2, \ldots$, such that all the Z_i have the same distribution. Such a data sequence is obtained, for example, by forming non-overlapping blocks of length d from a real-valued stationary process. To compress the data, each Z_i is mapped into a binary string b_i. Thus $Z_1, Z_2, \ldots$ is represented by the sequence of binary strings $b_1, b_2, \ldots$. Typically, Z_i can take a continuum of values, while b_i is always discrete, and so this representation is lossy (not invertible). The data is reconstructed by mapping each b_i into another d-dimensional vector $\widehat{Z}_i$, called the reproduction of Z_i.

The compactness of the representation is measured by its *rate*, defined as the average (expected) number of binary digits needed to obtain $\widehat{Z}_i$, i.e., the average length of b_i. Note that this number is the same for all i since the Z_i have the same distribution. If b_i has fixed length, the compression procedure is called *fixed-rate vector quantization*. If the length of b_i is not fixed (i.e., it depends on the value of Z_i), we talk about *variable-rate* vector quantization. The composition of mappings $Z_i \to b_i \to \widehat{Z}_i$ is called a fixed-rate vector quantizer in the first case, and a variable-rate vector quantizer in the second case.

Since $Z_i \neq \widehat{Z}_i$ in general, we need a way to measure how well $\widehat{Z}_i$ approximates Z_i. For this reason, we are given a nonnegative function $d(\cdot,\cdot)$ of two vector variables, called a *distortion measure*, and we use the quantity $d(Z_i,\widehat{Z}_i)$ to measure the reconstruction error in representing Z_i by $\widehat{Z}_i$. In these notes, we will use the popular mean squared error given by the squared Euclidean distance between Z_i and $\widehat{Z}_i$. The *distortion* of the scheme is characterized by a single number, the average (expected) value of $d(Z_i,\widehat{Z}_i)$. Again, this quantity is the same for all i since the Z_i have the same distribution.

The fact that the rate and distortion do not depend on the particular index i allows us to focus on the problem of quantizing (compressing) a generic random vector X, called the *source*, which has the common distribution of the Z_i. The goal is to make the distortion in quantizing X as small as possible, while keeping the rate at a given threshold. Quantizers which are optimal in the sense of achieving minimum distortion under a rate constraint depend on the distribution of the source. If the source distribution is known, then the problem of optimal quantization under a rate constraint can be posed as a (typically rather hard) optimization problem.

On the other hand, if the source distribution is unknown (as is often the case in practice), then an approximation to an optimal quantizer must be constructed (learned) on the basis of a finite number of training samples drawn from the source distribution. Questions of a different flavor arise in this situation. For example, a fundamental problem is whether an optimal quantizer can be learned as the number of training samples increases without bound. We would also like to know how many training samples are needed and what methods to use to construct a quantizer whose performance is close to the optimum. Our main goal in these notes is to demonstrate how tools and techniques from nonparametric statistics and statistical learning theory can be used to tackle these and related problems concerning learning vector quantizers from empirical data.

Notes

The model of data compression we consider is not the most general possible. For more general models and the information-theoretic framework for lossy data compression, see, e.g., Gray (1990). Fundamentals of vector quantization are given in Gersho and Gray (1992). A very thorough review of the history of quantization and an extensive survey of its literature can be found in Gray and Neuhoff (1998).

4.2 The fixed-rate quantization problem

A *fixed-rate N-point vector quantizer* ($N \geq 1$ is an integer) is a Borel measurable mapping $q : \mathcal{R}^d \to C$, where the *codebook* $C = \{y_1, \ldots, y_N\}$ is an ordered collection of N distinct points in $\mathcal{R}^d$, called the *codevectors*.

In our model, the source is an $\mathcal{R}^d$-valued random vector X. In order to compress the source, the quantizer q represents the "input" X by the "output" $\widehat{X} = q(X)$. Since $q(X)$ can only take N distinct values, it is possible to uniquely describe $q(X)$ using only a finite number of bits, while such a description is in general not possible for X. In fact, for values of N such that $\log_2 N$ is an integer, exactly $\log_2 N$ bits are necessary and sufficient to uniquely specify the values of $q(X)$ in a description using binary strings of a fixed length. In other words, for every realization of X, one needs to transmit or store $\log_2 N$ bits so that the value of $q(X)$ can be reconstructed. For this reason, the the *rate* of q is defined by

$$R(q) \triangleq \log_2 N.$$

In general, it is convenient (and customary) to define the rate by $\log_2 N$ even if this number is not an integer.

To measure the reconstruction error in representing X by $q(X)$, we use the quantity $d(X,q(X))$, where $d(x,y) \geq 0$, $x,y \in \mathcal{R}^d$, is a measurable function called a *distortion measure*. The *distortion* of q in quantizing X is the expected reconstruction error

$$D(\mu,q) \triangleq \mathbb{E}d(X,q(X)) = \int_{\mathcal{R}^d} d(x,q(x))\,\mu(dx)$$

where μ denotes the distribution of X. For simplicity, we assume mean squared distortion, i.e., $d(x,y) = \|x-y\|^2$, where $\|\cdot\|$ denotes the Euclidean norm in $\mathcal{R}^d$, so that

$$D(\mu,q) \triangleq \mathbb{E}\|X - q(X)\|^2.$$

Throughout we assume that $\mathbb{E}\|X\|^2 < \infty$, which implies $D(\mu,q) < \infty$.

Two quantizers have the same rate if they have the same number of codevectors. The primary goal of quantization is to find quantizers that have minimum distortion subject to a constraint on the rate, or equivalently, on the number of codevectors. Consequently, we define the optimal performance over N-point quantizers by

$$D_N^*(\mu) \triangleq \inf_{q \in \mathcal{Q}_N} D(\mu,q)$$

where $\mathcal{Q}_N$ denotes the set of all N-point quantizers. A quantizer $q^* \in \mathcal{Q}_N$ is called *optimal* if $D(\mu,q^*) = D_N^*(\mu)$.

It is useful to note that any N-point quantizer q is completely characterized by its codebook $\{y_i\}_{i=1}^N$ and the *cells* $S_i = \{x : q(x) = y_i\}$, $i = 1,\ldots,N$, via the rule

$$q(x) = y_i \quad \text{if and only if} \quad x \in S_i.$$

Note that $\{S_1,\ldots,S_N\}$ is a partition of $\mathcal{R}^d$. In what follows we will often define quantizers by specifying their codebooks and partition cells.

The next lemma shows how to choose the partition cells optimally for a fixed codebook, and how to choose the codebook optimally for fixed partition cells.

Lemma 4.1.

(NEAREST NEIGHBOR CONDITION) *Let q be an arbitrary N-point quantizer with codebook $C = \{y_i\}_{i=1}^N$, and let q' be an N-point quantizer which has the same codebook and is defined by*

$$q'(x) = \operatorname*{argmin}_{y_i \in C} \|x - y_i\|^2 \tag{4.1}$$

where ties are broken arbitrarily. Then

$$D(\mu, q') \leq D(\mu, q).$$

(CENTROID CONDITION) *Let q be an arbitrary quantizer with partition cells $\{S_i\}_{i=1}^N$. If q' is defined to have the same partition cells and codevectors given by*

$$y_i' = \operatorname*{argmin}_{y \in \mathcal{R}^d} \mathbb{E}[\|X - y\|^2 | X \in S_i] = \mathbb{E}[X | X \in S_i], \quad i = 1, \ldots, N$$

then

$$D(\mu, q') \leq D(\mu, q).$$

PROOF. To prove the nearest neighbor condition, note that (4.1) is equivalent to

$$\|x - q'(x)\|^2 = \min_{1 \leq i \leq N} \|x - y_i\|^2$$

i.e, $q(x)$ is the nearest neighbor of x among $\{y_i\}_{i=1}^N$. Thus for any q with codebook C and arbitrary partition cells $\{S_i\}_{i=1}^N$,

$$\begin{aligned}
\mathbb{E}\|X - q(X)\|^2 &= \sum_{j=1}^N \int_{S_j} \|x - y_j\|^2 \mu(dx) \\
&\geq \sum_{j=1}^N \int_{S_j} \min_{1 \leq i \leq N} \|x - y_i\|^2 \mu(dx) \\
&= \int_{\mathcal{R}^d} \min_{1 \leq i \leq N} \|x - y_i\|^2 \mu(dx) \\
&= \mathbb{E}\|X - q'(X)\|^2.
\end{aligned}$$

To prove the centroid condition, note that for any measurable $S \subset \mathcal{R}^d$ such that $\mu(S) > 0$, if $y' = \mathbb{E}[X | X \in S]$ and $y \in \mathcal{R}^d$ is arbitrary, then

$$\mathbb{E}[\|X - y\|^2 | X \in S] = \mathbb{E}[\|X - y'\|^2 | X \in S] + \|y - y'\|^2$$

since $\mathbb{E}[(y-y')\cdot(X-y')|X\in S]=0$, where $a\cdot b$ denotes the usual inner product of $a,b\in\mathcal{R}^d$. (This optimizing y' is often called the *centroid* of S'.) Hence y' is the unique point satisfying

$$\mathbb{E}[\|X-y'\|^2|X\in S]=\inf_{y\in\mathcal{R}^d}\mathbb{E}[\|X-y\|^2|X\in S].$$

Thus if q has partition cells $\{S_i\}_{i=1}^N$ and arbitrary codebook $\{y_i\}_{i=1}^N$, then

$$\begin{aligned}\mathbb{E}\|X-q(X)\|^2 &= \sum_{i=1}^N \mathbb{E}[\|X-y_i\|^2|X\in S_i]\mu(S_i)\\ &\geq \sum_{i=1}^N \mathbb{E}[\|X-y_i'\|^2|X\in S_i]\mu(S_i)\\ &= \mathbb{E}\|X-q'(X)\|^2.\end{aligned}$$

□

A quantizer q with codebook $C=\{y_i\}_{i=1}^N$ is called a *nearest neighbor* quantizer if for all $x\in\mathcal{R}^d$,

$$\|x-q(x)\|^2=\min_{y_i\in C}\|x-y_i\|^2.$$

The nearest neighbor condition of Lemma 4.1 implies that it suffices to consider nearest neighbor quantizers when searching for an optimal quantizer. Equivalently,

$$D_N^*(\mu)=\inf_{C:|C|=N}\mathbb{E}\min_{y_i\in C}\|X-y_i\|^2. \tag{4.2}$$

Note that (4.2) clearly implies

$$D_N^*(\mu)\leq D_{N+1}^*(\mu) \tag{4.3}$$

for any $N\geq 1$.

Although the partition cells of a nearest neighbor quantizer are not uniquely determined by the codebook $C=\{y_i\}_{i=1}^N$, one can make the definition unique via a fixed tie-breaking rule. For example, a tie-breaking rule that favors smaller indices gives

$$S_1=\{x:\|x-y_1\|\leq\|x-y_j\|,\ j=1,\ldots,N\}$$

and for $i=2,\ldots,N$,

$$S_i=\{x:\|x-y_i\|\leq\|x-y_j\|,\ j=1,\ldots,N\}\setminus\bigcup_{k=1}^{i-1}S_k.$$

Any $\{S_i\}_{i=1}^N$ obtained as the partition associated with a nearest neighbor quantizer with codebook C is called a *Voronoi* (or nearest neighbor) *partition* of $\mathcal{R}^d$ with respect to C.

In view of (4.2), it is not hard to show that an optimal N-point quantizer always exists.

Theorem 4.1. *There exists a nearest neighbor quantizer $q^* \in \mathcal{Q}_N$ such that $D(\mu, q^*) = D_N^*(\mu)$.*

PROOF. For a positive integer m and $(y_1, \ldots, y_m) \in (\mathcal{R}^d)^m$ define

$$g_m(y_1, \ldots, y_m) \triangleq \int_{\mathcal{R}^d} \min_{1 \le i \le m} \|x - y_i\|^2 \mu(dx).$$

Note that g_m is the distortion of a nearest neighbor quantizer with codevectors $y_1, \ldots, y_m$ (this quantizer may have less than m codevectors since the y_i are not necessarily distinct). Hence by (4.3),

$$\inf_{(y_1, \ldots, y_m) \in (\mathcal{R}^d)^m} g_m(y_1, \ldots, y_m) = D_m^*(\mu). \tag{4.4}$$

We can assume that $N \geq 2$ since for $N = 1$ the claim of the theorem follows from the centroid condition of Lemma 4.1. Also, to exclude the trivial case when μ is concentrated on a single point, we assume that the support of μ contains at least two distinct points. Then it is easy to show that

$$D_2^*(\mu) < D_1^*(\mu)$$

(to see this, note that by the proof of the centroid condition, if q is an arbitrary two-point nearest-neighbor quantizer with cells S_1, S_2 and codepoints y_1, y_2 such that $\mu(S_i) > 0$ and $y_i = \mathbb{E}[X | X \in S_i]$ for $i = 1, 2$, then $D(\mu, q) < \mathbb{E}\|X - \mathbb{E}X\|^2 = D_1^*(\mu)$). Hence there is a unique integer $2 \le k \le N$ such that

$$D_N^*(\mu) = \cdots = D_k^*(\mu) < D_{k-1}^*(\mu). \tag{4.5}$$

(In fact, one can prove that if the support of μ contains at least N points, then $D_N^*(\mu) < D_{N-1}^*(\mu)$, and so $k = N$).

Let $B_r \triangleq \{x : \|x\| \le r\}$ denote the closed ball of radius $r > 0$ centered at the origin. Fix $\varepsilon > 0$ such that

$$\varepsilon < \frac{1}{2}\left(D_{k-1}^*(\mu) - D_k^*(\mu)\right) \tag{4.6}$$

and pick $0 < r < R$ such that

$$(R - r)^2 \mu(B_r) > D_k^*(\mu) + \varepsilon, \qquad 4 \int_{B_{2R}^c} \|x\|^2 \mu(dx) < \varepsilon. \tag{4.7}$$

Choose $(y_1, \ldots, y_k)$ satisfying $g_k(y_1, \ldots, y_k) < D_k^*(\mu) + \varepsilon$, and suppose without loss of generality that the codevectors are indexed so that $\|y_1\| \le \cdots \le \|y_k\|$. Then $\|y_1\| \le R$, since otherwise, by the triangle inequality, $\min_{1 \le i \le k} \|x - y_i\|^2 \ge (R - r)^2$ for all $x \in B_r$, and so

$$D_k^*(\mu) + \varepsilon > \int_{B_r} \min_{1 \le i \le k} \|x - y_i\|^2 \mu(dx) \ge (R - r)^2 \mu(B_r)$$

contradicting (4.7). We will show that $\|y_j\| \le 5R$ for all j. Assume to the contrary that $\|y_k\| > 5R$. Then by the triangle inequality, for all $x \in \mathcal{R}^d$,

$$\|x - y_1\| \le \|x - y_k\| I_{\{x \in B_{2R}\}} + 2\|x\| I_{\{x \in B_{2R}^c\}} \tag{4.8}$$

where I_A denotes the indicator of the set A. Then letting $\{S_i\}_{i=1}^k$ be a Voronoi partition with respect to $\{y_i\}_{i=1}^k$, we obtain

$$\begin{aligned}
g_{k-1}(y_1, \ldots, y_{k-1}) &= \sum_{j=1}^{k} \int_{S_j} \min_{1 \le i \le k-1} \|x - y_i\|^2 \mu(dx) \\
&\le \sum_{j=1}^{k-1} \int_{S_j} \|x - y_j\|^2 \mu(dx) + \int_{S_k} \|x - y_1\|^2 \mu(dx) \\
&\le \sum_{j=1}^{k} \int_{S_j} \|x - y_j\|^2 \mu(dx) + 4 \int_{B_{2R}^c} \|x\|^2 \mu(dx) \\
&\le g_k(y_1, \ldots, y_k) + \varepsilon \le D_k^*(\mu) + 2\varepsilon < D_{k-1}^*(\mu)
\end{aligned}$$

where the second inequality follows from (4.8), the third from (4.7), and the last one from (4.6). This contradicts (4.4), so we obtain that $g_k(y_1, \ldots, y_k) < D_k^*(\mu) + \varepsilon$ implies $(y_1, \ldots, y_k) \in (B_{5R})^k$. Therefore

$$D_k^*(\mu) = \inf_{(y_1, \ldots, y_k) \in (B_{5R})^k} g_k(y_1, \ldots, y_k).$$

Since $(B_{5R})^k \subset (\mathcal{R}^d)^k$ is compact and g_k is continuous (as can be seen by an application of the dominated convergence theorem), there exists $(y_1^*, \ldots, y_k^*)$ in $(B_{5R})^k$ with $g_k(y_1^*, \ldots, y_k^*) = D_k^*(\mu)$. Thus there is a nearest neighbor quantizer with at most k codevectors achieving $D_k^*(\mu)$. Since $k \le N$ and $D_k^*(\mu) = D_N^*(\mu)$, this implies that there must exist an N-point nearest neighbor quantizer q^* that achieves $D_N^*(\mu)$. □

Remark Lemma 4.1 gives rise to an iterative algorithm for designing N-point quantizers. Start with an arbitrary N-point quantizer q_0 with codebook $\mathcal{C}_0$ and partition $\mathcal{S}_0$. In the mth iteration ($m = 1, 2, \ldots$), first let $\mathcal{S}_m = \{S_i^{(m)}\}_{i=1}^N$ be the Voronoi partition with respect to $\mathcal{C}_{m-1}$, and then set $\mathcal{C}_m = \{y_i^{(m)}\}_{i=1}^N$, where $y_i^{(m)} = \mathbb{E}[X | X \in S_i^{(m)}]$ for $i = 1, \ldots, N$. If q_m denotes the quantizer defined by $\mathcal{C}_m$ and $\mathcal{S}_m$, and we set $D_m = D(\mu, q_m)$, then Lemma 4.1 implies

$$D_m \le D_{m-1}$$

and so $\lim_{m \to \infty}(D_{m-1} - D_m) = 0$. The algorithm stops (after a finite number of iterations) when the drop in distortion falls below a given threshold. The distortion D_m is not guaranteed to converge to the minimum distortion $D_N^*(\mu)$, but quantizers obtained by this method or its variants yield sufficiently low distortion for practical applications.

Unless the dimension d is very small, computing D_m and the conditional expectations $\mathbb{E}[X|X \in S_i^{(m)}]$, $i = 1,\dots,N$, is hard for a general source distribution (given, e.g., by its probability density function). In practice, the algorithm is usually run with μ replaced by the empirical distribution of a finite number of samples drawn according to μ. As explained at the end of the next section, the implementation becomes straightforward in this case.

Notes

The optimality conditions of Lemma 4.1 were first derived by Lloyd (1957) (for the scalar $d = 1$ case) and by Steinhaus (1956) (who considered a problem equivalent to three-dimensional fixed-rate quantization). Theorem 4.1 is due to Pollard (1982a); the existence of optimal quantizers for more general distortion measures was shown, for example, by Pollard (1981), Abaya and Wise (1982), and Sabin (1984). Except for trivial cases, optimal quantizers and the minimum distortion $D_N^*(\mu)$ are very hard to determine analytically, but approximations that become tight as $N \to \infty$ can be derived for a large class of source distributions. We refer to Gray and Neuhoff (1998) and Graf and Luschgy (2000) for such asymptotic (high-rate) results. The design algorithm sketched above is basically also due to Lloyd (1957) and Steinhaus (1956). An extension to the vector case and to more general distortion measures was given by Linde, Buzo, and Gray (1980), and the algorithm is often referred to as "the LBG algorithm." For more details and other methods of vector quantizer design, see Gersho and Gray (1992).

4.3 Consistency of empirical design

In most situations, the distribution μ of the source X to be quantized is unknown, and the only available information about μ is in the form of *training data*, a finite sequence of vectors drawn according to μ. More formally, the training data $X_1^n \triangleq X_1,\dots,X_n$ consists of n independent and identically distributed (i.i.d.) copies of X. It is assumed that X_1^n and X are also independent. The training data is used to construct an N-point quantizer

$$q_n(\cdot) = q_n(\cdot, X_1,\dots,X_n).$$

Such a q_n is called an *empirically designed* quantizer. The goal is to "learn" the optimal quantizer from the data, i.e., to produce empirically designed quantizers with performance approaching (as n gets large) the performance of a quantizer optimal for X. We assume, as before, that $\mathbb{E}\|X\|^2 < \infty$.

We call the quantity

$$D(\mu, q_n) = \mathbb{E}\big[\|X - q_n(X)\|^2 \,|X_1^n\big]$$

the *test distortion* of q_n. Thus $D(\mu, q_n)$ measures the distortion resulting when q_n is applied to X;

it is the "true" distortion of the empirically designed quantizer. Note that $D(\mu, q_n)$ is a random variable since q_n depends on X_1^n.

Also of interest is the *training distortion* (or empirical distortion) of q_n, defined as the average distortion of q_n on the training data:

$$\frac{1}{n}\sum_{k=1}^{n} \|X_k - q_n(X_k)\|^2.$$

The empirical distribution μ_n of the training data is defined by

$$\mu_n(A) = \frac{1}{n}\sum_{k=1}^{n} I_{\{X_k \in A\}}$$

for every Borel measurable $A \subset \mathcal{R}^d$, i.e, μ_n places weight $1/n$ at each point X_k, $k = 1, \ldots, n$. Thus the training distortion becomes

$$\frac{1}{n}\sum_{k=1}^{n} \|X_k - q_n(X_k)\|^2 = D(\mu_n, q_n).$$

Intuitively, if q_n performs well on the training set, it should also have good performance on the source, assuming the training set is sufficiently large. We define an *empirically optimal* quantizer as an N-point quantizer q_n^* that minimizes the training distortion:

$$q_n^* \triangleq \operatorname*{argmin}_{q \in Q_N} \frac{1}{n}\sum_{k=1}^{n} \|X_k - q(X_k)\|^2.$$

In other words, $q_n^* \in Q_N$ satisfies

$$D(\mu_n, q_n^*) = \inf_{q \in Q_N} D(\mu_n, q) = D_N^*(\mu_n).$$

Note that by Theorem 4.1, q_n^* always exists. In fact, since μ_n is supported on at most n points, the existence of q_n^* is easy to show without resorting to Theorem 4.1. Note also that the definition of q_n^* outside the support of μ_n does not affect the training distortion, so q_n^* is not uniquely defined even if its codebook happens to be unique. We will resolve this problem by always requiring (as we may by Lemma 4.1) that q_n^* be a nearest neighbor quantizer.

Our goal is to show that the design based on empirical distortion minimization is *consistent* in the following sense: As the size of the training data grows, the sequence of test distortions converges almost surely (i.e., for almost every realization of the training sequence) to the minimum distortion achieved by an optimal quantizer. If this is the case, then for n large enough, q_n^* can effectively replace an optimal quantizer q^* in quantizing X.

THEOREM 4.2 (CONSISTENCY OF EMPIRICAL DESIGN) For any $N \geq 1$ the sequence of empirically optimal N-point nearest neighbor quantizers q_n^*, $n = 1, 2, \ldots$, satisfies

$$\lim_{n \to \infty} D(\mu, q_n^*) = D_N^*(\mu) \quad \text{a.s.}$$

To prove the theorem we need some intermediate results. The basic idea is that for large n the empirical distribution μ_n is a good estimate of μ, so the optimal quantizer for μ_n should provide a good approximation to the optimal quantizer for μ. In order to formalize this idea, we need a measure of closeness for probability distributions that is appropriate in quantization arguments.

Let μ and ν be probability distributions on $\mathcal{R}^d$ with finite second moment. The L_2 Wasserstein distance between μ and ν is defined by

$$\rho(\mu,\nu) \triangleq \inf_{X\sim\mu, Y\sim\nu} (\mathbb{E}\|X-Y\|^2)^{1/2}$$

where the infimum is taken over all joint distributions of two random vectors X and Y such that X has distribution μ, and Y has distribution ν (denoted by $X \sim \mu$ and $Y \sim \nu$, respectively). It is easy to see that $\rho(\mu,\nu)$ is finite; in fact, $\rho(\mu,\nu) \le (\mathbb{E}\|X\|^2)^{1/2} + (\mathbb{E}\|Y\|^2)^{1/2}$ by the triangle inequality for the L_2 norm.

Lemma 4.2. *The infimum defining $\rho(\mu,\nu)$ is a minimum. Moreover, $\rho(\mu,\nu)$ is a metric on the space of probability distributions on $\mathcal{R}^d$ with finite second moment.*

SKETCH OF PROOF. Let $\mathcal{P}(\mu,\nu)$ denote the family of probability distributions on $\mathcal{R}^d \times \mathcal{R}^d$ with marginals μ and ν. Then

$$\rho(\mu,\nu) = \inf_{P\in\mathcal{P}(\mu,\nu)} \left(\int \|x-y\|^2 P(dx,dy) \right)^{1/2}.$$

First we show that the infimum is in fact a minimum. Fix $\varepsilon > 0$ and pick a closed ball $B \subset \mathcal{R}^d$ with large enough radius such that $\mu(B) \ge 1-\varepsilon$ and $\nu(B) \ge 1-\varepsilon$. Then any $P \in \mathcal{P}(\mu,\nu)$ has $P(B\times B) \ge 1-2\varepsilon$. Since $B\times B$ is compact, we obtain that the set of distributions $\mathcal{P}(\mu,\nu)$ is tight (see, e.g., Ash (2000)). If $P_k \in \mathcal{P}(\mu,\nu)$, $k = 1,2,\ldots$, is a sequence such that

$$\int \|x-y\|^2 P_k(dx,dy) < \rho(\mu,\nu)^2 + \frac{1}{k}$$

then by Prokhorov's theorem (see, e.g., Theorem 7.2.4 in Ash (2000)) there is a subsequence of $\{P_k\}$, say $\{P_k'\}$, such that, as $k \to \infty$, P_k' converges weakly to some probability distribution P'. Clearly, $P' \in \mathcal{P}(\mu,\nu)$, and it is easy to show using a truncation argument that $\int \|x-y\|^2 P'(dx,dy) = \rho(\mu,\nu)^2$.

To show that ρ is a metric, note that $\rho(\mu,\nu) = \rho(\nu,\mu) \ge 0$, and if $\rho(\mu,\nu) = 0$, then by the preceding argument there exist $X \sim \mu$ and $Y \sim \nu$ such that $\mathbb{E}\|X-Y\|^2 = 0$, implying $\mu = \nu$. Thus it only remains to verify that ρ satisfies the triangle inequality.

Let μ, ν, and λ be probability distributions on $\mathcal{R}^d$ having finite second moment. Assume $P \in \mathcal{P}(\mu,\nu)$ achieves $\rho(\mu,\nu)$ and $P' \in \mathcal{P}(\nu,\lambda)$ achieves $\rho(\nu,\lambda)$. Construct a jointly distributed triplet (X,Y,Z) by specifying that $(X,Y) \sim P$, $(Y,Z) \sim P'$, and that X and Z are conditionally independent given Y (i.e., X, Y, Z form a Markov chain in this order). Since $X \sim \mu$ and $Z \sim \lambda$, by the triangle inequality for the L_2 norm

$$\begin{aligned} \rho(\mu,\lambda) &\le (\mathbb{E}\|X-Z\|^2)^{1/2} \\ &\le (\mathbb{E}\|X-Y\|^2)^{1/2} + (\mathbb{E}\|Y-Z\|^2)^{1/2} \\ &= \rho(\mu,\nu)+\rho(\nu,\lambda). \end{aligned}$$

□

The next lemma justifies the choice of ρ by showing that if two distributions are close in ρ metric, then any nearest neighbor quantizer will quantize these distributions with similar distortion. This fact also implies the stability of optimal quantizer performance with respect to the ρ metric.

Lemma 4.3. *If q is a nearest neighbor quantizer, then*

$$|D(\mu,q)^{1/2} - D(\nu,q)^{1/2}| \le \rho(\mu,\nu).$$

Consequently,

$$|D_N^*(\mu)^{1/2} - D_N^*(\nu)^{1/2}| \le \rho(\mu,\nu).$$

PROOF. To prove the first bound let $\{y_1,\ldots,y_N\}$ denote the codebook of q, and let $X \sim \mu$ and $Y \sim \nu$ achieve the minimum defining $\rho(\mu,\nu)$. Then

$$\begin{aligned} D(\mu,q)^{1/2} &= \{\mathbb{E} \min_{1\le i\le N} \|X-y_i\|^2\}^{1/2} \\ &= \{\mathbb{E}(\min_{1\le i\le N} \|X-y_i\|)^2\}^{1/2} \\ &\le \{\mathbb{E}(\min_{1\le i\le N}(\|X-Y\|+\|Y-y_i\|))^2\}^{1/2} \\ &= \{\mathbb{E}(\|X-Y\| + \min_{1\le i\le N}\|Y-y_i\|)^2\}^{1/2} \\ &\le \{\mathbb{E}\|X-Y\|^2\}^{1/2} + \{\mathbb{E}\min_{1\le i\le N}\|Y-y_i\|^2\}^{1/2} \\ &= \rho(\mu,\nu)+D(\nu,q)^{1/2}. \end{aligned}$$

The inequality $D(\nu,q)^{1/2} - D(\mu,q)^{1/2} \le \rho(\mu,\nu)$ is proved similarly.

To prove the second bound, assume q^* is an optimal N-point (nearest neighbor) quantizer for ν. Then

$$\begin{aligned} D_N^*(\mu)^{1/2} - D_N^*(\nu)^{1/2} &= D_N^*(\mu)^{1/2} - D(\nu,q^*)^{1/2} \\ &\le D(\mu,q^*)^{1/2} - D(\nu,q^*)^{1/2} \\ &\le \rho(\mu,\nu) \end{aligned}$$

by the first bound of the lemma. The inequality $D_N^*(\nu)^{1/2} - D_N^*(\mu)^{1/2} \le \rho(\mu,\nu)$ is proved in a similar fashion by considering an N-point optimal quantizer for μ. □

The following corollary relates the test distortion of the empirically optimal quantizer to the distortion of the optimal quantizer in terms of the ρ distance between the the empirical distribution and the true source distribution.

COROLLARY 4.1. *The test distortion of the empirically optimal N-point quantizer q_n^* is upper bounded as*

$$D(\mu,q_n^*)^{1/2} - D_N^*(\mu)^{1/2} \le 2\rho(\mu,\mu_n).$$

PROOF. Let q^* be an optimal N-point (nearest neighbor) quantizer for μ. Recall that we specified q_n^* to be a nearest neighbor quantizer. Then

$$\begin{aligned} & D(\mu,q_n^*)^{1/2} - D_N^*(\mu)^{1/2} \\ &= D(\mu,q_n^*)^{1/2} - D(\mu,q^*)^{1/2} \\ &= D(\mu,q_n^*)^{1/2} - D(\mu_n,q_n^*)^{1/2} + D(\mu_n,q_n^*)^{1/2} - D(\mu,q^*)^{1/2} \\ &\le D(\mu,q_n^*)^{1/2} - D(\mu_n,q_n^*)^{1/2} + D(\mu_n,q^*)^{1/2} - D(\mu,q^*)^{1/2} \\ &\le 2\rho(\mu,\mu_n) \end{aligned}$$

where the last inequality follows from Lemma 4.3. □

The consistency theorem is a consequence of Corollary 4.1 and the following lemma.

Lemma 4.4. *Given* ν *and a sequence* $\{\nu_n\}$, $\lim_{n\to\infty}\rho(\nu,\nu_n) = 0$ *if and only if* $\nu_n \to \nu$ *weakly and* $\int \|x\|^2\nu_n(dx) \to \int \|x\|^2\nu(dx)$ *as* $n \to \infty$.

PROOF. Suppose $\lim_{n\to\infty}\rho(\nu,\nu_n) = 0$. For each n let $Y \sim \nu$, $Y_n \sim \nu_n$ have joint distribution such that $\rho(\nu,\nu_n) = (\mathbb{E}\|Y - Y_n\|^2)^{1/2}$. Then $\lim_{n\to\infty}\mathbb{E}\|Y - Y_n\|^2 = 0$, implying $Y_n \to Y$ in distribution (i.e., $\nu_n \to \nu$ weakly) and $\mathbb{E}\|Y_n\|^2 \to \mathbb{E}\|Y\|^2$ (i.e., $\int \|x\|^2\nu_n(dx) \to \int \|x\|^2\nu(dx)$).

Conversely, suppose ν_n converges weakly to ν. Then by Skorohod's theorem (see, e.g., Theorem 11.7.2 in Dudley (1989)) there exist $Y_n \sim \nu_n$ and $Y \sim \nu$ jointly distributed such that $Y_n \to Y$ a.s. By the triangle inequality, $2\|Y_n\|^2 + 2\|Y\|^2 - \|Y_n - Y\|^2 \ge \|Y_n\|^2 + \|Y\|^2 - 2\|Y_n\|\|Y\| \ge 0$; hence Fatou's lemma implies

$$\begin{aligned} & \liminf_{n\to 0}\mathbb{E}\{2\|Y_n\|^2 + 2\|Y\|^2 - \|Y_n - Y\|^2\} \\ &\ge \mathbb{E}\{\liminf_{n\to 0}(2\|Y_n\|^2 + 2\|Y\|^2 - \|Y_n - Y\|^2)\} \\ &= 4\mathbb{E}\|Y\|^2. \end{aligned}$$

Thus, if $E\|Y_n\|^2 \to E\|Y\|^2$, we obtain that $\mathbb{E}\|Y_n - Y\|^2 \to 0$, implying $\rho(\nu,\nu_n) \to 0$. □

PROOF OF THEOREM 4.2. Clearly, we have $D(\mu,q_n^*) \geq D_N^*(\mu)$ for each n. Thus by Corollary 4.1 the theorem holds if the empirical distributions converge to the true source distribution in ρ metric almost surely, i.e., if

$$\lim_{n\to\infty} \rho(\mu,\mu_n) = 0 \quad \text{a.s.} \tag{4.9}$$

By Lemma 4.4 this happens if, almost surely, $\mu_n \to \mu$ weakly and $\int \|x\|^2 \mu_n(dx) \to \int \|x\|^2 \mu(dx)$. A basic result (due to Varadayan) concerning empirical distributions is that

$$\mathbb{P}\{\mu_n \to \mu \text{ weakly}\} = 1$$

(see, e.g., Theorem 11.4.1 in Dudley (1989)). Also, by the strong law of large numbers

$$\begin{aligned}\lim_{n\to\infty} \int \|x\|^2 \mu_n(dx) &= \lim_{n\to\infty} \frac{1}{n}\sum_{i=1}^{n} \|X_i\|^2 \\ &= \mathbb{E}\|X\|^2 = \int \|x\|^2 \mu(dx) \quad \text{a.s.}\end{aligned}$$

Thus

$$\mathbb{P}\left\{\mu_n \to \mu \text{ weakly}, \int \|x\|^2 \mu_n(dx) \to \int \|x\|^2 \mu(dx)\right\} = 1$$

completing the proof. □

Remarks (i) The consistency theorem says that when trained on a long enough training sequence, empirically optimal quantizers perform almost as well as optimal ones. In a similar manner, we can also show that the training distortion of the empirically optimal quantizer is a strongly consistent estimate of the optimal distortion. Indeed, applying the second bound of Lemma 4.3 with $\nu = \mu_n$,

$$|D(\mu_n,q_n^*)^{1/2} - D_N^*(\mu)^{1/2}| = |D_N^*(\mu_n)^{1/2} - D_N^*(\mu)^{1/2}| \leq \rho(\mu,\mu_n)$$

and so (4.9) implies

$$\lim_{n\to\infty} D(\mu_n,q_n^*) = D_N^*(\mu) \quad \text{a.s.}$$

However, for finite n the training distortion is optimistically biased: If q^* is an optimal N-level quantizer for μ, then $D(\mu_n,q_n^*) \leq D(\mu_n,q^*)$, so

$$\begin{aligned}\mathbb{E}D(\mu_n,q_n^*) &\leq \mathbb{E}D(\mu_n,q^*) \\ &= \mathbb{E}\left\{\frac{1}{n}\sum_{k=1}^{n} \|X_k - q^*(X_k)\|^2\right\} \\ &= D(\mu,q^*) = D_N^*(\mu).\end{aligned}$$

In fact, it is not hard to see that the above inequality is always strict unless μ is a discrete distribution concentrating on N points or less.

(ii) Finding an empirically optimal quantizer q_n^* for a given realization $x_1,\ldots,x_n$ of the training sequence is a computationally hard problem for quantizer dimensions $d \geq 2$. However, the iterative algorithm sketched at the end of Section 4.2 can be used with the empirical distribution of the training sequence replacing μ to find a suboptimal, but usually good enough approximation to q_n^*. The implementation of the algorithm is straightforward in this case: Let μ_n denote the empirical distribution of the samples $x_1,\ldots,x_n$, and let $Y \sim \mu_n$. Then the computation of the new codevectors in the mth iteration reduces to

$$y_i^{(m)} = \mathbb{E}[Y | Y \in S_i^{(m)}] = \frac{\frac{1}{n}\sum_{k=1}^n x_k I_{\{x_k \in S_i^{(m)}\}}}{\frac{1}{n}\sum_{k=1}^n I_{\{x_k \in S_i^{(m)}\}}}, \quad i = 1,\ldots,N.$$

Calculating the distortion D_m has similar complexity. Note that it is easy to decide whether or not $x_k \in S_i^{(m)}$ since $\{S_j^{(m)}\}_{j=1}^N$ is a Voronoi partition with respect to $\{y_j^{(m-1)}\}_{j=1}^N$. Indeed, we have $x_k \in S_i^{(m)}$ if $y_i^{(m-1)}$ is the nearest neighbor of x_k in the codebook $\{y_j^{(m-1)}\}_{j=1}^N$.

Notes

The proof of Theorem 4.2 is based on Pollard (1982a); see also Pollard (1981) and (1982b), Abaya and Wise (1984), and Graf and Luschgy (1994). See Graf and Luschgy (2000) for more recent stability and consistency results related to fixed-rate quantization. The L_2 Wasserstein metric ρ is a special case of the $\bar{\rho}$ ("ρ-bar") distance between random processes introduced by Gray, Neuhoff, and Shields (1975). Lemma 4.3 is due to Gray and Davisson (1975) who were the first to use the L_2 Wasserstein metric in quantization arguments. The problem of constructing an empirically optimal quantizer is often referred to as the k-means clustering problem in the statistical literature (see MacQueen (1967)). Convergence results for the empirical version of the iterative quantizer design algorithm (the LBG algorithm) are given in Sabin and Gray (1986).

4.4 Finite sample upper bounds

We know from the consistency theorem (Theorem 4.2) that as the training data size increases without bound, the performance of the empirically optimal quantizer will approximate the optimal performance with arbitrary accuracy. But large amounts of training data may be "expensive" to acquire, and the computational cost of finding a quantizer that minimizes (at least approximately) the training distortion may become prohibitive for large training sets. It is therefore of interest to quantify how fast the test distortion of the empirically optimal quantizer converges to the optimal distortion with increasing training set size.

In this section we develop finite sample bounds on the expected test (and training) distortion in empirical design. Such bounds are of both theoretical and practical interest. For example, an upper bound on the test distortion, if sufficiently tight, can give a useful bound on the minimum number of training samples sufficient to guarantee a preassigned level of performance for the designed quantizer.

To simplify matters, we only consider source distribution with bounded support. For $T > 0$, let $\mathcal{P}(T)$ denote the set of probability distributions on $\mathcal{R}^d$ supported on B_T, the closed ball of radius T centered at the origin. Throughout this section we require that $\mu \in \mathcal{P}(T)$, that is,

$$\mathbb{P}\{\|X\| \le T\} = 1.$$

Let $Q_N(T)$ denote the collection of all nearest neighbor quantizers with all their codevectors inside B_T. We start with a basic lemma.

Lemma 4.5. *For any $\mu \in \mathcal{P}(T)$,*

$$D(\mu, q_n^*) - D_N^*(\mu) \le 2 \sup_{q \in Q_N(T)} |D(\mu_n, q) - D(\mu, q)|.$$

PROOF. Let q^* be an optimal N-point nearest neighbor quantizer for μ. We can repeat the argument in the proof of Corollary 4.1 without taking square roots to obtain

$$\begin{aligned} &D(\mu, q_n^*) - D_N^*(\mu) \\ &\quad \le \; D(\mu, q_n^*) - D(\mu_n, q_n^*) + D(\mu_n, q^*) - D(\mu, q^*). \end{aligned} \tag{4.10}$$

Now observe that since B_T is a convex set, if an arbitrary N-point quantizer q has a codevector y_i outside B_T, we can replace y_i by its projection y_i' on B_T. The new codevector y_i' is then closer to all $x \in B_T$ than y_i, i.e., $\|x - y_i'\| < \|x - y_i\|$ if $\|x\| \le T$, $\|y_i\| > T$, and $y_i' = Ty_i/\|y_i\|$. Replacing all codevectors of q outside B_T by their projections, we obtain q' such that $\|x - q'(x)\| \le \|x - q(x)\|$ for all $x \in B_T$. Thus for any $\nu \in \mathcal{P}(T)$, we have $D(\nu, q') \le D(\nu, q)$. Since $\mu \in \mathcal{P}(T)$, we also have $\mu_n \in \mathcal{P}(T)$ a.s., and it follows that both q^* and q_n^* can be assumed to have all their codevectors inside B_T. Since both q^* and q_n^* were assumed to be nearest neighbor quantizers, we obtain $q^*, q_n^* \in Q_N(T)$. This and bound (4.10) imply the lemma. □

The following result is the key to the finite sample bounds in this section.

Theorem 4.3. *There is a constant C, depending only on d, N, and T, such that for all $n \ge 1$ and $\mu \in \mathcal{P}(T)$,*

$$\mathbb{E}\left\{ \sup_{q \in Q_N(T)} |D(\mu_n, q) - D(\mu, q)| \right\} \le \frac{C}{\sqrt{n}}.$$

PROOF. There are several (not fundamentally different) ways to prove the bound of the theorem, resulting in somewhat different values for the constant C. These methods are all based on bounds on uniform deviations of empirical averages given in terms of combinatorial properties of certain classes of sets or functions. The proof given below is perhaps the simplest; it is based on a sharpened version of the basic Vapnik-Chervonenkis inequality. In what follows we will use notions and results of Vapnik-Chervonenkis theory given in Section 1.4 (see also Devroye, Györfi, and Lugosi (1996), Vapnik (1998), or Anthony and Bartlett (1999)).

For $q \in \mathcal{Q}_N(T)$ define the distortion function

$$f_q(x) \triangleq \|x - q(x)\|^2.$$

Note that $q \in \mathcal{Q}_N(T)$ implies $0 \leq f_q(x) \leq 4T^2$ for all $x \in B_T$. Thus

$$D(\mu, q) = \mathbb{E} f_q(X) = \int_0^{4T^2} \mathbb{P}\{f_q(X) > u\}\, du$$

and

$$D(\mu_n, q) = \frac{1}{n} \sum_{k=1}^{n} f_q(X_k) = \int_0^{4T^2} \frac{1}{n} \sum_{k=1}^{n} I_{\{f_q(X_k) > u\}}\, du \quad \text{a.s.}$$

Hence

$$\begin{aligned}
&\sup_{q \in \mathcal{Q}_N(T)} \left| D(\mu_n, q) - D(\mu, q) \right| \\
&\quad = \sup_{q \in \mathcal{Q}_N(T)} \left| \frac{1}{n} \sum_{k=1}^{n} f_q(X_k) - \mathbb{E} f_q(X) \right| \\
&\quad = \sup_{q \in \mathcal{Q}_N(T)} \left| \int_0^{4T^2} \left(\frac{1}{n} \sum_{k=1}^{n} I_{\{f_q(X_k) > u\}} - \mathbb{P}\{f_q(X) > u\} \right) du \right| \\
&\quad \leq 4T^2 \sup_{q \in \mathcal{Q}_N(T), u > 0} \left| \frac{1}{n} \sum_{k=1}^{n} I_{\{f_q(X_k) > u\}} - \mathbb{P}\{f_q(X) > u\} \right| \\
&\quad = 4T^2 \sup_{A \in \mathcal{A}_N} |\mu_n(A) - \mu(A)| \quad \text{a.s.}
\end{aligned} \tag{4.11}$$

where $\mathcal{A}_N$ is the family of sets in $\mathcal{R}^d$ defined by

$$\mathcal{A}_N \triangleq \Big\{ \{x : f_q(x) > u\} : q \in \mathcal{Q}_N(T), u > 0 \Big\}.$$

Denoting by $V(\mathcal{A}_N)$ the VC dimension of $\mathcal{A}_N$, the sharpened version of the Vapnik-Chervonenkis inequality in Section 1.4.5 gives

$$\mathbb{E}\left\{ \sup_{A \in \mathcal{A}_N} |\mu_n(A) - \mu(A)| \right\} \leq c \sqrt{\frac{V(\mathcal{A}_N)}{n}} \tag{4.12}$$

where c is a universal constant. Therefore

$$\mathbb{E}\left\{\sup_{q\in\mathcal{Q}_N(T)}\left|D(\mu_n,q)-D(\mu,q)\right|\right\}\leq 4T^2c\sqrt{\frac{V(\mathcal{A}_N)}{n}}. \tag{4.13}$$

Of course, the bound is meaningful only if $V(\mathcal{A}_N)<\infty$, which we show to be the case next.

For a nearest neighbor quantizer q with codevectors $\{y_i\}_{i=1}^N$, $f_q(x)>u$ if and only if $\|x-y_i\|>\sqrt{u}$ for all $i=1,\ldots,N$. Thus each $A\in\mathcal{A}_N$ is the intersection of the complements of N closed balls of equal radii in $\mathcal{R}^d$. Letting $\mathcal{A}$ denote the collection of all complements of closed balls in $\mathcal{R}^d$, we therefore have

$$\mathcal{A}_N\subset\bar{\mathcal{A}}_N\triangleq\{A_1\cap\cdots\cap A_N: A_i\in\mathcal{A}, i=1,\ldots,N\}. \tag{4.14}$$

Note that $\bar{\mathcal{A}}_1=\mathcal{A}$, and $\bar{\mathcal{A}}_N=\{A\cap B: A\in\mathcal{A}, B\in\bar{\mathcal{A}}_{N-1}\}$ for $N\geq 2$. Thus Theorem 1.12(4) implies that for $N\geq 2$ the mth shatter coefficient of $\bar{\mathcal{A}}_N$ is upper bounded as

$$\mathbb{S}_{\bar{\mathcal{A}}_N}(m)\leq\mathbb{S}_{\mathcal{A}}(m)\mathbb{S}_{\bar{\mathcal{A}}_{N-1}}(m).$$

By induction on N, we obtain that for all $N\geq 1$ and $m\geq 1$,

$$\mathbb{S}_{\bar{\mathcal{A}}_N}(m)\leq\mathbb{S}_{\mathcal{A}}(m)^N. \tag{4.15}$$

Define $\mathcal{B}=\{B^c: B\in\mathcal{A}\}$. From Theorem 1.12(3), we have $\mathbb{S}_{\mathcal{A}}(m)=\mathbb{S}_{\mathcal{B}}(m)$. Hence (4.14) and (4.15) yield

$$\mathbb{S}_{\mathcal{A}_N}(m)\leq\mathbb{S}_{\bar{\mathcal{A}}_N}(m)\leq\mathbb{S}_{\mathcal{B}}(m)^N. \tag{4.16}$$

Since $\mathcal{B}$ is the collection of all closed balls in $\mathcal{R}^d$, we have $V(\mathcal{B})=d+1$ by the remark following Corollary 1.4. Thus a consequence of Sauer's lemma, Corollary 1.3, implies that for all $m\geq d+1$,

$$\mathbb{S}_{\mathcal{B}}(m)\leq\left(\frac{me}{d+1}\right)^{d+1}.$$

This and (4.16) imply that for all $m\geq d+1$,

$$\mathbb{S}_{\mathcal{A}_N}(m)\leq\left(\frac{me}{d+1}\right)^{N(d+1)}. \tag{4.17}$$

An upper bound to $V(\mathcal{A}_N)$ can now be obtained by finding an m for which the right side is less than 2^m. It is easy to check that if $d\geq 2$, then $m=4N(d+1)\ln(N(d+1))$ satisfies this requirement. Since for $d=1$ we obviously have $V(\mathcal{A}_N)\leq 2N$, we obtain that for all $N,d\geq 1$,

$$V(\mathcal{A}_N)\leq 4N(d+1)\ln(N(d+1)).$$

This and the sharpened Vapnik-Chervonenkis inequality (4.13) imply the theorem with

$$C=4T^2c\sqrt{4N(d+1)\ln(N(d+1))}.$$

□

Combining Lemma 4.5 with the bound of Theorem 4.3, we obtain the main result of this section, a finite sample bound on the expected test distortion of the empirically optimal quantizer.

Theorem 4.4. *There is a constant C_1, depending only on d, N, and T, such that for all $n \geq 1$ and $\mu \in \mathcal{P}(T)$,*

$$\mathbb{E}D(\mu, q_n^*) - D_N^*(\mu) \leq \frac{C_1}{\sqrt{n}}.$$

We have seen in Section 4.3 that the training distortion of the empirically optimal quantizer is a strongly consistent, but optimistically biased estimate of the optimal distortion. Theorem 4.3 immediately provides an upper bound on the size of this bias.

Theorem 4.5. *For all $n \geq 1$ and $\mu \in \mathcal{P}(T)$,*

$$D_N^*(\mu) - \mathbb{E}D(\mu_n, q_n^*) \leq \frac{C}{\sqrt{n}}.$$

where C is the constant in Theorem 4.3.

PROOF. As before, let q^* denote an optimal N-point quantizer for μ. Then we have

$$\begin{aligned}
D_N^*(\mu) - \mathbb{E}D(\mu_n, q_n^*) &= \mathbb{E}\left\{D(\mu, q^*) - D(\mu_n, q_n^*)\right\} \\
&\leq \mathbb{E}\left\{D(\mu, q_n^*) - D(\mu_n, q_n^*)\right\} \\
&\leq \mathbb{E}\left\{\sup_{q \in Q_N(T)} |D(\mu_n, q) - D(\mu, q)|\right\}
\end{aligned}$$

so the statement follows from Theorem 4.3. □

Combined with an appropriate concentration inequality (namely, the bounded difference inequality (see Section 1.3), Theorem 4.3 also provides a convergence rate in the consistency theorem for sources with bounded support.

Theorem 4.6. *For every $\mu \in \mathcal{P}(T)$, as $n \to \infty$,*

$$D(\mu, q_n^*) - D_N^*(\mu) = O\left(\sqrt{\frac{\ln n}{n}}\right) \quad a.s.$$

PROOF. Let

$$Y_n \triangleq \sup_{q \in Q_N(T)} |D(\mu_n, q) - D(\mu, q)|.$$

From Lemma 4.5 and Theorem 4.3,

$$D(\mu,q_n^*) - D_N^*(\mu) \leq 2Y_n \leq 2(Y_n - \mathbb{E}Y_n) + O\left(\frac{1}{\sqrt{n}}\right). \tag{4.18}$$

We can view Y_n as a function $Y_n = g(X_1,\ldots,X_n)$ of the independent random vectors $X_1,\ldots,X_n$. For any $q \in Q_N(T)$, and any $x_1,\ldots,x_n \in B_T$, $\hat{x}_1,\ldots,\hat{x}_n \in B_T$ such that $x_k = \hat{x}_k$ for all $k \neq i$,

$$\begin{aligned}
&\left|\frac{1}{n}\sum_{k=1}^{n}\|x_k - q(x_k)\|^2 - \frac{1}{n}\sum_{k=1}^{n}\|\hat{x}_k - q(\hat{x}_k)\|^2\right| \\
&\quad = \frac{1}{n}\left|\|x_i - q(x_i)\|^2 - \|\hat{x}_i - q(\hat{x}_i)\|^2\right| \\
&\quad \leq \frac{1}{n}4T^2.
\end{aligned}$$

It follows that $g : (B_T)^n \to \mathcal{R}$ satisfies the assumptions of the bounded difference inequality (Theorem 1.8) with $c_i = 4T^2/n$, and we obtain for all $t > 0$,

$$\mathbb{P}\{Y_n - \mathbb{E}Y_n \geq t\} \leq e^{-nt^2/(8T^4)}.$$

Choosing $t = \sqrt{\hat{c}\ln n/n}$ with $\hat{c} > 8T^4$, the right side is summable in n, and so the Borel-Cantelli lemma implies $Y_n - \mathbb{E}Y_n = O(\sqrt{\ln n/n})$ a.s. In view of (4.18) this proves the theorem. □

Notes

Theorem 4.4 is due to Linder, Lugosi, and Zeger (1994). The constant can be improved by using covering numbers and metric entropy bounds from empirical process theory (see, e.g., Dudley (1978) or Pollard (1990)) instead of the Vapnik-Chervonenkis inequality; see Linder (2000). Related bounds are given in Linder, Lugosi, and Zeger (1997) for quantization of sources corrupted by noise, and for combined quantization and transmission over noisy channels. A generalization of Theorem 4.4 to dependent (mixing) training data is given in Zeevi (1998). Graf and Luschgy (1999) proved almost sure convergence rates for the training distortion. The sample behavior of the test distortion for a class of sources with smooth densities is given in Chou (1994). The dependence of the test distortion on the size of the training data was empirically investigated by Cosman *et al.* (1991) and Cohn, Riskin, and Ladner (1992) in the context of image coding.

4.5 Minimax lower bounds

We showed in the previous section that for source distributions with bounded support, the expected test and training distortions of an empirically optimal quantizer trained on n data samples are bounded as

$$D_N^*(\mu) \le \mathbb{E}D(\mu, q_n^*) \le D_N^*(\mu) + \frac{C_1}{\sqrt{n}} \tag{4.19}$$

and

$$D_N^*(\mu) - \frac{C}{\sqrt{n}} \le \mathbb{E}D(\mu_n, q_n^*) \le D_N^*(\mu). \tag{4.20}$$

In these bounds the positive constants C_1 and C depend only on the dimension, the number of codevectors, and the diameter of the support.

Unfortunately, the proofs give no indication whether the $O(1/\sqrt{n})$ terms can be tightened. More generally, we don't know if there exists a method, perhaps different from empirical error minimization, which provides an empirically designed quantizer with substantially smaller test distortion.

Let us examine the simple case of quantizers with $N = 1$ codevector. In this case, as the centroid condition of Lemma 4.1 implies, the optimal quantizer q^* has a unique codepoint $y_1 = \mathbb{E}X$ and its distortion is

$$D_1^*(\mu) = \mathbb{E}\|X - \mathbb{E}X\|^2.$$

The empirically optimal 1-point quantizer q_n^* is also unique with codepoint $y_1^{(n)} = \frac{1}{n}\sum_{j=1}^n X_j$, and its expected test distortion is easily seen to be

$$\begin{aligned}
\mathbb{E}D(\mu, q_n^*) &= \mathbb{E}\left\| X - \frac{1}{n}\sum_{k=1}^n X_k \right\|^2 \\
&= \left(1 + \frac{1}{n}\right)\mathbb{E}\|X - \mathbb{E}X\|^2 \\
&= D_1^*(\mu) + \frac{D_1^*(\mu)}{n}.
\end{aligned}$$

Similarly, the expected training distortion of q_n^* is

$$\begin{aligned}
\mathbb{E}D(\mu_n, q_n^*) &= \mathbb{E}\left\{ \frac{1}{n}\sum_{k=1}^n \left\| X_k - \frac{1}{n}\sum_{j=1}^n X_j \right\|^2 \right\} \\
&= \left(1 - \frac{1}{n}\right)\mathbb{E}\|X - \mathbb{E}X\|^2 \\
&= D_1^*(\mu) - \frac{D_1^*(\mu)}{n}.
\end{aligned}$$

Thus the convergence rate in both cases is $O(1/n)$, which is substantially faster than the $O(1/\sqrt{n})$ rate in (4.19) and (4.20). However, perhaps surprisingly, the main results of this section show that the case $N = 1$ is something of an anomaly, and for $N \ge 3$, the $O(1/\sqrt{n})$ convergence rate of Theorems 4.4 and 4.5 cannot be improved upon in the minimax sense.

To formalize the problem, recall that an empirically designed N-point quantizer is a (measurable) function $q_n : (\mathcal{R}^d)^{n+1} \to \mathcal{R}^d$ such that for any fixed $x_1, \ldots, x_n \in \mathcal{R}^d$, $q_n(\cdot, x_1, \ldots, x_n)$ is an N-point quantizer. Thus an empirically designed quantizer consists of a family of quantizers and an "algorithm" which chooses one of these quantizers for each value of the training data $x_1, \ldots, x_n$.

We are interested in the *minimax distortion redundancy* defined by

$$\inf_{q_n} \sup_{\mu \in \mathcal{P}(T)} \mathbb{E}D(\mu, q_n) - D_N^*(\mu)$$

where the supremum is taken over all source distributions supported in B_T, and the infimum is over all empirically designed N-point quantizers. As the next theorem shows, for all $N \geq 3$ the minimax distortion redundancy is lower bounded by a quantity proportional to $1/\sqrt{n}$. This means that no matter what method of empirical design we use, there always exists a "bad" source distribution in $\mathcal{P}(T)$ such that the expected test distortion exceeds the optimal distortion by constant times $1/\sqrt{n}$.

Theorem 4.7. *If $N \geq 3$, then for any empirically designed N-point quantizer q_n trained on $n \geq n_0$ samples,*

$$\sup_{\mu \in \mathcal{P}(T)} \mathbb{E}D(\mu, q_n) - D_N^*(\mu) \geq \frac{C_2}{\sqrt{n}}$$

where the threshold n_0 depends only on N, and C_2 is a positive constant that depends only on d, N, and T.

The idea behind the proof is best illustrated by the special case $d = 1, N = 3$. Assume that μ is concentrated on four points: $-1, -1+\Delta, 1-\Delta$, and 1, such that either $\mu(-1) = \mu(-1+\Delta) = (1-\delta)/4$ and $\mu(1-\Delta) = \mu(1) = (1+\delta)/4$, or $\mu(-1) = \mu(-1+\Delta) = (1+\delta)/4$ and $\mu(1-\Delta) = \mu(1) = (1-\delta)/4$. Then if Δ is sufficiently small, the codepoints of the 3-point optimal quantizer are $-1+\Delta/2, 1-\Delta, 1$ in the first case, and $-1, -1+\Delta, 1-\Delta/2$ in the second case. Therefore, an empirical quantizer should "learn" from the data which of the two distributions generates the data. This leads to a hypothesis testing problem whose error may be estimated by appropriate inequalities for the binomial distribution. Proper choice of the parameters Δ and δ yields the desired $1/\sqrt{n}$-type lower bound for the minimax expected distortion redundancy.

The proof of the general case is more complicated, but the basic idea is the same. A complete proof is given only for the special case; the main steps of the proof of the general case are sketched in the Appendix at the end of the section.

PROOF OF THEOREM 4.7 (CASE $d = 1, N = 3$).

For simplicity we assume $T = 1$ (i.e., we consider distributions supported in the interval

$[-1,1]$); the result is generalized for arbitrary $T > 0$ by straightforward scaling. Define

$$J(\mu, q_n) \triangleq \mathbb{E}D(\mu, q_n) - D_N^*(\mu)$$

and note that

$$\sup_{\mu \in \mathcal{P}(1)} J(\mu, q_n) \geq \sup_{\mu \in \mathcal{D}} J(\mu, q_n)$$

for any restricted class of distributions $\mathcal{D}$ supported in $[-1,1]$. In particular, we let $\mathcal{D}$ contain two discrete distributions ν_1 and ν_2 concentrated on four points $\{-1, -1+\Delta, 1-\Delta, 1\}$ with probabilities

$$\nu_1(-1) = \nu_1(-1+\Delta) = \frac{1-\delta}{4}, \qquad \nu_1(1-\Delta) = \nu_1(1) = \frac{1+\delta}{4}$$

and

$$\nu_2(-1) = \nu_2(-1+\Delta) = \frac{1+\delta}{4}, \qquad \nu_2(1-\Delta) = \nu_2(1) = \frac{1-\delta}{4}$$

where the parameters $0 < \delta < 1$ and $0 < \Delta \leq 1/2$ are to be specified later.

The optimal 3-point quantizer $q^{(i)}$ for ν_i is easy to find. Since $q^{(i)}$ is a nearest neighbor quantizer, its partition cells are intervals, and optimality requires that each of these three intervals contain at least one of the four atoms of ν_i. There are only three such partitions of the four points, and the centroid condition (Lemma 4.1) will uniquely determine the codepoints in each case. Thus there are only three possible candidates for an optimal quantizer for ν_i.

It is easy to check that if $0 < \delta < 1$ and $0 < \Delta \leq 1/2$, then the unique codebooks C_i of the optimal quantizers $q^{(i)}$, $i = 1,2$, are given by

$$C_1 = \{-1+\Delta/2, 1-\Delta, 1\}, \qquad C_2 = \{-1, -1+\Delta, 1-\Delta/2\}$$

with equal minimum distortions

$$D_3^*(\nu_1) = D_3^*(\nu_2) = \frac{\Delta^2}{8}(1-\delta).$$

It is also easy to check that if $\delta \leq 1/2$ and $\Delta \leq 1/2$, then for any 3-point quantizer q there is a $q' \in \{q^{(1)}, q^{(2)}\}$, such that

$$D(\nu_i, q') \leq D(\nu_i, q), \qquad i = 1,2.$$

(Clearly, we need only to check that under both ν_1 and ν_2, the distortion of say $q^{(1)}$ is less than that of any q that partitions the four points as $\{-1\}, \{-1+\Delta, 1-\Delta\}$ and $\{1\}$).

Let $Q^{(n)}$ denote the family of empirically designed quantizers q_n such that for every $x_1, \ldots, x_n$, we have $q_n(\cdot, x_1, \ldots, x_n) \in \{q^{(1)}, q^{(2)}\}$. Then by the preceding discussion,

$$\begin{aligned} \inf_{q_n} \sup_{\mu \in \mathcal{P}(1)} J(\mu, q_n) &\geq \inf_{q_n} \max_{\mu \in \{\nu_1, \nu_2\}} J(\mu, q_n) \\ &= \inf_{q_n \in Q^{(n)}} \max_{\mu \in \{\nu_1, \nu_2\}} J(\mu, q_n). \end{aligned} \tag{4.21}$$

The maximum can be lower bounded by an average: If Z is a random variable with distribution $\mathbb{P}\{Z=1\}=\mathbb{P}\{Z=2\}=1/2$, then for any q_n,

$$\max_{\mu\in\{\nu_1,\nu_2\}} J(\mu,q_n) \geq \frac{1}{2}\sum_{i=1}^{2} J(\nu_i,q_n) = \mathbb{E}J(\nu_Z,q_n). \tag{4.22}$$

Define

$$M \triangleq |\{k : x_k \in \{-1,-1+\Delta\}, k=1\dots n\}|$$

i.e., M is the number of training samples falling in $\{-1,-1+\Delta\}$, and let $Q_n^* \in \mathcal{Q}^{(n)}$ be the "maximum likelihood" quantizer defined by

$$Q_n^*(\cdot,x_1,\dots,x_n) = \begin{cases} q^{(1)} & \text{if } M < n/2 \\ q^{(2)} & \text{otherwise.} \end{cases}$$

The idea is that $M < n/2$ indicates that the training data was drawn from ν_1, in which case $q^{(1)}$ is better than $q^{(2)}$. Next we show that this is indeed the optimal strategy, i.e.,

$$\inf_{q_n\in\mathcal{Q}^{(n)}} \mathbb{E}J(\nu_Z,q_n) = \mathbb{E}J(\nu_Z,Q_n^*). \tag{4.23}$$

To prove (4.23), note that

$$\begin{aligned}\mathbb{E}J(\nu_Z,q_n) &= \mathbb{E}\|Y-q_n(Y,Y_1,\dots,Y_n)\|^2 - \mathbb{E}D_N^*(\mu_Z)\\ &= \mathbb{E}\{\mathbb{E}(\|Y-q_n(Y,Y_1,\dots,Y_n)\|^2|Y_1,\dots,Y_n)\} - \mathbb{E}D_N^*(\mu_Z)\end{aligned}$$

where, under the condition $Z=i$, the sequence $Y,Y_1,\dots,Y_n$ is conditionally i.i.d. with $Y\sim\nu_i$. Thus for any given $x_1,\dots,x_n$, a q_n achieving the infimum on the left side of (4.23) must pick a quantizer that is optimal for the conditional distribution of Y given $Y_i=x_i$, $i=1,\dots,n$. Note that by conditional independence,

$$\begin{aligned}&\mathbb{P}\{Y=x|Y_1=x_1,\dots,Y_n=x_n\}\\ &= \sum_{i=1}^{2}\mathbb{P}\{Y=x|Y_1=x_1,\dots,Y_n=x_n,Z=i\}\mathbb{P}\{Z=i|Y_1=x_1,\dots,Y_n=x_n\}\\ &= \sum_{i=1}^{2}\mathbb{P}\{Y=x|Z=i\}\mathbb{P}\{Z=i|Y_1=x_1,\dots,Y_n=x_n\}\end{aligned} \tag{4.24}$$

for all $x,x_1,\dots,x_n \in \{-1,-1+\Delta,1-\Delta,1\}$. Since Z is uniformly distributed,

$$\mathbb{P}\{Z=1|Y_1=x_1,\dots,Y_n=x_n\} > \mathbb{P}\{Z=2|Y_1=x_1,\dots,Y_n=x_n\} \tag{4.25}$$

if and only if

$$\mathbb{P}\{Y_1=x_1,\dots,Y_n=x_n|Z=1\} > \mathbb{P}\{Y_1=x_1,\dots,Y_n=x_n|Z=2\}. \tag{4.26}$$

We have

$$\mathbb{P}\{Y_1 = x_1, \ldots, Y_n = x_n | Z = i\} = \left(\frac{1-\delta_i}{4}\right)^M \left(\frac{1+\delta_i}{4}\right)^{n-M} \tag{4.27}$$

where $\delta_1 = \delta$ and $\delta_2 = -\delta$. Thus (4.25) holds if and only if $M < n/2$. Introducing the notation

$$p_n(x) \triangleq \mathbb{P}\{Y = x | Y_1 = x_1, \ldots, Y_n = x_n\}$$

and noting that $\mathbb{P}\{Y = x | Z = i\} = \nu_i(x)$, we obtain from (4.24)-(4.27) that

$$p_n(-1) = p_n(-1+\Delta) < p_n(1-\Delta) = p_n(1)$$

if $M < n/2$, and

$$p_n(-1) = p_n(-1+\Delta) \geq p_n(1-\Delta) = p_n(1)$$

otherwise. To avoid the asymmetry caused by tie-breaking if n is even, we assume here that n is odd (this assumption is clearly insignificant). It follows that the optimal q_n must pick $q^{(1)}$ if $M < n/2$, and $q^{(2)}$ otherwise, i.e., $q_n = Q_n^*$ as claimed.

By symmetry we have

$$\mathbb{E}J(\nu_Z, Q_n^*) = J(\nu_1, Q_n^*). \tag{4.28}$$

With a slight abuse of notation, let now M denote the number of training samples, drawn independently under ν_1, falling in $\{-1, -1+\Delta\}$. Since $D(\nu_1, q^{(1)}) = \Delta^2(1-\delta)/8$ and $D(\nu_1, q^{(2)}) = \Delta^2(1+\delta)/8$, it is easy to see that

$$\begin{aligned} J(\nu_1, Q_n^*) & \\ &= \frac{\Delta^2}{8}(1-\delta)\mathbb{P}\{M < n/2\} + \frac{\Delta^2}{8}(1+\delta)\mathbb{P}\{M \geq n/2\} - \frac{\Delta^2}{8}(1-\delta) \\ &= \frac{\Delta^2}{4}\delta\mathbb{P}\{M \geq n/2\}. \end{aligned}$$

Note that M has binomial distribution with parameters n and $p = (1-\delta)/2$. From (4.21), (4.22), (4.23), and (4.28) we conclude that for all $\Delta, \delta \leq 1/2$,

$$\inf_{q_n} \sup_{\mu \in \mathcal{P}(1)} J(\mu, q_n) \geq \frac{\Delta^2}{4}\delta\mathbb{P}\{M \geq n/2\}. \tag{4.29}$$

The above binomial probability can be lower bounded via the standard method of normal approximation. However, it is more convenient to use a non-asymptotic inequality by Slud (1977) which states that for all $np \leq k \leq n(1-p)$,

$$\mathbb{P}\{M \geq k\} \geq \Phi\left(-\frac{k-np}{\sqrt{np(1-p)}}\right)$$

where Φ is the standard normal distribution function. Letting

$$\delta = \frac{1}{\sqrt{n}}$$

for $n \geq 4$, the choice $k = \lceil n/2 \rceil$ satisfies the conditions of Slud's inequality, and we obtain

$$\begin{aligned}\mathbb{P}\{M \geq n/2\} &\geq \Phi\left(-\frac{n/2+1-n(1-1/\sqrt{n})/2}{\sqrt{n(1-1/\sqrt{n})(1+1/\sqrt{n})/4}} \right) \\ &\geq \Phi(-2)\end{aligned}$$

where the second inequality holds for all $n \geq 6$. Combining this with (4.29) and setting $\Delta = 1/2$, we obtain for all $n \geq 6$,

$$\inf_{q_n} \sup_{\mu \in \mathcal{P}(1)} J(\mu, q_n) \geq \frac{\Phi(-2)/16}{\sqrt{n}}.$$

If we consider $\mathcal{P}(T)$ rather than $\mathcal{P}(1)$, then Δ becomes $T/2$, and so the theorem holds with $C_2 = T^2\Phi(-2)/16$ and $n_0 = 6$. □

The next result provides a similar lower bound on the bias of the training distortion of empirically optimal quantizers in estimating the optimal distortion. The theorem shows that there exist "bad" distributions for which the bias is on the order of $1/\sqrt{n}$. This lower bound matches, at least in order of magnitude, the upper bound of Theorem 4.5. Again, we prove only the special case $d = 1, N = 3$. In the general case the argument is similar, but the details are more involved.

Theorem 4.8. *If $N \geq 3$, then for the empirically optimal N-point quantizer q_n^* trained on $n \geq n_0$ samples,*

$$\sup_{\mu \in \mathcal{P}(T)} D_N^*(\mu) - \mathbb{E}D(\mu_n, q_n^*) \geq \frac{C_3}{\sqrt{n}}$$

where n_0 depends only on N, and C_3 is a positive constant that depends only on d, N, and T.

PROOF (case $d = 1$, $N = 3$). Suppose $T = 1$ and let the discrete random variable $X \sim \mu$ be uniformly distributed on $\{-1, -1+\Delta, 1-\Delta, 1\}$. It is easy to see that if $0 < \Delta < 2/3$, then there are exactly two optimal 3-point quantizers for X; one with codebook

$$C_1 = \{-1+\Delta/2, 1-\Delta, 1\}$$

the other with codebook

$$C_2 = \{-1, -1+\Delta, 1-\Delta/2\}.$$

Let q^* be the nearest neighbor quantizer with codebook C_1. The training data $X_1,\ldots,X_n$ consist of i.i.d. copies of X. Let M be the number of X_k, $k=1,\ldots,n$ such that $X_k \in \{-1,-1+\Delta\}$. Then

$$\begin{aligned} D_3^*(\mu) &= \mathbb{E}(X-q^*(X))^2 = \mathbb{E}\left\{\frac{1}{n}\sum_{k=1}^{n}(X_k-q^*(X_k))^2\right\} \\ &= \frac{1}{n}\frac{\Delta^2}{4}\mathbb{E}M \end{aligned} \tag{4.30}$$

where the second equality holds because $(q^*(X_k)-X_k)^2 = \Delta^2/4$ if X_k takes value in $\{-1,-1+\Delta\}$ and $(q^*(X_k)-X_k)^2=0$ otherwise.

Let the training set dependent nearest neighbor quantizer q_n be the following: the codebook of q_n is C_1 if $M<n/2$ and the codebook is C_2 if $M \geq n/2$. Then the expected training distortion of q_n is

$$\begin{aligned} \mathbb{E}D(\mu_n,q_n) &= \mathbb{E}\left\{\frac{1}{n}\sum_{k=1}^{n}(X_k-q_n(X_k))^2\right\} \\ &= \frac{1}{n}\frac{\Delta^2}{4}\mathbb{E}\min(M,n-M). \end{aligned} \tag{4.31}$$

Since the empirically optimal 3-point quantizer q_n^* minimizes the training distortion, we have

$$\mathbb{E}D(\mu_n,q_n) \geq \mathbb{E}D(\mu_n,q_n^*).$$

Hence (4.30) and (4.31) imply

$$D_3^*(\mu)-\mathbb{E}D(\mu_n,q_n^*) \geq \frac{1}{n}\frac{\Delta^2}{4}\mathbb{E}\{M-\min(M,n-M)\}.$$

Since M is binomial with parameters $(n,1/2)$, its distribution is symmetric about $n/2$. Thus

$$\begin{aligned} \mathbb{E}\{M-\min(M,n-M)\} &= \mathbb{E}\{(2M-n)^+\} \\ &= \mathbb{E}\left\{2\left(M-\frac{n}{2}\right)^+\right\} \\ &= \mathbb{E}\left|M-\frac{n}{2}\right|. \end{aligned}$$

Now we can apply a special case of Khintchine's inequality (see Szarek (1976)) stating that if M has binomial distribution with parameters $(n,1/2)$, then

$$\mathbb{E}\left|M-\frac{n}{2}\right| \geq \sqrt{\frac{n}{8}}.$$

We conclude that for all $\Delta < 2/3$ and $n \geq 1$,

$$\sup_{\mu \in \mathcal{P}(1)} D_3^*(\mu) - \mathbb{E}D(\mu, q_n^*) \geq \frac{\Delta^2}{8\sqrt{2}} \frac{1}{\sqrt{n}}.$$

If we consider distributions in $\mathcal{P}(T)$ rather than in $\mathcal{P}(1)$, then the constraint on Δ becomes $\Delta < 2T/3$, and we obtain

$$\sup_{\mu \in \mathcal{P}(T)} D_3^*(\mu) - \mathbb{E}D(\mu, q_n^*) \geq \frac{T^2}{18\sqrt{2}} \frac{1}{\sqrt{n}}.$$

Notes

Theorem 4.7 is from Bartlett, Linder, and Lugosi (1998). A related lower bound in an information-theoretic setting was proved by Merhav and Ziv (1997). Theorem 4.8 is based on Linder (2000).

Appendix

SKETCH OF PROOF OF THEOREM 4.7 (CASE $d \geq 1, N \geq 3$).

STEP 1: Define the restricted class of distributions $\mathcal{D}$ as follows: each member of $\mathcal{D}$ is concentrated on the set of $2m = 4N/3$ fixed points $\{z_i, z_i + w : i = 1\ldots, m\}$, where $w = (\Delta, 0, 0, \ldots, 0)$ is a fixed d-vector, and Δ is a small positive number to be determined later. The positions of $z_1, \ldots, z_m \in B_T$ satisfy the property that the distance between any two of them is greater than 3Δ. For the sake of simplicity, we assume that N is divisible by 3. (This assumption is clearly insignificant.) For $0 < \delta \leq 1/2$ and $i = 1, \ldots, m$, set

$$\nu(z_i) = \nu(z_i + w) = \begin{cases} \text{either} & \dfrac{1-\delta}{2m} \\ \text{or} & \dfrac{1+\delta}{2m} \end{cases}$$

such that exactly half of the pairs $\{z_i, z_i + w\}$ have mass $(1-\delta)/m$, and the other half of the pairs have mass $(1+\delta)/m$, so that the total mass adds up to one. Let $\mathcal{D}$ contain all such distributions. The cardinality of $\mathcal{D}$ is $K = \binom{m}{m/2}$. Denote the members of $\mathcal{D}$ by $\nu_1, \ldots, \nu_K$.

STEP 2: Let $\mathcal{Q}$ denote the collection of N-point quantizers q such that for $m/2$ values of $i \in \{1, \ldots, m\}$, q has codepoints at both z_i and $z_i + w$, and for the remaining $m/2$ values of i, q has a single codepoint at $z_i + w/2$. Then for each ν_i the unique optimal N-point quantizer is in $\mathcal{Q}$. It is easy to see that for all i,

$$D_N^*(\nu_i) = \min_{q \in \mathcal{Q}} D(\nu_i, q) = \frac{\Delta^2}{8}(1-\delta).$$

STEP 3: One can show that for any N-point quantizer q there exists a $q' \in \mathcal{Q}$ such that, for all ν in $\mathcal{D}$, $D(\nu, q') \le D(\nu, q)$. Thus if $\mathcal{Q}^{(n)}$ denotes the family of empirically designed quantizers q_n such that for every fixed $x_1, \ldots, x_n$, we have $q_n(\cdot, x_1, \ldots, x_n) \in \mathcal{Q}$, then

$$\begin{aligned}\inf_{q_n} \sup_{\mu \in \mathcal{P}(T)} J(\mu, q_n) &\ge \inf_{q_n} \max_{\mu \in \mathcal{D}} J(\mu, q_n) \\ &= \inf_{q_n \in \mathcal{Q}^{(n)}} \max_{\mu \in \mathcal{D}} J(\mu, q_n).\end{aligned}$$

STEP 4: Let Z be a random variable which is uniformly distributed on the set of integers $\{1, 2, \ldots, K\}$. Then, for any q_n, we have

$$\max_{\nu \in \mathcal{D}} J(\mu, q_n) \ge \frac{1}{K} \sum_{i=1}^{K} J(\nu_i, q_n) = \mathbb{E} J(\nu_Z, q_n).$$

Let $\mathcal{Q}_n^*$ denote the "maximum-likelihood" quantizer from $\mathcal{Q}$, that is, if M_i denotes the number of training samples falling in $\{z_i, z_i + w\}$, then $\mathcal{Q}_n^*$ has a codepoint at both z_i and $z_i + w$ if the corresponding M_i is one of the $m/2$ largest values. For the other i's (i.e., those with the $m/2$ smallest M_i's) $\mathcal{Q}_n^*$ has a codepoint at $z_i + w/2$. Then it can be proved that

$$\inf_{q_n \in \mathcal{Q}^{(n)}} \mathbb{E} J(\nu_Z, q_n) = \mathbb{E} J(\nu_Z, \mathcal{Q}_n^*).$$

STEP 5: By symmetry

$$\mathbb{E} J(\nu_Z, \mathcal{Q}_n^*) = J(\nu_1, \mathcal{Q}_n^*).$$

Under ν_1, the vector of random integers $(M_1, \ldots, M_m)$ is multinomially distributed with parameters $(n, p_1, \ldots, p_m)$, where $p_1 = p_2 = \cdots = p_{m/2} = (1-\delta)/m$, and $p_{m/2+1} = \cdots = p_m = (1+\delta)/m$. Let $M_{\sigma(1)}, \ldots, M_{\sigma(m)}$ be a reordering of the M_i's such that $M_{\sigma(1)} \le M_{\sigma(2)} \le \cdots \le M_{\sigma(m)}$. (In case of equal values, break ties using a random ordering.) Let P_j, $j = 1, \ldots, m/2$ be the probability of the event that among $M_{\sigma(1)}, \ldots, M_{\sigma(m/2)}$ there are exactly j of the M_i's with $i \ge m/2$ (i.e., the "maximum likelihood" estimate makes j mistakes). Then it is easy to see that

$$J(\nu_1, \mathcal{Q}_n^*) = \frac{\Delta^2 \delta}{2m} \sum_{j=1}^{m/2} j P_j$$

since one "mistake" increases the distortion by $\Delta^2 \delta / (2m)$.

STEP 6: Using properties of the multinomial distribution, the above sum can be lower bounded in terms of the binomial probability $\mathbb{P}\{M_1 > n/m\}$. Using normal approximation to binomials and choosing $\delta = \sqrt{m/n}$, we obtain for all $n \ge 8m/\Phi(-2)^2$,

$$J(\nu_1, \mathcal{Q}_n^*) \ge \frac{\Delta^2 \Phi(-2)^4}{512} \sqrt{\frac{m}{n}}.$$

STEP 7: It remains to choose Δ as large as possible such that m pairs of points $\{z_i, z_i + w\}$ can be placed in B_T so that the distance between any two of the z_i's is at least 3Δ. A standard argument relating packings and coverings shows that $\Delta = T/(4m^{1/d})$ satisfies this property, and so we obtain for all $n \geq n_0 = 16N/(3\Phi(-2)^2)$,

$$\inf_{q_n} \sup_{\mu \in \mathcal{P}(T)} J(\mu, q_n) \geq \frac{C_2}{\sqrt{n}}$$

where $C_2 = T^2 \Phi(-2)^4 2^{-13} \sqrt{(2N/3)^{1-4/d}}$. □

4.6 Fundamentals of variable-rate quantization

In fixed-rate quantization, the possible outputs of a quantizer are represented by distinct binary strings of equal length. One can make a quantizer more efficient by using a variable-length representation. The idea is to assign shorter binary strings to quantizer outputs that occur more frequently (i.e., have higher probability), and longer binary strings to outputs that occur less frequently (i.e., have lower probability). This way the average rate, defined as the expected number of bits per quantizer output, can be substantially reduced. In this case, the number of codevectors no longer determines the rate, and in fact the average rate can be finite even if there are infinitely many codevectors.

Formally, a *variable-rate vector quantizer* q is described by an *encoder* $\alpha : \mathcal{R}^d \to I$, where I is a countable index set, a *decoder* $\beta : I \to \mathcal{R}^d$, and an *index coder* $\psi : I \to \{0,1\}^*$, where $\{0,1\}^*$ denotes the collection of all finite-length binary strings. If I is finite with N elements, without loss of generality we always take $I = \{1, \ldots, N\}$; otherwise I is taken to be the set of all positive integers.

In variable-rate quantization, an input $x \in \mathcal{R}^d$ is encoded into an index $i = \alpha(x)$, which is represented by the binary string $\psi(i)$ for purposes of storage or transmission. We require that ψ be invertible (in fact, as explained below, we require more) and so i can be recovered from $\psi(i)$, and the decoder can output $\beta(i)$. Thus q maps any point $x \in \mathcal{R}^d$ into one of the codevectors in the codebook $\{\beta(i); i \in I\}$, via the rule

$$q(x) = \beta(\alpha(x)).$$

Letting $S_i = \{x : \alpha(x) = i\}$ and $y_i = \beta(i)$ for all $i \in I$, we have, just as in the fixed-rate case,

$$q(x) = y_i \quad \text{if and only if} \quad x \in S_i.$$

As before, we assume that the d-dimensional random vector $X \sim \mu$ has finite second moment $\mathbb{E}\|X\|^2 < \infty$, and define the distortion of q in the usual way:

$$D(\mu, q) = \mathbb{E}\|X - q(X)\|^2.$$

We require that the index coder ψ have the *prefix-free* property: If $i \neq j$, then the string $\psi(i)$ is not a prefix of the string $\psi(j)$. In particular, the prefix-free property implies that the binary *codewords* $\psi(i)$, $i \in I$ are all distinct, so ψ is invertible. More importantly, the prefix-free property also makes sure that if q is successively applied to a sequence of source outputs $x_1, x_2, \ldots, x_k$ of arbitrary length k, then from the binary string

$$\psi(\alpha(x_1))\psi(\alpha(x_2))\ldots\psi(\alpha(x_k))$$

obtained by concatenating the codewords $\psi(\alpha(x_j))$, $j = 1, \ldots, k$, one can uniquely recover the sequence of indices $\alpha(x_1), \alpha(x_2), \ldots, \alpha(x_k)$ and thus the quantizer outputs $q(x_1), q(x_2), \ldots, q(x_k)$.

The *length function* $\ell : I \to \{0, 1, 2, \ldots\}$ associates with each index i the length of the corresponding codeword $\psi(i)$, i.e., $\ell(i) = \text{length}(\psi(i))$. The rate of the variable-rate quantizer q is defined as the expected codeword length:

$$r(\mu, q) \triangleq \mathbb{E}\ell(\alpha(X)) = \sum_{i \in I} \ell(i)\mathbb{P}\{q(X) = y_i\}.$$

The following fundamental lemma gives a characterization of the set of codelengths for index coders that have the prefix-free property. The proof can be found, for example, in Cover and Thomas (1991).

LEMMA 4.6 (KRAFT'S INEQUALITY) If the binary codewords $\psi(i)$, $i \in I$ have the prefix-free property, then their lengths $\ell(i)$, $i \in I$ must satisfy the inequality

$$\sum_{i \in I} 2^{-\ell(i)} \leq 1.$$

Conversely, if the nonnegative integers $\ell(i)$, $i \in I$, satisfy this inequality, then there exists a set of codewords with these lengths that has the prefix-free property.

A length function ℓ is called *admissible* if it satisfies Kraft's inequality. Note that the distortion of q depends only on α and β, and that the rate of q depends on ψ only through ℓ. By Kraft's inequality, a prefix-free index coder ψ exists for a given length function ℓ if and only if ℓ is admissible. Thus for our purposes, it is enough to specify a quantizer q by its encoder α, decoder β, and admissible length function ℓ. In this case, we write $q \equiv (\alpha, \beta, \ell)$.

We are interested in the minimum distortion achievable for a given rate $R \geq 0$:

$$\delta_R^*(\mu) \triangleq \inf_{q: r(\mu,q) \leq R} D(\mu, q).$$

To give some insight into the advantage of variable-rate quantization, assume $\log_2 N$ is an integer and q^* is an optimal N-level fixed-rate quantizer. Then the lengths $\ell(i) = \log_2 N$, $i = 1, \ldots, N$, are clearly admissible, so we can view q^* as a special variable-rate quantizer with (constant) length function ℓ and rate $r(\mu, q^*) = \log_2 N$. Hence,

$$D(\mu, q^*) \geq \inf_{q: r(\mu,q) \leq \log_2 N} D(\mu, q)$$

i.e., variable-rate quantizers always perform at least as well as fixed-rate ones.

To assess the advantage quantitatively, suppose $\{S_i\}_{i=1}^N$ is the partition of q^* and denote $p_i = \mathbb{P}\{X \in S_i\}$. The *entropy* of the discrete random variable $q^*(X)$ is the nonnegative quantity

$$H(q^*(X)) \triangleq -\sum_{i=1}^{N} p_i \log_2 p_i.$$

(Here we use the convention that $0\log_2 0 = 0$.) Using the inequality $\log_2 t \le (t-1)\log_2 e$, valid for all $t > 0$, we see that

$$\begin{aligned} H(q^*(X)) - \log_2 N &= \sum_{i=1}^{N} p_i \log_2 \frac{1/N}{p_i} \\ &\le \log_2 e \sum_{i:\, p_i>0} p_i \left(\frac{1}{Np_i} - 1\right) \\ &\le 0 \end{aligned}$$

that is,

$$H(q^*(X)) \le \log_2 N. \tag{4.32}$$

It is easy to see that the inequality is strict unless $p_i = 1/N$ for all i. It can be shown in a similar manner that for any admissible ℓ,

$$\sum_{i\in I} \ell(i) p_i \ge H(q^*(X)).$$

Hence the entropy of a variable-rate quantizer is an absolute lower bound on its rate. This lower bound can be approached by the *Shannon-Fano codelengths*

$$\ell^*(i) \triangleq \lceil -\log_2 p_i \rceil, \qquad i = 1, \ldots N$$

(ℓ^* is admissible since $\sum_{i=1}^N 2^{-\lceil -\log_2 p_i \rceil} \le \sum_{i=1}^N 2^{\log_2 p_i} = 1$). Then ℓ^* achieves the lower bound within one bit since

$$\sum_{i=1}^{N} p_i \ell^*(i) \le \sum_{i=1}^{N} p_i(-\log_2 p_i + 1) = H(q^*(X)) + 1. \tag{4.33}$$

When the probabilities p_i, $i = 1, \ldots, N$, are highly nonuniform, the entropy $H(q^*(X))$ can be much less than $\log_2 N$. In this case, as (4.32) and (4.33) show, the rate of the variable-rate quantizer obtained from q^* using the Shannon-Fano codelengths can be significantly less than the rate $R(q^*) = \log_2 N$ of the original fixed-rate quantizer.

The discussion above illustrates how the performance of an optimal fixed-rate quantizer can be improved by appropriate variable-rate encoding. Note, however, that even when q^* is equipped with a length function ℓ that minimizes the expected codeword length, the resulting quantizer is not necessarily an optimal variable-rate quantizer. In general, optimality issues in variable-rate quantization are more difficult than in the fixed-rate case.

Notes

Fundamentals of the theory of lossless coding are given in Cover and Thomas (1991). Gersho and Gray (1992) and Sayood (2000) discuss several methods of variable-length lossless coding used in lossy data compression. To facilitate analyses, it is customary in the quantization literature to approximate the average rate of a variable-rate quantizer by the entropy of the quantizer output. In this "entropy-constrained" setting the optimal quantizer performance is still hard to determine analytically (nontrivial examples are known only in the scalar $d = 1$ case; see Berger (1972) and György and Linder (2000)), but this approach makes it possible to find approximations to the optimal performance that become tight as the rate increases; see, e.g., Zador (1966) and (1982), Gish and Pierce (1968), and Gray, Linder, and Li (2001). For a survey of results in this area, see Gray and Neuhoff (1998).

4.7 The Lagrangian formulation

At the core of our results on learning fixed-rate quantizers from empirical data was the observation that we could restrict attention to the parametric class of N-point nearest neighbor quantizers instead of having to deal with the much larger class of all N-point quantizers. Unfortunately, for $d \geq 2$ there is very little known concerning the structure of variable-rate quantizers achieving minimum distortion $\delta_R^*(\mu)$ under a rate constraint R; nor is it known whether an optimal variable-rate quantizer always exists.

In this section we recast the problem of optimal distortion-rate tradeoff for variable-rate quantization in a Lagrangian formulation that will resolve most of these difficulties. For a variable-rate quantizer $q \equiv (\alpha, \beta, \ell)$, and for $\lambda > 0$, define the *Lagrangian distortion* by

$$\Delta_\lambda(\mu, q) \stackrel{\triangle}{=} D(\mu, q) + \lambda r(\mu, q) = \mathbb{E}\{\|X - q(X)\|^2 + \lambda \ell(\alpha(X))\}$$

and the optimal Lagrangian performance by

$$\Delta_\lambda^*(\mu) \stackrel{\triangle}{=} \inf_q \Delta_\lambda(\mu, q)$$

where the infimum is taken over all variable-rate quantizers.

To see the connection between the original and the Lagrangian formulation of optimal variable-rate quantization, suppose that q_λ^* achieves $\Delta_\lambda^*(\mu)$, i.e.,

$$\Delta_\lambda(\mu, q_\lambda^*) = \inf_q D(\mu, q) + \lambda r(\mu, q).$$

Consider any quantizer q' with rate $r(\mu,q') \le r(\mu,q^*_\lambda)$. Since $D(\mu,q') + \lambda r(\mu,q') \ge D(\mu,q^*_\lambda) + \lambda r(\mu,q^*_\lambda)$, we have

$$\begin{aligned} D(\mu,q') &\ge D(\mu,q^*_\lambda) + \lambda(r(\mu,q^*_\lambda) - r(\mu,q')) \\ &\ge D(\mu,q^*_\lambda). \end{aligned}$$

Thus q^*_λ is an optimal variable-rate quantizer for the rate constraint $R = r(\mu,q^*_\lambda)$, i.e,

$$D(\mu,q^*_\lambda) = \inf_{q:r(\mu,q)\le R} D(\mu,q), \qquad r(\mu,q^*_\lambda) \le R.$$

Unfortunately, the converse statement does not hold in general: For a given R there may not exist $\lambda > 0$ such that q^*_λ achieves $\delta^*_R(\mu)$. In other words, we may not be able to find an optimal quantizer for an arbitrary rate constraint R by minimizing $\Delta_\lambda(\mu,q)$ for some value of λ.

One can characterize the rates for which an optimal variable-rate quantizer can be obtained by the Lagrangian design by considering the *convex hull* of $\delta^*_R(\mu)$, defined as the largest convex function $\hat{\delta}^*_R(\mu)$, $R \ge 0$, which is majorized by $\delta^*_R(\mu)$ (see Rockafellar (1970)). One can show that for a given rate R, $\delta^*_R(\mu)$ is achievable by q^*_λ for some $\lambda > 0$ if and only if $\delta^*_R(\mu)$ coincides with its convex hull at this rate, i.e, $\delta^*_R(\mu) = \hat{\delta}^*_R(\mu)$. (Here the Lagrange multiplier λ is the slope of the line supporting the convex hull.) Thus by minimizing $\Delta_\lambda(\mu,q)$ for all values of λ, one can obtain all variable-rate quantizers that achieve the convex hull of $\delta^*_R(\mu)$. For values of R such that $\hat{\delta}^*_R(\mu)$ is strictly less than $\delta^*_R(\mu)$ (such R exist if and only if $\delta^*_R(\mu)$ is not convex), optimal variable-rate quantizers cannot be obtained by the Lagrangian method. However, this is not a serious limitation in practical applications since any rate and distortion pair $(R, \hat{\delta}^*_R(\mu))$ on the convex hull can be achieved by "timesharing" between two quantizers that achieve the convex hull, i.e., two quantizers that can be obtained by Lagrangian minimization.

The Lagrangian formulation yields a set of useful necessary conditions for quantizer optimality. The following result is the variable-rate counterpart of Lemma 4.1.

Lemma 4.7. *Suppose $q \equiv (\alpha,\beta,\ell)$ is an arbitrary variable-rate quantizer. Then in each of the following three cases the variable-rate quantizer q' defined there satisfies*

$$\Delta_\lambda(\mu,q') \le \Delta_\lambda(\mu,q).$$

(a) $q' \equiv (\alpha',\beta,\ell)$, *where the encoder α' is defined by*

$$\alpha'(x) = \operatorname*{argmin}_{i\in I}\big(\|x-\beta(i)\|^2 + \lambda\ell(i)\big), \qquad x \in \mathcal{R}^d$$

(ties are broken arbitrarily).

(b) $q' \equiv (\alpha,\beta',\ell)$, *where the decoder β' is defined by*

$$\beta'(i) = \operatorname*{argmin}_{y\in\mathcal{R}^d} \mathbb{E}[\|X-y\|^2|\alpha(X)=i] = \mathbb{E}[X|\alpha(X)=i], \quad i \in I.$$

(c) $q' \equiv (\alpha, \beta, \ell')$, *where the codelength function* ℓ' *minimizes*

$$\sum_{i\in I} \ell(i)\mathbb{P}\{\alpha(X) = i\}$$

over all admissible codelengths ℓ.

PROOF. To prove (a), let $S_i = \{x : \alpha(x) = i\}$, $i \in I$ denote the partition cells of q. Note that the equation defining α' is equivalent to

$$\|x - \beta(\alpha'(x))\|^2 + \lambda\ell(\alpha'(x)) = \min_{i\in I}\big(\|x - \beta(i)\|^2 + \lambda\ell(i)\big). \tag{4.34}$$

To see that the minimum (and so α') is well defined for all x even when I is not finite, note that in this case the admissibility of ℓ implies that $\lim_{i\to\infty} \ell(i) = \infty$. Thus for each x it suffices to take the minimum over a finite subset of I. Hence (4.34) is always well defined, and we obtain

$$\begin{aligned}
\Delta_\lambda(\mu, q) &= \sum_{j\in I} \int_{S_j} \big(\|x - \beta(j)\|^2 + \lambda\ell(j)\big)\,\mu(dx) \\
&\geq \sum_{j\in I} \int_{S_j} \min_{i\in I}\big(\|x - \beta(i)\|^2 + \lambda\ell(i)\big)\,\mu(dx) \\
&= \Delta_\lambda(\mu, q').
\end{aligned}$$

To prove (b), notice that the choice of the decoder only affects the term $\|x - q(x)\|^2$ in the Lagrangian expression. Therefore (b) follows directly from the centroid condition of Lemma 4.1.

Finally, write

$$\Delta_\lambda(\mu, q) = \mathbb{E}\|X - \beta(\alpha(X))\|^2 + \lambda \sum_{i\in I} \ell(i) p_i$$

where $p_i = \mathbb{P}\{\alpha(X) = i\}$. Now (c) follows since for α and β fixed, the admissible length function that minimizes $\sum_{i\in I} \ell(i) p_i$ is the one that minimizes the overall Lagrangian distortion. □

Remarks (i) Lemma 4.7 (a) is analogous to the nearest neighbor condition of fixed-rate quantization. An optimal α' for a given β and ℓ is called a *modified nearest neighbor encoder*.

(ii) For a finite index set I, optimal codelengths in part (c) of the lemma can be obtained, for example, as the codelengths of the binary Huffman code for the probabilities $p_i = \mathbb{P}\{\alpha(X) = i\}$, $i \in I$ (see, e.g., Cover and Thomas (1991)). For infinite index sets the existence of an optimal prefix-free code and an associated optimal admissible length function is shown in Linder, Tarokh, and Zeger (1997) under the condition that the entropy $-\sum_{i\in I} p_i \log_2 p_i$ is finite.

The next result shows the existence of quantizers minimizing the Lagrangian distortion. The proof, which we omit here, relies on the optimality criteria of Lemma 4.7, but is somewhat more involved than the proof of the analogous Theorem 4.1. The difficulty is posed by the fact that optimal variable-rate quantizers can have an infinite number of codevectors. Of course, if a quantizer minimizes the Lagrangian distortion, it can be assumed to have a modified nearest neighbor encoder by Lemma 4.7.

Theorem 4.9. *For any μ with finite second moment and $\lambda > 0$ there is a variable-rate quantizer q^*_λ with a modified nearest neighbor encoder such that*

$$\Delta_\lambda(\mu, q^*_\lambda) = \Delta^*_\lambda(\mu).$$

Remark The necessary conditions for optimality in Lemma 4.7 suggest an iterative algorithm for designing variable-rate quantizers in a manner analogous to the fixed-rate case. Let $\Delta_\lambda(\alpha,\beta,\ell)$ stand for $\Delta_\lambda(\mu,q)$ if $q \equiv (\alpha,\beta,\ell)$. Start with an arbitrary quantizer $q_0 \equiv (\alpha^{(0)},\beta^{(0)},\ell^{(0)})$ with a finite index set. In the mth iteration ($m = 1,2,\ldots$), first choose $\alpha^{(m)}$ to minimize $\Delta_\lambda(\alpha,\beta^{(m-1)},\ell^{(m-1)})$ for fixed $\beta^{(m-1)}$ and $\ell^{(m-1)}$, then choose $\beta^{(m)}$ to minimize $\Delta_\lambda(\alpha^{(m)},\beta,\ell^{(m-1)})$ for fixed $\alpha^{(m)}$ and $\ell^{(m-1)}$, and then choose an admissible $\ell^{(m)}$ to minimize $\Delta_\lambda(\alpha^{(m)},\beta^{(m)},\ell)$ for fixed $\alpha^{(m)}$ and $\beta^{(m)}$. Since the Lagrangian distortion is decreasing (or at least not increasing) in each step, setting

$$q^{(m)} \equiv (\alpha^{(m)},\beta^{(m)},\ell^{(m)})$$

we obtain

$$\Delta_\lambda(\mu,q^{(m)}) \le \Delta_\lambda(\mu,q^{(m-1)})$$

so $\lim_{m\to\infty} \Delta_\lambda(\mu,q^{(m-1)}) - \Delta_\lambda(\mu,q^{(m)}) = 0$. The algorithm stops (after a finite number of iterations) when the drop in distortion falls below a given threshold. It may be necessary to repeat this procedure several times with different values of λ to obtain a quantizer with rate that is close enough to the desired rate.

As in the fixed-rate case, the algorithm is most often used with the empirical distribution μ_n in place of μ. Although the sequence of distortions converges as $m \to \infty$, there is no guarantee that the limit is the optimum distortion. However, the quantizers designed using this algorithm have very favorable performance in general.

Notes

The optimality conditions of Lemma 4.7 are due to Chou, Lookabaugh, and Gray (1989) who also introduced the Lagrangian formulation of variable-rate vector quantization discussed in this section. Theorem 4.9 is proved in György and Linder (2001b) (see György and Linder (2001a) for an existence result that does not assume the Lagrangian formulation). Gray, Linder, and Li (2001) used the Lagrangian formulation for the asymptotic (high-rate) analysis of optimal entropy-constrained vector quantizer performance. The algorithm sketched above is the well-known entropy-constrained vector quantizer design algorithm of Chou *et al.* (1989).

4.8 Consistency of Lagrangian empirical design

We are interested in the performance of quantizers learned from a finite training sequence. As before, let $X_1^n = X_1, \ldots, X_n$ be i.i.d. copies of X such that X_1^n and X are also independent, and let μ_n denote the empirical distribution of X_1^n. Fix $\lambda > 0$ and assume that the empirically optimal variable-rate quantizer q_n^* is one that minimizes the empirical Lagrangian distortion:

$$\Delta_\lambda(\mu_n, q_n^*) = \inf_q \Delta_\lambda(\mu_n, q) = \Delta_\lambda^*(\mu_n)$$

i.e.,

$$q_n^* = \operatorname*{argmin}_{q\equiv(\alpha,\beta,\ell)} \frac{1}{n} \sum_{k=1}^{n} \|X_k - \beta(\alpha(X_k))\|^2 + \lambda \ell(\alpha(X_k)).$$

We will always assume that q_n^* has a modified nearest neighbor encoder (see Theorem 4.9).

As before, the performance of the empirically optimal variable-rate quantizer $q_n^* = (\alpha_n^*, \beta_n^*, \ell_n^*)$ is measured by its Lagrangian test distortion, given by

$$\Delta_\lambda(\mu, q_n^*) = \mathbb{E}\big[\|X - \beta_n^*(\alpha_n^*(X))\|^2 + \lambda \ell_n^*(\alpha_n^*(X)) | X_1^n\big].$$

The following theorem, the variable-rate counterpart of Theorem 4.2, shows that the design based on the minimization of the empirical Lagrangian distortion is consistent.

THEOREM 4.10 (CONSISTENCY OF LAGRANGIAN EMPIRICAL DESIGN) For any $\lambda > 0$ the sequence of variable-rate quantizers q_n^*, $n = 1, 2, \ldots$, minimizing the empirical Lagrangian distortion satisfies

$$\lim_{n\to\infty} \Delta_\lambda(\mu, q_n^*) = \Delta_\lambda^*(\mu) \quad \text{a.s.}$$

The proof is based on the properties of the metric $\rho(\mu, \nu)$ introduced in Section 4.3, but we need some additional definitions and auxiliary results.

Let $\mathcal{D}$ denote the set of all discrete distributions on $\mathcal{R}^d$ with finite second moment and finite entropy. That is, $\nu \in \mathcal{D}$ if and only if ν is concentrated on a finite or countably infinite set $\{x_i; i \in I_\nu\} \subset \mathcal{R}^d$, and satisfies

$$\sum_{i\in I_\nu} \|x_i\|^2 \nu(x_i) < \infty, \qquad -\sum_{i\in I_\nu} \nu(x_i)\log_2 \nu(x_i) < \infty.$$

For any $\nu \in \mathcal{D}$ let L_ν denote the minimum expected codelength over all admissible codelength functions $\ell : I_\nu \to \{0, 1, \ldots\}$,

$$L_\nu = \min_\ell \sum_{i\in I_\nu} \ell(i)\nu(x_i). \tag{4.35}$$

Note that by the remark after Lemma 4.7, a minimizing admissible ℓ always exists, and that $L_\nu < \infty$ by the Shannon-Fano bound (4.33).

For $\lambda > 0$, μ with finite second moment, and $\nu \in \mathcal{D}$ define

$$\rho_\lambda(\mu,\nu) \triangleq \left(\rho(\mu,\nu)^2 + \lambda L_\nu\right)^{1/2}.$$

To interpret $\rho_\lambda(\mu,\nu)$, suppose $X \sim \mu$ and $Y \sim \nu$ achieve $\rho(\mu,\nu)$, and let ℓ_ν be an admissible codelength achieving L_ν in (4.35). Then Y can be viewed as the output of a variable-rate "random quantizer" that, to each x, assigns the reproduction vector x_i and a binary codeword of length $\ell_\nu(i)$ with probability $\mathbb{P}\{Y = x_i | X = x\}$. The quantity $\rho_\lambda(\mu,\nu)^2$ is the Lagrangian distortion of this random quantizer.

In this interpretation, the next lemma states that deterministic quantizers always outperform random quantizers in the Lagrangian sense.

Lemma 4.8.

$$\Delta_\lambda^*(\mu) = \inf_{\nu \in \mathcal{D}} \rho_\lambda(\mu,\nu)^2.$$

PROOF. Suppose $q \equiv (\alpha, \beta, \ell)$ is a variable-rate quantizer such that $\Delta_\lambda(\mu,q) < \infty$. Let ν_q denote the distribution of the discrete random variable $q(X)$ and note that $\nu_q \in \mathcal{D}$. Since $X \sim \mu$ and $q(X) \sim \nu_q$,

$$\begin{aligned}\Delta_\lambda(\mu,q) &= \mathbb{E}\|X - q(X)\|^2 + \lambda \mathbb{E}\ell(\alpha(X)) \\ &\geq \rho(\mu,\nu_q)^2 + \lambda L_{\nu_q} \\ &= \rho_\lambda(\mu,\nu_q)^2\end{aligned}$$

and hence

$$\Delta_\lambda^*(\mu) = \inf_q \Delta_\lambda(\mu,q) \geq \inf_{\nu \in \mathcal{D}} \rho_\lambda(\mu,\nu)^2.$$

To show the reverse inequality, assume $X \sim \mu$ and $Y \sim \nu$ achieve $\rho(\mu,\nu)$, where $\nu \in \mathcal{D}$ is concentrated on a countable set of points $\{y_i; i \in I_\nu\}$. Define the variable-rate quantizer q with index set I_ν to have decoder $\beta(i) = y_i$, $i \in I_\nu$, codelength ℓ_ν such that

$$L_\nu = \sum_{i \in I_\nu} \ell_\nu(i) \nu(y_i)$$

and encoder α that is optimized for β and ℓ_ν, i.e.,

$$\alpha(x) = \operatorname*{argmin}_{i \in I_\nu} \left(\|x - y_i\|^2 + \lambda \ell_\nu(i) \right).$$

Then

$$\Delta_\lambda(\mu,q) = \mathbb{E} \min_{i \in I_\nu} \left(\|X - y_i\|^2 + \lambda \ell_\nu(i) \right).$$

Since X and Y achieve $\rho(\mu,\nu)$, and Y takes values in $\{y_i; i \in I_\nu\}$,

$$\begin{aligned}\rho_\lambda(\mu,\nu)^2 &= \mathbb{E}\{\|X-Y\|^2+\lambda L_\nu\} \\ &= \int_{\mathcal{R}^d}\sum_{i\in I_\nu}(\|x-y_i\|^2+\lambda\ell_\nu(i))\mathbb{P}\{Y=y_i|X=x\}\mu(dx) \\ &\geq \int_{\mathcal{R}^d}\min_{i\in I_\nu}\big(\|x-y_i\|^2+\lambda\ell_\nu(i)\big)\,\mu(dx) \\ &= \Delta_\lambda(\mu,q)\end{aligned}$$

and so we obtain

$$\inf_{\nu\in\mathcal{D}}\rho_\lambda(\mu,\nu)^2 \geq \inf_q \Delta_\lambda(\mu,q).$$

□

As a consequence of the previous lemma, we obtain a stability result for the optimal Lagrangian performance which is the variable-rate counterpart of Lemma 4.3.

Lemma 4.9. *For any μ and μ' with finite second moment,*

$$\left|\Delta^*_\lambda(\mu)^{1/2}-\Delta^*_\lambda(\mu')^{1/2}\right| \leq \rho(\mu,\mu').$$

PROOF. Assume $\Delta^*_\lambda(\mu)\geq\Delta^*_\lambda(\mu')$. Fix $\varepsilon>0$ and let $\nu'\in\mathcal{D}$ be such that

$$\rho_\lambda(\mu',\nu') \leq \inf_{\nu\in\mathcal{D}}\rho_\lambda(\mu',\nu)+\varepsilon.$$

Then by Lemma 4.8,

$$\begin{aligned}\Delta^*_\lambda(\mu)^{1/2}-\Delta^*_\lambda(\mu')^{1/2} &= \inf_{\nu\in\mathcal{D}}\rho_\lambda(\mu,\nu)-\inf_{\nu\in\mathcal{D}}\rho_\lambda(\mu',\nu) \\ &\leq \inf_{\nu\in\mathcal{D}}\rho_\lambda(\mu,\nu)-\rho_\lambda(\mu',\nu')+\varepsilon \\ &\leq \rho_\lambda(\mu,\nu')-\rho_\lambda(\mu',\nu')+\varepsilon \\ &= \big(\rho(\mu,\nu')^2+\lambda L_{\nu'}\big)^{1/2}-\big(\rho(\mu',\nu')^2+\lambda L_{\nu'}\big)^{1/2}+\varepsilon \\ &\leq |\rho(\mu,\nu')-\rho(\mu',\nu')|+\varepsilon \\ &\leq \rho(\mu,\mu')+\varepsilon\end{aligned}$$

where the third inequality holds because $(a+c)^{1/2}-(b+c)^{1/2}\leq a^{1/2}-b^{1/2}$ for all $a\geq b\geq 0$, $c\geq 0$ by the concavity of the square root, and the last inequality follows from the triangle inequality since ρ is a metric (see Lemma 4.2). Since $\varepsilon>0$ was arbitrary, we obtain $\Delta^*_\lambda(\mu)^{1/2}-\Delta^*_\lambda(\mu')^{1/2}\leq\rho(\mu,\mu')$. The case $\Delta^*_\lambda(\mu')\geq\Delta^*_\lambda(\mu)$ is handled similarly. □

The preceding lemma immediately implies that the Lagrangian training distortion $\Delta_\lambda(\mu_n,q^*_n)$ is a strongly consistent estimate of the optimal Lagrangian distortion $\Delta^*_\lambda(\mu)$.

Theorem 4.11.

$$\lim_{n\to\infty} \Delta_\lambda(\mu_n, q_n^*) = \Delta_\lambda^*(\mu) \quad a.s.$$

PROOF. Since $\Delta_\lambda(\mu_n, q_n^*) = \Delta_\lambda^*(\mu_n)$, Lemma 4.9 with $\mu' = \mu_n$ gives

$$\left|\Delta_\lambda^*(\mu)^{1/2} - \Delta_\lambda^*(\mu_n)^{1/2}\right| \le \rho(\mu, \mu_n).$$

The statement follows since we know from the proof of Theorem 4.2 that

$$\lim_{n\to\infty} \rho(\mu, \mu_n) = 0 \quad \text{a.s.} \tag{4.36}$$

□

Now we are ready to prove the consistency theorem.

PROOF OF THEOREM 4.10. We have

$$\begin{aligned} &\Delta_\lambda(\mu, q_n^*) - \Delta_\lambda^*(\mu) \\ &\quad = \Delta_\lambda(\mu, q_n^*) - \Delta_\lambda^*(\mu_n) + \Delta_\lambda^*(\mu_n) - \Delta_\lambda^*(\mu). \end{aligned} \tag{4.37}$$

The second difference on the right side converges to zero a.s. by Theorem 4.11.

To bound the first difference, recall that by assumption, the encoder of $q_n^* \equiv (\alpha_n^*, \beta_n^*, \ell_n^*)$ uses the modified nearest neighbor rule. Thus for any $x, y \in \mathcal{R}^d$,

$$\|x - q_n^*(x)\|^2 + \lambda \ell_n^*(\alpha_n^*(x)) \le \|x - q_n^*(y)\|^2 + \lambda \ell_n^*(\alpha_n^*(y)). \tag{4.38}$$

Now let $X \sim \mu$ and $Y \sim \mu_n$ achieve $\rho(\mu, \mu_n)$, and in addition suppose that the pair X, Y is independent of the training sequence X_1^n. Letting $\mathbb{E}_n$ denote conditional expectation with respect to X_1^n, (4.38) implies

$$\begin{aligned} &\Delta_\lambda(\mu, q_n^*) \\ &\quad = \mathbb{E}_n\left\{\|X - q_n^*(X)\|^2 + \lambda \ell_n^*(\alpha_n^*(X))\right\} \\ &\quad \le \mathbb{E}_n\left\{\|X - q_n^*(Y)\|^2 + \lambda \ell_n^*(\alpha_n^*(Y))\right\} \\ &\quad \le \mathbb{E}_n\|X - Y\|^2 + \mathbb{E}_n\left\{\|Y - q_n^*(Y)\|^2 + \lambda \ell_n^*(\alpha_n^*(Y))\right\} \\ &\qquad + 2\mathbb{E}_n\left\{\|X - Y\| \|Y - q_n^*(Y)\|\right\} \\ &\quad = \rho(\mu, \mu_n)^2 + \Delta_\lambda(\mu_n, q_n^*) + 2\mathbb{E}_n\left\{\|X - Y\| \|Y - q_n^*(Y)\|\right\} \\ &\quad \le \rho(\mu, \mu_n)^2 + \Delta_\lambda(\mu_n, q_n^*) + 2\left(\mathbb{E}_n\|X - Y\|^2\right)^{1/2}\left(\mathbb{E}_n\|Y - q_n^*(Y)\|^2\right)^{1/2} \\ &\quad = \rho(\mu, \mu_n)^2 + \Delta_\lambda^*(\mu_n) + 2\rho(\mu, \mu_n) D(\mu_n, q_n^*)^{1/2} \end{aligned}$$

where the last inequality follows from the Cauchy-Schwarz inequality. Since $D(\mu_n, q_n^*) \le \Delta_\lambda(\mu_n, q_n^*) = \Delta_\lambda^*(\mu_n)$, using Lemma 4.9 again, we obtain

$$D(\mu_n, q_n^*)^{1/2} \le \Delta_\lambda^*(\mu)^{1/2} + \rho(\mu, \mu_n)$$

and so

$$\Delta_\lambda(\mu, q_n^*) \le \rho(\mu, \mu_n)^2 + \Delta_\lambda^*(\mu_n) + 2\rho(\mu, \mu_n)\big(\Delta_\lambda^*(\mu)^{1/2} + \rho(\mu, \mu_n)\big)$$

or

$$\Delta_\lambda(\mu, q_n^*) - \Delta_\lambda^*(\mu_n) \le \rho(\mu, \mu_n)^2 + 2\rho(\mu, \mu_n)\big(\Delta_\lambda^*(\mu)^{1/2} + \rho(\mu, \mu_n)\big).$$

Therefore, by (4.36), the first difference in (4.37) converges to zero a.s. as $n \to \infty$, completing the proof of the theorem. □

Notes

The material of this section is new. Lemma 4.8 has a counterpart in fixed-rate quantization; see Pollard (1982a). Zador (1966) was probably the first to use random quantization as a tool for analyzing the performance of optimal variable-rate quantizers.

4.9 Finite sample bounds in Lagrangian design

In analogy to the fixed-rate case, we would like to establish a connection between the performance of the empirically designed variable-rate quantizer and the number of training samples used in the design. The methods used in proving such results in the fixed-rate case will turn out to be applicable once we have established a variable-rate counterpart of Lemma 4.5.

As we did in Section 4.4, we assume here that the source distribution is an element of $\mathcal{P}(T)$, the set of probability distributions on $\mathcal{R}^d$ supported on the closed ball B_T of radius T centered at the origin.

In the next lemma we show that for sources with a given bounded support, variable-rate quantizers that are optimal in the Lagrangian sense can be assumed not to have too many codevectors or very large codeword lengths. The strength of the Lagrangian approach is evident here; no such general result is known for variable-rate quantizers that minimize the distortion for a given rate constraint.

For $T > 0$ and positive integers N and L, let $Q_{N,L}(T)$ denote the collection of all variable-rate quantizers $q \equiv (\alpha, \beta, \ell)$ with index set I such that

(i) $\|\beta(i)\| \le T$ for all $i \in I$;

(ii) α is a modified nearest neighbor encoder;

(iii) $\ell(i) \le L$ for all $i \in I$, and I is finite with cardinality $|I| \le N$.

Lemma 4.10. *For any $\mu \in \mathcal{P}(T)$ and $\lambda > 0$,*

$$\min_q \Delta_\lambda(\nu, q) = \min_{q \in Q_{N,L}(T)} \Delta_\lambda(\nu, q) \tag{4.39}$$

where $N = \lfloor 2^{5T^2/\lambda} \rfloor$ and $L = \lfloor 5T^2/\lambda \rfloor$. Thus there exists $q_n^* \in Q_{N,L}(T)$, and for this q_n^* we have

$$\Delta_\lambda(\mu, q_n^*) - \Delta_\lambda^*(\mu) \le 2 \sup_{q \in Q_{N,L}(T)} |\Delta_\lambda(\mu_n, q) - \Delta_\lambda(\mu, q)| \tag{4.40}$$

PROOF. The second statement is an easy consequence of the first one. Let q^* denote a variable-rate quantizer achieving the minimum Lagrangian distortion $\Delta_\lambda^*(\mu)$. Using the same argument as in Lemma 4.5, we obtain the basic inequality

$$\begin{aligned} &\Delta_\lambda(\mu, q_n^*) - \Delta_\lambda^*(\mu) \\ &\quad \le \Delta_\lambda(\mu, q_n^*) - \Delta_\lambda(\mu_n, q_n^*) + \Delta_\lambda(\mu_n, q^*) - \Delta_\lambda(\mu, q^*). \end{aligned}$$

Since $\mu, \mu_n \in \mathcal{P}(T)$, (4.39) implies that there exist $q^*, q_n^* \in Q_{N,L}(T)$, and we obtain (4.40).

To prove (4.39), suppose $q \equiv (\alpha, \beta, \ell)$ with index set I achieves $\Delta_\lambda^*(\nu)$. Using the argument of Lemma 4.5, we see that any codevector of q outside B_T can be replaced by its projection to the surface of B_T without increasing the Lagrangian distortion, and so we can assume that $\|\beta(i)\| \le T$ for all $i \in I$.

Next note that by Lemma 4.7 (a) we can assume that α is a modified nearest neighbor encoder. Also, we can assume that for each $S_i = \{x : \alpha(x) = i\}$, $i \in I$, we have $S_i \cap B_T \ne \emptyset$; otherwise, since $\nu \in \mathcal{P}(T)$, we can discard i from I without affecting the performance. Let $i_0 \in I$ be an index with minimum codeword length, i.e.,

$$\ell(i_0) = \min_{i \in I} \ell(i).$$

Since α is a modified nearest neighbor encoder, for any $i \in I$ and $x \in S_i$,

$$\|x - \beta(i)\|^2 + \lambda \ell(i) \le \|x - \beta(i_0)\|^2 + \lambda \ell(i_0).$$

Since $\|\beta(i_0)\| \le T$, we have $\|x - \beta(i_0)\|^2 \le 4T^2$ for all $x \in B_T$, and since $S_i \cap B_T$ is nonempty, the previous inequality implies that for all $i \in I$,

$$\ell(i) \le \frac{4T^2}{\lambda} + \ell(i_0). \tag{4.41}$$

Now let q_1 denote the quantizer with a single codepoint $y = 0$ and rate $r(\nu, q_1) = 0$ (formally, the single binary codeword of q_1 is the empty string of length zero). Then since $\nu \in \mathcal{P}(T)$,

$$\Delta_\lambda(\nu, q_1) = D(\nu, q_1) + \lambda r(\nu, q_1) \le T^2. \tag{4.42}$$

On the other hand,

$$\Delta_\lambda(\nu,q) \geq \lambda r(\nu,q) \geq \lambda \ell(i_0)$$

which, together with (4.42) and the fact that $\Delta_\lambda(\nu,q) \leq \Delta_\lambda(\nu,q_1)$ (since q minimizes the Lagrangian distortion for ν), implies that

$$\ell(i_0) \leq \frac{T^2}{\lambda}.$$

Hence by (4.41), we have for all $i \in I$,

$$\ell(i) \leq \frac{5T^2}{\lambda}.$$

The admissibility of ℓ then yields

$$1 \geq \sum_{i \in I} 2^{-\ell(i)} \geq |I| 2^{-5T^2/\lambda}$$

and so $|I| \leq 2^{5T^2/\lambda}$. Setting $N = \lfloor 2^{5T^2/\lambda} \rfloor$ and $L = \lfloor 5T^2/\lambda \rfloor$, we obtain $q \in \mathcal{Q}_{N,L}(T)$, which completes the proof. □

Lemma 4.10 allows us to adapt the proof of Theorem 4.3 to the Lagrangian case. The following finite sample bound on the Lagrangian performance on an empirically optimal variable-rate quantizer is the main result of this section. In this result we assume (as we may by Lemma 4.10) that $q_n^* \in \mathcal{Q}_{N,L}(T)$.

Theorem 4.12. *There is a constant C_4, depending only on d, λ, and T, such that for all $n \geq 1$ and $\mu \in \mathcal{P}(T)$,*

$$\mathbb{E}\Delta_\lambda(\mu,q_n^*) - \Delta_\lambda^*(\mu) \leq \frac{C_4}{\sqrt{n}}.$$

PROOF. By Lemma 4.10, it suffices to give an appropriate upper bound on the expected value of $\sup_{q \in \mathcal{Q}_{N,L}(T)} |\Delta_\lambda(\mu_n,q) - \Delta_\lambda(\mu,q)|$.

For N and L as in Lemma 4.10 and $q \equiv (\alpha,\beta,\ell) \in \mathcal{Q}_{N,L}(T)$, define the distortion function

$$f_{\lambda,q}(x) \triangleq \|x - q(x)\|^2 + \lambda \ell(\alpha(x)).$$

Then for all $x \in B_T$,

$$0 \leq f_{\lambda,q}(x) \leq 4T^2 + \lambda L \leq 4T^2 + \lambda \lfloor 5T^2/\lambda \rfloor = 9T^2.$$

Hence, we can repeat the steps leading to (4.11) in the proof of Theorem 4.3 to obtain

$$\begin{aligned}\sup_{q\in \mathcal{Q}_{N,L}(T)} &\left|\Delta_\lambda(\mu_n,q)-\Delta_\lambda(\mu,q)\right| \\ &\le 9T^2 \sup_{q\in \mathcal{Q}_{N,L}(T),u>0}\left|\frac{1}{n}\sum_{k=1}^{n} I_{\{f_{\lambda,q}(X_k)>u\}} - \mathbb{P}\{f_{\lambda,q}(X)>u\}\right| \\ &= 9T^2 \sup_{A\in\tilde{\mathcal{A}}_N}|\mu_n(A)-\mu(A)| \quad \text{a.s.}\end{aligned}$$

where now $\tilde{\mathcal{A}}_N$ is the family of subsets of $\mathcal{R}^d$ defined by

$$\tilde{\mathcal{A}}_N \triangleq \Big\{\{x: f_{\lambda,q}(x)>u\}: q\in \mathcal{Q}_{N,L}(T), u>0\Big\}.$$

Denoting by $V(\tilde{A}_N)$ the VC dimension of $\tilde{A}_N$ and using the sharpened version of the Vapnik-Chervonenkis inequality as in the proof of Theorem 4.3, we obtain

$$\mathbb{E}\left\{\sup_{q\in \mathcal{Q}_{N,L}(T)}\left|\Delta_\lambda(\mu_n,q)-\Delta_\lambda(\mu,q)\right|\right\} \le 9T^2 c\sqrt{\frac{V(\tilde{A}_N)}{n}}. \tag{4.43}$$

Since the encoder of each $q\in \mathcal{Q}_{N,L}(T)$ uses the modified nearest neighbor rule, $f_{\lambda,q}(x)>u$ if and only if

$$\|x-\beta(i)\|^2+\lambda\ell(i)>u \quad \text{for all } i\in I$$

i.e.,

$$\{x: f_{\lambda,q}(x)>u\} = \bigcap_{i\in I}\{x: \|x-\beta(i)\|^2 > u-\lambda\ell(i)\}.$$

Since $|I|\le N$, we obtain that either $\{x: f_{\lambda,q}(x)>u\}$ is an intersection of the complements of at most N closed balls of possibly different radii in $\mathcal{R}^d$ (if $u-\lambda\ell(i)\ge 0$ for some i), or $\{x: f_{\lambda,q}(x)>u\}=\mathcal{R}^d$ (this is the case if $u-\lambda\ell(i)<0$ for all i). As in the proof of Theorem 4.3, let $\bar{\mathcal{A}}_N$ denote the family of all intersections of complements of N closed balls in $\mathcal{R}^d$. Then we have $\tilde{\mathcal{A}}_N \subset \bar{\mathcal{A}}_N \cup \{\mathcal{R}^d\}$, and so $V(\tilde{\mathcal{A}}_N)\le V(\bar{\mathcal{A}}_N)$. Thus we can use the upper bound $V(\bar{\mathcal{A}}_N)\le 4N(d+1)\ln(N(d+1))$ derived in the proof of Theorem 4.3 to obtain

$$V(\tilde{\mathcal{A}}_N)\le 4N(d+1)\ln(N(d+1)).$$

Thus (4.43) and Lemma 4.10 imply the theorem with

$$C_4 = 18T^2 c\sqrt{4N(d+1)\ln(N(d+1))}$$

where $N=\lfloor 2^{5T^2/\lambda}\rfloor$. □

Remark A variable-rate versions of Theorem 4.5 and Theorem 4.6 can also be derived from Lemma 4.10 and the inequality (4.43). In particular, the variable-rate counterpart of Theorem 4.5 states that for every $\mu \in \mathcal{P}(T)$ the bias of the expected Lagrangian training distortion is upper bounded as

$$\Delta_\lambda^*(\mu) - \mathbb{E}\Delta_\lambda(\mu_n, q_n^*) \le \frac{C_4/2}{\sqrt{n}}$$

where C_4 is the constant in Theorem 4.12. Also, the bounded difference inequality can be used to show that for every $\mu \in \mathcal{P}(T)$,

$$\Delta_\lambda(\mu, q_n^*) - \Delta_\lambda^*(\mu) = O\left(\sqrt{\frac{\ln n}{n}}\right) \quad \text{a.s.}$$

which is the variable-rate counterpart of Theorem 4.6. The details are left as an exercise.

Notes

The proof of Lemma 4.10 is based on an idea of Chou and Betts (1998). The rest of the material in this section is new.

4.10 References

E. A. Abaya and G. L. Wise. On the existence of optimal quantizers. *IEEE Trans. Inform. Theory*, 28:937 – 940, Nov. 1982.

E. A. Abaya and G. L. Wise. Convergence of vector quantizers with applications to optimal quantization. *SIAM Journal on Applied Mathematics*, 44:183–189, 1984.

M. Anthony and P. L. Bartlett. *Neural Network Learning: Theoretical Foundations*. Cambridge University Press, 1999.

R. B. Ash. *Probability and Measure Theory*. Academic Press, New York, 2000.

P. Bartlett, T. Linder, and G. Lugosi. The minimax distortion redundancy in empirical quantizer design. *IEEE Trans. Inform. Theory*, IT-44(5):1802–1813, Sep. 1998.

T. Berger. Optimum quantizers and permutation codes. *IEEE Trans. Inform. Theory*, IT-18:759–765, Nov. 1972.

P. A. Chou. The distortion of vector quantizers trained on n vectors decreases to the optimum as $O_p(1/n)$. in *Proc. IEEE Int. Symp. Information Theory* (Trondheim, Norway, Jun. 27-Jul. 1, 1994), p. 457.

P. A. Chou and B. J. Betts. When optimal entropy-constrained quantizers have only a finite number of codewords. in *Proc. IEEE Int. Symp. Information Theory* (Cambridge, MA, USA, Aug. 16-21, 1998), p. 97.

P. A. Chou, T. Lookabaugh, and R. M. Gray. Entropy-constrained vector quantization. *IEEE Trans. Acoust. Speech, Signal Processing*, ASSP-37:31–42, Jan. 1989.

D. Cohn, E. Riskin, and R. Ladner. Theory and practice of vector quantizers trained on small training sets. *IEEE Trans. on Pattern Analysis and Machine Intelligence*, 16:54–65, Jan. 1994.

P. C. Cosman, K. O. Perlmutter, S. M. Perlmutter, R. A. Olshen, and R. M. Gray. Training sequence size and vector quantizer performance. In *Proceedings of Asilomar Conference on Signals, Systems, and Computers*, pages 434–438, 1991.

T. Cover and J. A. Thomas. *Elements of Information Theory*. Wiley, New York, 1991.

L. Devroye, L. Györfi, and G. Lugosi. *A Probabilistic Theory of Pattern Recognition*. Springer, New York, 1996.

R. M. Dudley. *Real Analysis and Probability*. Chapman & Hall, New York, 1989.

R.M. Dudley. Central limit theorems for empirical measures. *Annals of Probability*, 6:899–929, 1978.

A Gersho and R. M. Gray. *Vector Quantization and Signal Compression*. Kluwer, Boston, 1992.

H. Gish and J. N. Pierce. Asymptotically efficient quantizing. *IEEE Trans. Inform. Theory*, IT-14:676–683, Sep. 1968.

S. Graf and H. Luschgy. Consistent estimation in the quantization problem for random vectors. In *Transactions of the Twelfth Prague Conference on Information Theory, Statistical Decision Functions, Random Processes*, pages 84–87, 1994.

S. Graf and H. Luschgy. Rates of convergence for the empirical quantization error. preprint, 1999.

S. Graf and H. Luschgy. *Foundations of Quantization for Probability Distributions*. Springer Verlag, Berlin, Heidelberg, 2000.

R. M. Gray. *Source Coding Theory*. Kluwer, Boston, 1990.

R. M. Gray and L. D. Davisson. Quantizer mismatch. *IEEE Trans. Communications*, 23:439–443, 1975.

R. M. Gray, T. Linder, and J. Li. A Lagrangian formulation of Zador's entropy-constrained quantization theorem. *IEEE Trans. Inform. Theory*, 2001 (to appear).

R. M. Gray and D. L. Neuhoff. Quantization. *IEEE Trans. Inform. Theory,* (Special Commemorative Issue), IT-44(6):2325–2383, Oct. 1998.

R. M. Gray, D. L. Neuhoff, and P. C. Shields. A generalization of Orstein's $\bar{d}$-distance with applications to information theory. *Annals of Probability*, 3:315–328, 1975.

A. György and T. Linder. Optimal entropy-constrained scalar quantization of a uniform source. *IEEE Trans. Inform. Theory*, IT-46:pp. 2704–2711, Nov. 2000.

A. György and T. Linder. On the structure of optimal entropy-constrained scalar quantizers. *IEEE Trans. Inform. Theory*, 2001 (to appear).

A. György and T. Linder. On optimal Lagrangian entropy-constrained vector quantization. preprint, 2001.

Y. Linde, A. Buzo, and R. M. Gray. An algorithm for vector quantizer design. *IEEE Transactions on Communications*, COM-28:84–95, Jan. 1980.

T. Linder. On the training distortion of vector quantizers. *IEEE Trans. Inform. Theory*, IT-46:1617–1623, Jul. 2000.

T. Linder, G. Lugosi, and K. Zeger. Rates of convergence in the source coding theorem, in empirical quantizer design, and in universal lossy source coding. *IEEE Trans. Inform. Theory*, 40:1728–1740, Nov. 1994.

T. Linder, G. Lugosi, and K. Zeger. Empirical quantizer design in the presence of source noise or channel noise. *IEEE Trans. Inform. Theory*, IT-43:612–623, Mar. 1997.

T. Linder, V. Tarokh, and K. Zeger. Existence of optimal prefix codes for infinite source alphabets. *IEEE Trans. Inform. Theory*, IT-43:2026–2028, Nov. 1997.

S. P. Lloyd. Least squared quantization in PCM. unpublished memorandum, Bell Lab., 1957; Also, *IEEE Trans. Inform. Theory,* vol. IT-28, no. 2, pp. 129-137., Mar. 1982.

J. MacQueen. Some methods for classification and and analysis of multivariate observations. in *Proc. 5th Berkeley Symp. on Mathematical Statistics and Probability*, vol. 1, pp. 281–296, 1967.

N. Merhav and J. Ziv. On the amount of side information required for lossy data compression. *IEEE Trans. Inform. Theory*, IT-43:1112–1121, July 1997.

D. Pollard. Strong consistency of k-means clustering. *Annals of Statistics*, 9, no. 1:135–140, 1981.

D. Pollard. Quantization and the method of k-means. *IEEE Trans. Inform. Theory*, IT-28:199–205, Mar. 1982.

D. Pollard. A central limit theorem for k-means clustering. *Annals of Probability*, vol. 10, no. 4:919–926, 1982.

D. Pollard. *Empirical Processes: Theory and Applications*. NSF-CBMS Regional Conference Series in Probability and Statistics, Institute of Mathematical Statistics, Hayward, CA, 1990.

R. T. Rockafellar. *Convex Analysis*. Princeton University Press, Princeton, NJ, 1970.

M. J. Sabin. *Global convergence and empirical consistency of the generalized Lloyd algorithm*. PhD thesis, Stanford Univ., 1984.

M. J. Sabin and R. M. Gray. Global convergence and empirical consistency of the generalized Lloyd algorithm. *IEEE Trans. Inform. Theory*, IT-32:148–155, Mar. 1986.

K. Sayood. *Introduction to Data Compression*. Morgan Kaufmann Publishers, San Francisco, 2nd edition, 2001.

E.V. Slud. Distribution inequalities for the binomial law. *Annals of Probability*, 5:404–412, 1977.

H. Steinhaus. Sur la division des corp materiels en parties. *Bull. Acad. Polon. Sci*, IV:801–804, May 1956.

S. Szarek. On the best constants in the Khintchine inequality. *Studia Mathematica*, 63:197–208, 1976.

V. N. Vapnik. *Statistical Learning Theory*. Wiley, New York, 1998.

P. Zador. Topics in the asymptotic quantization of continuous random variables. unpublished memorandum, Bell Laboratories, Murray Hill, NJ, Feb. 1966.

P. Zador. Asymptotic quantization error of continuous signals and the quantization dimension. *IEEE Trans. Inform. Theory*, IT-28:139–149, Mar. 1982.

A. J. Zeevi. On the performance of vector quantizers empirically designed from dependent sources. in *Proceedings of Data Compression Conference, DCC'98*, (J. Storer, M. Cohn, ed.) pp. 73–82, IEEE Computer Society Press, Los Alamitos, California, 1998.

5

DISTRIBUTION AND DENSITY ESTIMATION

L. Devroye
McGill University, Montreal, Canada

L. Györfi
Budapest University of Technology and Economics, Budapest, Hungary

5.1 Distribution estimation

The classical nonparametric example is the problem estimating a distribution function $F(x)$ from i.i.d. samples $X_1, X_2, \ldots X_n$ taking values in $\mathcal{R}^d$ $(d \geq 1)$. Here on the one hand the construction of the empirical distribution function $F_n(x)$ is distribution-free, and on the other hand its uniform convergence, the Glivenko-Cantelli Theorem holds for all $F(x)$:

$$\lim_{n\to\infty} \sup_x |F_n(x) - F(x)| = 0 \text{ a.s.}$$

The Glivenko-Cantelli Theorem is really distribution-free, and the convergence in Kolmogorov-Smirnov distance means uniform convergence, so virtually it seems that there is no need to go further. However, if, for example, in a decision problem one wants to use empirical distribution functions for two unknown continuous distribution functions for creating a kind of likelihood then these estimates are useless. It turns out that we should look for stronger error criteria.

For this purpose it is obvious to consider the total variation: if μ and ν are probability measures on $\mathcal{R}^d$ then the total variation of μ and ν is defined by

$$V(\mu, \nu) = \sup_A |\mu(A) - \nu(A)|,$$

where the supremum is taken over all Borel sets A.

However, if μ stands for the common distribution of $\{X_i\}$ and μ_n denotes the empirical distribution then for nonatomic μ

$$V(\mu, \mu_n) = 1$$

a.s., so the empirical distribution is a bad estimate in total varition.

One may expect to find a more sophisticated sequence $\{\mu_n^*\}$ of distribution estimates of μ which is consistent in total variation:

$$\lim_{n\to\infty} V(\mu,\mu_n^*) = 0 \text{ a.s.}$$

Theorem 5.1. (DEVROYE AND GYÖRFI (1990)) *Given any sequence of distribution estimators $\{\mu_n^*\}$ there always exists a probability measure μ for which*

$$V(\mu,\mu_n^*) > 1/4 \textit{ for all } n \textit{ a.s.}$$

PROOF. This negative finding means that the total variation is a much stronger error criterion than the Kolmogorov-Smirnov distance such that it is impossible to construct a distribution estimate with distribution-free consistency in total variation. The proof borrows some arguments from Devroye (1983) and Rényi (1959). First, we need a rich family of singular continuous probability measures. The family of probability measures considered here is parametrized by a number $b \in [0,1]$ with binary expansion $b = 0.b_{(1)}b_{(2)}b_{(3)}\dots$, $b_{(i)} \in \{0,1\}$. Let the binary random variables $Y_{(1)}, Y_{(2)}, \dots$ be i.i.d. and uniformly distributed on $\{0,1\}$. We define the random variable $X = X(Y,b)$ by setting $X = 0.X_{(1)}X_{(2)}X_{(3)}\dots$ in the ternary radix system used for $Y = 0.Y_{(1)}Y_{(2)}Y_{(3)}\dots$, where

$$X_{(k)} = \begin{cases} 0, & \text{if } b_{(k)} = 0, \\ Y_{(k)}, & \text{if } b_{(k)} = 1. \end{cases}$$

Let μ_b denote the probability measure of $X = X(Y,b)$. If in the binary expansion of b there are finitely many (L) zeros, then μ_b is absolutely continuous and distributes its mass uniformly on a set of Lebesgue measure 2^{-L}.If in the binary expansion of b there are finitely many (L) ones, then μ_b is discrete and puts its mass uniformly on a set of cardinality 2^L. In other cases, μ_b is singular.

We write $X(Y_1,b),\dots,X(Y_n,b)$ to denote a sample drawn from the distribution of $X(Y,b)$. We will replace b at a crucial step in the argument by a uniform $[0,1]$ random variable B, which is independent of $Y_1,\dots Y_n$. Put

$$A_k = \{0.x_{(1)}x_{(2)}\cdots : x_{(i)} \in \{0,1\} \text{ for all } i; x_{(k)} = 0\}.$$

Then

$$\mu_b(A_k) = \begin{cases} 1, & \text{if } b_{(k)} = 0, \\ 1/2, & \text{if } b_{(k)} = 1. \end{cases}$$

Let μ_n^* be an arbitrary distribution estimate based upon $X(Y_1,b),\dots,X(Y_n,b)$. Let us now define the parameter estimate $b_n = 0.b_{n1}b_{n2}\dots$ by its binary expansion with bits

$$b_{nk} = \begin{cases} 0, & \text{if } \mu_n^*(A_k) > 3/4, \\ 1, & \text{otherwise.} \end{cases}$$

Then

$$|\mu_n^*(A_k) - \mu_b(A_k)| \geq 1/4 I_{\{b_{nk} \neq b_{(k)}\}}.$$

Therefore

$$\begin{aligned}\sup_b \inf_n V(\mu_n^*, \mu_b) &= \sup_b \inf_n \sup_A |\mu_n^*(A) - \mu_b(A)| \\ &\geq \sup_b \inf_n \sup_k |\mu_n^*(A_k) - \mu_b(A_k)| \\ &\geq \sup_b \inf_n \sup_k 1/4 I_{\{b_{nk} \neq b_{(k)}\}}.\end{aligned}$$

Replace b by B and resulting b_{nk} by B_{nk}. Then

$$\begin{aligned}\sup_b \inf_n V(\mu_n^*, \mu_b) &\geq \inf_n \sup_k 1/4 I_{\{B_{nk} \neq B_{(k)}\}} \\ &= 1/4 \inf_n Z_n.\end{aligned}$$

Our theorem is proved if we can show that $Z_n = 1$ almost surely for all n. Put $Z_{Nn} = I_{\{\cup_{k=1}^N [B_{nk} \neq B_{(k)}]\}}$. Then $Z_{Nn} \uparrow Z_n = I_{\{\cup_{k=1}^\infty [B_{nk} \neq B_{(k)}]\}}$. Therefore it suffices to show that

$$\lim_{N \to \infty} \mathbb{P}\{\cup_{k=1}^N [B_{nk} \neq B_{(k)}]\} = 1.$$

But $\mathbb{P}\{\cup_{k=1}^N [B_{nk} \neq B_{(k)}]\}$ is the error probability of the decision $(B_{n1}, \ldots, B_{nN})$ on $(B_{(1)}, \ldots B_{(N)})$ for the observations $X_1, \ldots, X_n$. For this decision problem the Bayes decision is

$$\tilde{B}_{nk} = \begin{cases} 0, & \text{if } X_{i(k)} = 0 \text{ for all } i = 1, \ldots, n, \\ 1, & \text{otherwise.} \end{cases}$$

Thus,

$$\begin{aligned}\mathbb{P}\{Z_{Nn} = 1\} &= \mathbb{P}\{\cup_{k=1}^N [B_{nk} \neq B_{(k)}]\} \\ &\geq \mathbb{P}\{\cup_{k=1}^N [\tilde{B}_{nk} \neq B_{(k)}]\} \\ &= 1 - \left(1 - \frac{1}{2 \cdot 2^n}\right)^N \\ &\uparrow 1.\end{aligned}$$

□

For distribution estimation we may consider other error criteria. Such error criteria can be derived from dissimilarity measures of probability measures, like f-divergences introduced by Csiszár (1967) (see also Liese, Vajda (1987) and Vajda (1989)). The three most important f-divergences in mathematical statistics are the total variation, the information divergence and the χ^2-divergence.

If μ and ν are probability measures on $\mathcal{R}^d$ then the information divergence (or I-divergence, relative entropy, Kullback-Leibler number) of μ and ν is defined by

$$I(\mu,\nu) = \sup_{\{A_j\}} \sum_j \mu(A_j) \log \frac{\mu(A_j)}{\nu(A_j)},$$

where the supremum is taken over all finite Borel measurable partition $\{A_j\}$ of $\mathcal{R}^d$.

If μ and ν are discrete distributions then

$$I(\mu,\nu) = \sum_j \mu(\{j\}) \log \frac{\mu(\{j\})}{\nu(\{j\})}.$$

The following inequality, also called Pinsker's inequality, upperbounds the total variation in terms of I-divergence (cf. Csiszár (1967), Kemperman (1969) and Kullback (1967)):

$$2\{V(\mu,\nu)\}^2 \le I(\mu,\nu) \tag{5.1}$$

If $\mu_n^* = \mu_n^*(.;X_1,\ldots,X_n)$ is a distribution estimate of μ, then $\{\mu_n^*\}$ is said to be consistent in information divergence if

$$\lim_{n\to\infty} I(\mu,\mu_n^*) = 0 \; a.s.$$

By Pinsker's inequality (5.1), the information divergence dominates the total variation, so it follows from Theorem 5.1 that given any sequence of distribution estimators $\{\mu_n^*\}$ there always exists a probability measure μ for which the sequence $\{\mu_n^*\}$ is not consistent in information divergence. The situation is even worse, a discrete distribution with known support cannot be estimated consistently in information divergence:

Theorem 5.2. (GYÖRFI, PÁLI AND VAN DER MEULEN (1994)) *Assume that μ is a probability measure on the set of positive integers. Given any sequence of distribution estimators $\{\mu_n^*\}$ there always exists a probability measure μ with finite Shannon entropy*

$$H(\mu) = -\sum_{j=1}^{\infty} \mu(\{i\}) \log \mu(\{i\})$$

for which

$$I(\mu,\mu_n^*) = \infty \; a.s.$$

5.2 The density estimation problem

HOW TO MEASURE CLOSENESS. A random variable X on $\mathcal{R}^d$ has a density f if, for all Borel sets A of $\mathcal{R}^d$, $\int_A f(x)\,dx = \mathbb{P}\{X \in A\}$. It thus serves as a tool for computing probabilities of sets. As it is a function that reveals the local concentration of probability mass, it may be used to visualize distributions of random variables. The statistician's problem, then, is to estimate f from an i.i.d. sample $X_1, \ldots, X_n$ drawn from f. A density estimate is simply a mapping $f_n : \mathcal{R}^d \times (\mathcal{R}^d)^n \to \mathcal{R}^d$ (we write $f_n(x; X_1, \ldots, X_n)$ or $f_n(x)$). It is the global closeness of f_n to f that interests us. The choice of a density estimate is governed by a number of factors, like consistency, smoothness, ease of computation, interpretability, flexibility, robustness, versatility, and optimality for certain criteria. The early work in the field approached the problem largely as a functional estimation problem: f was treated as any function, and tools from function approximation theory came to the rescue in the analysis of the performance of density estimates—Taylor series expansions played a key role, for example. The view we take in nonparametric density estimation is that f is largely unknown and that no assumptions can be made about its properties. Methods or properties that are valid for all densities are said to be universal. It is quite surprising that there are density estimates that can approximate any density f asymptotically in an appropriate sense. We see in section 5.3 that the histogram has this property. Other examples will follow.

The quality of a density estimate is measured by how well it performs the task at hand, estimating probabilities. In this respect, the total variation criterion is a natural distance:

$$\sup_{A \in \mathcal{B}} \left| \int_A f_n - \int_A f \right| ,$$

where $\mathcal{B}$ is the class of Borel sets of $\mathcal{R}^d$. If this is smaller than ε, then all probabilities will be estimated with errors not exceeding ε. We measure the closeness of two densities f and g by their L_1 distance $\int |f - g|$. There are many reasons for this, but all more or less follow from Scheffé's identity (Theorem 5.4 below): $\sup_{A \in \mathcal{B}} |\int_A f_n - \int_A f| = (1/2) \int |f - f_n|$. We can thus compare the performance of density estimates on an absolute scale, and L_1 distances indeed have a physical interpretation: if we know that $\int |f - g| < 0.04$, then we know that differences in probabilities are at most 0.02. In contrast, the interpretation of the inequalities involving other metrics such as the L_2 metric (example: $\int (f - g)^2 \leq 0.04$) or the Kullback-Leibler metric (example: $\int f \log(f/g) \leq 0.04$) in terms of probabilities is less obvious.

THE COUPLING DISTANCE. There is an interpretation of the L_1 distance in terms of samples that is interesting in its own right. Let $X_1, \ldots, X_n$ and $Y_1, \ldots, Y_n$ be independent samples of identically distributed random variables, the former having density f, and the latter having density g. It should be clear that with probability one, all $2n$ are different. However, if we allow the X and Y samples to be dependent (while maintaining independence within each sample), it is possible to have $X_i = Y_i$ for many i's. Let us introduce the Hamming distance

$$H_n = \sum_{i=1}^{n} \mathbb{I}_{[X_i \neq Y_i]} .$$

It is the number of X_i's we need to change to turn the X-sample into the Y-sample. If $f = g$, then we can obviously take identical samples and achieve $H_n = 0$. Let $\mathcal{F}$ denote all possible joint distributions of $X_1,\ldots,X_n,Y_1,\ldots,Y_n$ such that $X_1,\ldots,X_n$ is i.i.d. and drawn from f and $Y_1,\ldots,Y_n$ is i.i.d. and drawn from g. $\mathcal{F}$ is called a coupling. Define the coupling distance by

$$C_n = \inf_{\mathcal{F}} \mathbb{E}\{H_n\} .$$

The coupling distance measures the minimal expected Hamming distance over all possible dependencies between samples. It is interesting to note that

$$C_n = \frac{n}{2} \int |f - g| .$$

Thus, $\int |f - g|/2$, a number between 0 and 1, is the fraction of a sample we need to alter in order to turn it in a sample with the other density. To see why this is true, consider the functions $m = \min(f,g)$, $(f-g)_+$ and $(g-f)_+$, and note that $f = m + (f-g)_+$, $g = m + (g-f)_+$. Also, $\int (f-g)_+ = \int (g-f)_+ = (1/2)\int |f-g|$, and $\int m = 1 - (1/2)\int |f-g|$. We can generate our two samples in stages. First we generate a binomial $(n,(1/2)\int |f-g|)$ random variable N, which decides what fraction of each sample is drawn from $(f-g)_+$ or $(g-f)_+$. The remainder, $n-N$, is drawn from m, and as it is common, we may make those values identical in both samples. If the samples are generated in this way, we have $H_n = N$, and $\mathbb{E}\{H_n\} = (n/2)\int |f-g|$. This shows that there is a coupling that achieves C_n. To see that we cannot improve over it, observe that $(f-g)_+$ and $(g-f)_+$ are functions of disjoint support.

INVARIANCE TO TRANSFORMATIONS. The total variation is invariant with respect to monotone transformations of the axes. In fact, let $T : \mathcal{R}^d \to S \subseteq \mathcal{R}^d$ be a bijection and a Borel-measurable mapping. Let X and Y be random variables with densities f and g, respectively. Then the total variation distance is

$$\sup_B |\mathbb{P}\{X \in B\} - \mathbb{P}\{Y \in B\}|.$$

If T is a bijection, this is equal to

$$\sup_B |\mathbb{P}\{T(X) \in T(B)\} - \mathbb{P}\{T(Y) \in T(B)\}|,$$

which is nothing but the total variation distance between $T(X)$ and $T(Y)$, provided that $\{T(B) : B \in \mathcal{B}\}$ coincides with the Borel sets of $S = T(\mathcal{R}^d)$. We leave this easy verification as an exercise.

The implications of this fact are tremendous. For example, if we know that a certain density estimate performs very well when the data belong to a compact set such as $[-1,1]$, then we might apply the monotone mapping $T : x \to x/(1+|x|)$, estimate the density of $T(X)$, and obtain an estimate of the density of X by well-known back-transformation methods. Such estimates are called transformed density estimates. Here is another variation on the same theme: assume that we wish to visualize the densities f and g to get an idea of the size and the location of the error. Computer screens cannot show infinite tails, but by showing the graphs of the transformed densities (for a transformation to a compact set), we can make do, as the total variation distance remains unchanged.

THE LEBESGUE DENSITY THEOREM. Note that two densities f and g are identical if $\int_B f = \int_B g$ for all Borel sets B, that is, if and only if $\int |f-g| = 0$. Thus, we may alter f on a set of zero Lebesgue measure, and still be in the same equivalence class of densities. This immediately makes criteria such as $|f(x)-g(x)|$ (pointwise error) or $\sup_x |f(x)-g(x)|$ (the supremum error) suspect, unless we replace these criteria by quantities that are defined in terms of these equivalence classes. Some help in this respect comes from the following theorem.

Theorem 5.3. (THE LEBESGUE DENSITY THEOREM) *Let $\mathcal{Q}$ be a subclass of the Borel sets of $\mathcal{R}^d$ with the property that*

$$\sup_{Q \in \mathcal{Q}} \frac{\lambda(Q^*)}{\lambda(Q)} \leq c < \infty$$

for some constant c, where Q^* is the smallest cube centered at the origin that contains Q, and $\lambda(\cdot)$ denotes the volume of a set. (Here $a/0$ is treated as ∞ for all $a \geq 0$.) Let $\mathcal{Q}_r$ be the subclass of $\mathcal{Q}$ containing only sets Q with $\lambda(Q) \leq r$. Let f be any density on $\mathcal{R}^d$. Let $z+Q$ denote the translation of Q by z. Then for almost all x:

$$\lim_{r \downarrow 0} \sup_{Q \in \mathcal{Q}_r} \left| \frac{1}{\lambda(Q)} \int_{x+Q} f(y)\, dy - f(x) \right| = 0.$$

The points x at which this convergence takes place are called Lebesgue points for f. Classes that satisfy the condition are the classes of all cubes, or all balls of $\mathcal{R}^d$.

This theorem is typically applied to balls $B_{x,r}$ of radius r centered at x. Let us define $g(x) = \liminf_{r \downarrow 0} \int_{B_{x,r}} f/\lambda(B_{x,r})$. The Lebesgue density theorem says that g is in the equivalence class of f. And in fact, it is easy to see that all members of that equivalence class lead to the same g, and thus, g may be considered as the representative of that class. We will call it the Lebesgue representative, denoted temporarily by $L(f)$. Now, of course, we may return to the pointwise criterion, and use $|L(f)(x)-L(g)(x)|$ instead of $|f(x)-g(x)|$. And we may use $\sup_x |L(f)(x) - L(g)(x)|$ as the new definition of supremum distance.

The Lebesgue density theorem states that for every density, without exception, for almost all x, $f(x)$ is close to an integral over a small ball centered at x. But integrals can be approximated by empirical measures based on data, and thus, the Lebesgue density theorem permits us to construct estimates that converge for all densities in the L_1 sense. Differentiation of measures, and approximation of functions by convolutions, is dealt with at length by Stein (1970), de Guzmán (1975, 1981), and Wheeden and Zygmund (1977).

Because of the following theorem, a distribution μ can be consistently estimated in total variation if it has a density.

Theorem 5.4. (Scheffé (1947)) *If μ and ν are absolutely continuous with respect to a σ-finite measure λ with densities f and g respectively, then*

$$\|f-g\| := \int_{\mathcal{R}^d} |f(x)-g(x)|\lambda(dx) = 2V(\mu,\nu).$$

Proof. Note that

$$\begin{aligned} V(\mu,\nu) &= \sup_A |\mu(A)-\nu(A)| \\ &= \sup_A \left|\int_A f d\lambda - \int_A g d\lambda\right| \\ &= \sup_A \left|\int_A (f-g) d\lambda\right| \\ &= \int_{f>g} (f-g) d\lambda \\ &= \int_{g>f} (g-f) d\lambda \\ &= \frac{1}{2}\int |f-g| d\lambda. \end{aligned}$$

The Scheffé Theorem results in a way of distribution estimation consistent in total variation via L_1-consistent density estimation: assume that f_n is L_1-consistent, i.e.

$$\lim_{n\to\infty} \|f-f_n\| = 0 \text{ a.s.}$$

Introduce the distribution estimate induced by the density estimate f_n:

$$\mu_n^*(A) = \int_A f_n(x)\lambda(dx),$$

then

$$\lim_{n\to\infty} V(\mu,\mu_n^*) = 0 \text{ a.s.}$$

5.3 The histogram density estimate

The standard examples for L_1-consistent density estimates are the histogram and the kernel estimates.

Let $\mathcal{P}_n$ be a partition of $\mathcal{R}^d$ with cells $\{A_{n,j}\}$ of positive and finite Lebesgue measure. Then the **histogram** is as follows:

$$f_n(x) = \frac{\mu_n(A_n(x))}{\lambda(A_n(x))},$$

where $A_n(x) = A_{n,j}$ if $x \in A_{n,j}$.

Theorem 5.5. (ABOU-JAOUDE (1976)) *Assume that μ has a density f. If for each sphere S centered at the origin*

$$\lim_{n\to\infty} \max_{j:A_{n,j}\cap S\neq\emptyset} diam(A_{n,j}) = 0 \tag{5.2}$$

and

$$\lim_{n\to\infty} \frac{|\{j : A_{n,j}\cap S \neq \emptyset\}|}{n} = 0 \tag{5.3}$$

then

$$\lim_{n\to\infty} \mathbb{E}\|f - f_n\| = 0$$

and

$$\lim_{n\to\infty} \|f - f_n\| = 0 \ a.s.$$

The beauty of Theorem 5.5 is that L_1-consistency is possible without any condition on the density f, thus we can have distribution estimate consistent in total variation if μ is absolutely continuous with respect to the Lebesgue measure. The histogram can be extended to distribution estimates consistent in total variation if the nonatomic part of μ is absolutely continuous with respect to a known σ-finite measure λ (Barron, Györfi and van der Meulen (1992)).

The proof of Theorem 5.5 uses a denseness lemma:

Lemma 5.1. *The set of continuous functions of bounded support is dense in $L_1(\lambda)$.*

PROOF. We prove that for any $f \in L_1(\lambda)$ and $\varepsilon > 0$ there is a continuous function g with compact support such that

$$\int |f(x) - g(x)|\lambda(dx) \leq \varepsilon.$$

Without loss of generality assume that $f \geq 0$. The case for arbitrary f can be handled similarily as the case $f \geq 0$ by using the decomposition $f = f^+ - f^-$ where $f^+ = \max\{f, 0\}$ and $f^- = -\min\{f, 0\}$. $f \in L_1(\lambda)$ implies that there is a closed sphere $S = S_{0,R}$ centered at the origin with radius R and $K > 0$ such that

$$\int |f(x) - \min\{f(x), K\}I_{x\in S}|\lambda(dx) \leq \varepsilon/2,$$

therefore it suffices to show that there is a continuous function g with support in S such that

$$\int_S |\min\{f(x), K\} - g(x)|\lambda(dx) \leq \varepsilon/2.$$

Put

$$f^*(x) = \min\{f(x), K\} I_{x \in S}$$

then using the technique of Lusin's Theorem we show that there is a continuous function g with support in S such that $0 \le g \le K$ and

$$\lambda\{x \in S; f^*(x) \ne g(x)\} \le \varepsilon/(2K).$$

Obviously this will imply the lemma since

$$\int_S |f^*(x) - g(x)| \lambda(dx) \le K\lambda\{x \in S; f^*(x) \ne g(x)\} \le K\varepsilon/(2K) = \varepsilon/2.$$

We have now that $0 \le f^* \le K$. Without loss of generality we may assume that $K = 1$. A function s is called a *simple function* if its range consists of finitely many points in $[0, \infty)$. We can construct a monotonically increasing sequence of simple functions $s_1 \le s_2 \le \cdots \le f^*$ such that $s_n(x) \to f^*(x)$ as $n \to \infty$, for every $x \in S$. Indeed, for $n = 1, 2, \ldots$, and for $1 \le i \le 2^n$, define sets

$$E_{n,i} = \left\{ x \in S : \frac{i-1}{2^n} \le f^*(x) < \frac{i}{2^n} \right\}$$

and the simple function

$$s_n = \sum_{i=1}^{2^n} \frac{i-1}{2^n} I_{E_{n,i}}.$$

Sets $E_{n,i}$ are inverses images of half open intervals through a measurable function and thus are measurable. Clearly sequence s_n is monotonically increasing and $s_n \le f^*$.

Now define $t_1 = s_1$ and $t_n = s_n - s_{n-1}, n = 2, 3, \ldots$. Then $2^n t_n$ is the indicator function of a set $T_n \subset S$ and

$$f^*(x) = \sum_{n=1}^{\infty} t_n(x).$$

Then there exist compact sets U_n and open sets V_n such that $U_n \subset T_n \subset V_n \subset S$ and $\lambda(V_n - U_n) < 2^{-n}\varepsilon/4$. Next we need a special version of Urysohn's Lemma which states that for any compact set U and open set V such that $U \subset V$ there exists a continuous function $h : S \to [0, 1]$ such that $h(x) = 1$ for $x \in U$ and $h(x) = 0$ for $x \in V^c$, where V^c denotes the complement of V. In order to show this special case of Uryshon's Lemma for any set A introduce the function

$$d(x, A) = \inf_{z \in A} \|x - z\|.$$

Then $d(x, A)$ is continuous, and $d(x, A) = 0$ iff x lies in the closure of A. Such function h can be defined as

$$h(x) = \frac{d(x, V^c)}{d(x, U) + d(x, V^c)}.$$

Thus by Uryshon's Lemma there are continuous functions h_n that assume value 1 on U_n and are zero outside V_n. Let

$$g(x) = \sum_{n=1}^{\infty} 2^{-n} h_n(x).$$

Since h_n are bounded by 1 this series converges uniformly on S and thus g is continuous. Also its support lies in S. Since $2^{-n}h_n(x) = t_n(x)$ except in $V_n - U_n$, we obtain $g(x) = f^*(x)$ except in $\cup(V_n - U_n)$, but

$$\lambda(\cup(V_n - U_n)) \leq \sum_{n=1}^{\infty} \lambda(V_n - U_n) \leq \varepsilon/2 \sum_{n=1}^{\infty} 2^{-n} = \varepsilon/2.$$

□

PROOF OF THEOREM 5.5. By triangle inequality

$$\|f - f_n\| \leq \|f - \mathbb{E}f_n\| + \|\mathbb{E}f_n - f_n\|.$$

The first term of the right hand side is called bias, and the second one is the variation term.

$$\mathbb{E}f_n(x) = \mathbb{E}\frac{\mu_n(A_n(x))}{\lambda(A_n(x))} = \frac{\mu(A_n(x))}{\lambda(A_n(x))} = \frac{\int_{A_n(x)} f(z)\lambda(dz)}{\lambda(A_n(x))} =: T_n f(x).$$

If f is uniformly continuous with bounded support then because of (5.2)

$$\|T_n f - f\| \to 0.$$

For arbitrary f, according to Lemma 5.1 choose continuous $\tilde{f}$ with bounded support such that

$$\|f - \tilde{f}\| < \varepsilon.$$

Then by the choice of $\tilde{f}$

$$\|T_n\tilde{f} - \tilde{f}\| \to 0.$$

Moreover

$$\begin{aligned}
\|T_n\tilde{f} - T_n f\| &= \int \left| \frac{\int_{A_n(x)} \tilde{f}(z)\lambda(dz)}{\lambda(A_n(x))} - \frac{\int_{A_n(x)} f(z)\lambda(dz)}{\lambda(A_n(x))} \right| \lambda(dx) \\
&= \sum_j \int_{A_{n,j}} \left| \frac{\int_{A_n(x)} \tilde{f}(z)\lambda(dz)}{\lambda(A_n(x))} - \frac{\int_{A_n(x)} f(z)\lambda(dz)}{\lambda(A_n(x))} \right| \lambda(dx) \\
&= \sum_j \left| \int_{A_{n,j}} \tilde{f}(z)\lambda(dz) - \int_{A_{n,j}} f(z)\lambda(dz) \right| \\
&\leq \sum_j \int_{A_{n,j}} |\tilde{f}(z) - f(z)|\lambda(dz) \\
&= \|\tilde{f} - f\|.
\end{aligned}$$

Thus

$$
\begin{aligned}
\|f-\mathbb{E}f_n\| &= \|f-T_nf\| \\
&\le \|f-\tilde{f}\|+\|T_n\tilde{f}-\tilde{f}\|+\|T_n\tilde{f}-T_nf\| \\
&\le 2\|f-\tilde{f}\|+\|T_n\tilde{f}-\tilde{f}\| \\
&\to 2\|f-\tilde{f}\| \\
&\le 2\varepsilon,
\end{aligned}
$$

therefore the bias term tends to zero. Observe that

$$
\begin{aligned}
\|\mathbb{E}f_n-f_n\| &= \int\left|\frac{\mu(A_n(x))}{\lambda(A_n(x))}-\frac{\mu_n(A_n(x))}{\lambda(A_n(x))}\right|\lambda(dx) \\
&= \sum_j|\mu(A_{n,j})-\mu_n(A_{n,j})|.
\end{aligned}
$$

Fix a sphere S centered at the origin such that $\mu(S^c)<\varepsilon$, and without loss of generality assume that $\{A_{n,j}\}$ are indexed such that $A_{n,j}\cap S\neq 0$ for $j=1,2,\dots,m_n$ and $A_{n,j}\cap S=0$ otherwise. Put

$$
A_n=\cup_{j=m_n+1}^{\infty}A_{n,j}.
$$

By (5.3)

$$
\frac{m_n}{n}\to 0
$$

and $\mu(A_n)\le\mu(S^c)<\varepsilon$, thus

$$
\begin{aligned}
\|\mathbb{E}f_n-f_n\| &\le \sum_{j=1}^{m_n}|\mu(A_{n,j})-\mu_n(A_{n,j})|+\mu(A_n)+\mu_n(A_n) \\
&\le \sum_{j=1}^{m_n}|\mu(A_{n,j})-\mu_n(A_{n,j})|+|\mu(A_n)-\mu_n(A_n)|+2\mu(A_n) \\
&\le \sum_{j=1}^{m_n}|\mu(A_{n,j})-\mu_n(A_{n,j})|+|\mu(A_n)-\mu_n(A_n)|+2\varepsilon.
\end{aligned}
$$

By the Cauchy-Schwarz and Jensen inequalities

$$\begin{aligned}\mathbb{E}\|\mathbb{E}f_n - f_n\| &\le \sum_{j=1}^{m_n} \mathbb{E}|\mu(A_{n,j}) - \mu_n(A_{n,j})| + \mathbb{E}|\mu(A_n) - \mu_n(A_n)| + 2\varepsilon \\ &\le \sum_{j=1}^{m_n} \sqrt{\mathbb{E}|\mu(A_{n,j}) - \mu_n(A_{n,j})|^2} + \sqrt{\mathbb{E}|\mu(A_n) - \mu_n(A_n)|^2} + 2\varepsilon \\ &\le \sum_{j=1}^{m_n} \sqrt{\mu(A_{n,j})/n} + \sqrt{\mu(A_n)/n} + 2\varepsilon \\ &\le \sqrt{(m_n+1)/n} + 2\varepsilon \\ &\to 2\varepsilon.\end{aligned}$$

Concerning the a.s. convergence, let $\mathcal{A}$ be the family of sets whose elements are the unions of $A_{n,1}, \ldots, A_{n,m_n}$ and A_n then by the Scheffé Theorem and Hoeffding's inequality

$$\begin{aligned}\mathbb{P}\{\|\mathbb{E}f_n - f_n\| > 3\varepsilon\} &\le \mathbb{P}\left\{\sum_{j=1}^{m_n} |\mu(A_{n,j}) - \mu_n(A_{n,j})| + |\mu(A_n) - \mu_n(A_n)| > \varepsilon\right\} \\ &= \mathbb{P}\left\{2 \sup_{A \in \mathcal{A}} |\mu(A) - \mu_n(A)| > \varepsilon\right\} \\ &\le 2^{m_n+1} \sup_{A \in \mathcal{A}} \mathbb{P}\{|\mu(A) - \mu_n(A)| > \varepsilon/2\} \\ &\le 2^{m_n+1} 2 e^{-2n(\varepsilon/2)^2} \\ &= e^{-n(\varepsilon^2/2 - \ln 2 \frac{m_n+2}{n})},\end{aligned}$$

which is summable, therefore applying the Borel-Cantelli Lemma the variation term tends to zero a.s. □

In the proof of Theorem 5.5, in fact, we have shown that

$$\mathbb{P}\{\|f - f_n\| > \varepsilon\} \le 2e^{-n(\varepsilon^2/2 + o(1))}.$$

Using McDiarmid's inequality (Theorem 1.8) Devroye(1991) proved that for the histogram and the kernel density estimates (see Section 5.6)

$$\mathbb{P}\{|\|f - f_n\| - \mathbb{E}\|f - f_n\|| > \varepsilon\} \le 2e^{-n\varepsilon^2/2}. \tag{5.4}$$

Using large deviation techniques the L_1 error of the histogram can be characterized as follows:

Theorem 5.6. (BEIRLANT, DEVROYE, GYÖRFI, VAJDA (2001)) *Assume (5.2). If there is a sequence of spheres S_n centered at the origin such that $S_n \uparrow \mathcal{R}^d$ and*

$$\lim_{n\to\infty} \frac{|\{A_{n,j} \cap S_n \neq \emptyset\}|}{n} = 0 \tag{5.5}$$

then for all $0<\varepsilon<2$

$$\lim_{n\to\infty}\frac{1}{n}\log\mathbb{P}\{\|f-f_n\|>\varepsilon\}=-g(\varepsilon), \tag{5.6}$$

where

$$g(\varepsilon)=\min_{0<p<1-\varepsilon/2}\left(p\log\frac{p}{p+\varepsilon/2}+(1-p)\log\frac{1-p}{1-p-\varepsilon/2}\right).$$

Note that in Theorem 5.6 there is no condition on f, and so (5.6) holds for all f, and the rate function $g(\varepsilon)$ does not depend on f.

(5.4) implies that for consistent f_n

$$\limsup_{n\to\infty}\frac{1}{n}\log\mathbb{P}\{\|f-f_n\|>\varepsilon\}\le-\varepsilon^2/2,$$

thus

$$g(\varepsilon)\ge\varepsilon^2/2.$$

We can get upper bound $\hat{g}(\varepsilon)$ on $g(\varepsilon)$ if in the definition of $g(\varepsilon)$ we substitute p by $\frac{1-\varepsilon/2}{2}$. Then

$$\hat{g}(\varepsilon)=\frac{\varepsilon}{2}\log\frac{2+\varepsilon}{2-\varepsilon}\ge g(\varepsilon).$$

(Vajda (1970)). Further bounds can be found on p. 294-295 in Vajda (1989). For small ε the upper and the lower bounds on $g(\varepsilon)$ are close to each other, and are approximately equal to $\varepsilon^2/2$.

In the proof of Theorem 5.6 we apply some lemmas, where we shall use the function

$$D(\alpha\|\beta)=\alpha\log\frac{\alpha}{\beta}+(1-\alpha)\log\frac{1-\alpha}{1-\beta}. \tag{5.7}$$

Lemma 5.2. (SANOV (1957), SEE P. 16 IN DEMBO, ZEITOUNI (1992), OR PROBLEM 1.2.11 IN CSISZÁR AND KÖRNER (1981)). *Let* $\Sigma=\{1,2,\ldots,m\}$ *be a finite set (alphabet),* $\mathcal{L}_n$ *be a set of types (possible empirical distributions) on* Σ*, and let* Γ *be a subset of* $\mathcal{L}_n$*. If* $Z_1,\ldots Z_n$ *are* Σ*-valued i.i.d. random variables with empirical distribution* μ_n *then*

$$\left|\frac{1}{n}\log\mathbb{P}\{\mu_n\in\Gamma\}+\min_{\tau\in\Gamma}I(\tau,\mu)\right|\le\frac{\log|\mathcal{L}_n|}{n} \tag{5.8}$$

where $|\mathcal{L}_n|$ *denotes the cardinality of* $\mathcal{L}_n$*.*

PROOF. We shall prove that

$$\mathbb{P}\{\mu_n\in\Gamma\}\le|\mathcal{L}_n|e^{-n\min_{\tau\in\Gamma}I(\tau,\mu)}$$

and

$$\mathbb{P}\{\mu_n\in\Gamma\}\ge\frac{1}{|\mathcal{L}_n|}e^{-n\min_{\tau\in\Gamma}I(\tau,\mu)}.$$

Because of our assumptions

$$\begin{aligned}
\mathbb{P}\{Z_1 = z_1, \dots Z_n = z_n\} &= \prod_{i=1}^{n} \mathbb{P}\{Z_i = z_i\} \\
&= \prod_{i=1}^{n} \mu(z_i) \\
&= e^{\sum_{i=1}^{n} \log \mu(z_i)} \\
&= e^{\sum_{i=1}^{n} \sum_{j=1}^{m} I_{z_i = j} \log \mu(z_i)} \\
&= e^{\sum_{i=1}^{n} \sum_{j=1}^{m} I_{z_i = j} \log \mu(j)} \\
&= e^{\sum_{j=1}^{m} n\mu_n(j) \log \mu(j)} \\
&= e^{-n(H(\mu_n) + I(\mu_n, \mu))} \\
&=: \mathbb{P}_\mu(z_1^n).
\end{aligned}$$

For any probability distribution $\tau \in \mathcal{L}_n$ we can define a probability distribution $\mathbb{P}_\tau(z_1^n)$ in this way:

$$\mathbb{P}_\tau(z_1^n) := e^{-n(H(\mu_n) + I(\mu_n, \tau))}.$$

Put

$$T_n(\tau) = \{z_1^n : \mu_n(z_1^n) = \tau\},$$

then

$$1 \geq \mathbb{P}_\tau\{\mu_n = \tau\} = \mathbb{P}_\tau\{z_1^n \in T_n(\tau)\} = |T_n(\tau)| e^{-nH(\tau)}$$

therefore

$$|T_n(\tau)| \leq e^{nH(\tau)},$$

which implies the upper bound:

$$\begin{aligned}
\mathbb{P}\{\mu_n \in \Gamma\} &= \sum_{\tau \in \Gamma} \mathbb{P}_\mu\{\mu_n = \tau\} \\
&\leq |\mathcal{L}_n| \max_{\tau \in \Gamma} \mathbb{P}_\mu\{\mu_n = \tau\} \\
&= |\mathcal{L}_n| \max_{\tau \in \Gamma} |T_n(\tau)| e^{-n(H(\tau) + I(\tau, \mu))} \\
&\leq |\mathcal{L}_n| \max_{\tau \in \Gamma} e^{-nI(\tau, \mu)} \\
&= |\mathcal{L}_n| e^{-n \min_{\tau \in \Gamma} I(\tau, \mu)}.
\end{aligned}$$

Concerning the lower bound notice that for any probability distribution $\nu \in \mathcal{L}_n$

$$\begin{aligned}\frac{\mathbb{P}_\tau\{\mu_n = \tau\}}{\mathbb{P}_\tau\{\mu_n = \nu\}} &= \frac{|T_n(\tau)|\prod_{a\in\Sigma}\tau(a)^{n\tau(a)}}{|T_n(\nu)|\prod_{a\in\Sigma}\tau(a)^{n\nu(a)}} \\ &= \prod_{a\in\Sigma}\frac{(n\nu(a))!}{(n\tau(a))!}\tau(a)^{n(\tau(a)-\nu(a))} \\ &\geq 1.\end{aligned}$$

This last inequality can be seen as follows: the terms of the last product are of the forms $\frac{m!}{l!}\left(\frac{l}{n}\right)^{l-m}$. It is easy to check that $\frac{m!}{l!} \geq l^{m-l}$, therefore

$$\prod_{a\in\Sigma}\frac{(n\nu(a))!}{(n\tau(a))!}\tau(a)^{n(\tau(a)-\nu(a))} \geq \prod_{a\in\Sigma} n^{n(\tau(a)-\nu(a))} = n^{n(\sum_{a\in\Sigma}\tau(a)-\sum_{a\in\Sigma}\nu(a))} = 1.$$

It implies that

$$\mathbb{P}_\tau\{\mu_n = \tau\} \geq \mathbb{P}_\tau\{\mu_n = \nu\}$$

and thus

$$1 = \sum_\nu \mathbb{P}_\tau\{\mu_n = \nu\} \leq |\mathcal{L}_n|\mathbb{P}_\tau\{\mu_n = \tau\} = |\mathcal{L}_n||T_n(\tau)|e^{-nH(\tau)},$$

consequently

$$|T_n(\tau)| \geq \frac{1}{|\mathcal{L}_n|}e^{nH(\tau)}.$$

This implies the lower bound:

$$\begin{aligned}\mathbb{P}\{\mu_n \in \Gamma\} &= \sum_{\tau\in\Gamma}\mathbb{P}_\mu\{\mu_n = \tau\} \\ &\geq \max_{\tau\in\Gamma}\mathbb{P}_\mu\{\mu_n = \tau\} \\ &= \max_{\tau\in\Gamma}|T_n(\tau)|e^{-n(H(\tau)+I(\tau,\mu))} \\ &\geq \frac{1}{|\mathcal{L}_n|}\max_{\tau\in\Gamma}e^{-nI(\tau,\mu)} \\ &= \frac{1}{|\mathcal{L}_n|}e^{-n\min_{\tau\in\Gamma}I(\tau,\mu)}.\end{aligned}$$

Lemma 5.3. (BEIRLANT, DEVROYE, GYÖRFI, VAJDA (2001)) *Consider*

$$J_n = \sum_{j=1}^{m_n}|\mu(A_{n,j}) - \mu_n(A_{n,j})|,$$

based on a finite partition $\mathcal{P}_n = \{A_{n,1},\ldots,A_{n,m_n}\}$, $(n \geq 2)$, of $\mathcal{R}^d$. Assume

$$\lim_{n\to\infty} m_n = \infty \tag{5.9}$$

and

$$\lim_{n\to\infty} \frac{m_n}{n} = 0 \tag{5.10}$$

then for all $0 < \varepsilon < 2$

$$\lim_{n\to\infty} \frac{1}{n} \log \mathbb{P}\{J_n > \varepsilon\} = -g(\varepsilon).$$

PROOF. We apply (5.8) for

$$\Sigma = \{A_{n,1}, \ldots, A_{n,m_n}\}$$

such that

$$\Gamma = \{\tau : 2V(\bar{\mu}, \tau) \geq \varepsilon\}.$$

Then, according to (5.8),

$$\left| \frac{1}{n} \log \mathbb{P}\{J_n \geq \varepsilon\} + \inf_{\tau \in \Gamma \cap \mathcal{L}_n} I(\tau, \bar{\mu}) \right| \leq \frac{\log |\mathcal{L}_n|}{n}.$$

Barron (1989) observed that

$$|\mathcal{L}_n| = \binom{n + m_n - 1}{m_n - 1} = e^{n\delta_n},$$

where $\delta_n \to 0$, provided (5.10). Thus, under (5.10),

$$\lim_{n\to\infty} \frac{1}{n} \log \mathbb{P}\{J_n > \varepsilon\} = -\lim_{n\to\infty} \inf_{\tau \in \Gamma \cap \mathcal{L}_n} I(\tau, \bar{\mu}).$$

It now remains to show that

$$\lim_{n\to\infty} \inf_{\tau \in \Gamma \cap \mathcal{L}_n} I(\tau, \bar{\mu}) = g(\varepsilon).$$

The distributions in $\mathcal{L}_n$ are possible empirical distributions, having components of the form $\frac{r}{n}$, where r is integer. Because of (5.9) and because of the continuity of $V(\tau, \bar{\mu})$ and $I(\tau, \bar{\mu})$

$$\lim_{n\to\infty} \inf_{\tau \in \Gamma \cap \mathcal{L}_n} I(\tau, \bar{\mu}) = \lim_{n\to\infty} \inf_{2V(\tau, \bar{\mu}_n) \geq \varepsilon} I(\tau, \bar{\mu}).$$

Here

$$I(\tau, \bar{\mu}) = \sum_{j=1}^{m_n} \tau(A_{n,j}) \log \frac{\tau(A_{n,j})}{\mu(A_{n,j})}.$$

Put

$$L = \{j : \mu(A_{n,j}) > \tau(A_{n,j})\}$$

and

$$A_n = \cup_{j \in L} A_{n,j}.$$

Then

$$2V(\tau, \bar{\mu}) = 2(\mu(A_n) - \tau(A_n))$$

and, by the definition of I-divergence,

$$I(\tau, \bar{\mu}) \geq D(\tau(A_n)\|\mu(A_n)),$$

where the equality holds iff $\frac{\tau(A_{n,j})}{\mu(A_{n,j})}$ is constant both on L and L^c. Thus

$$\begin{aligned}
\lim_{n\to\infty} \inf_{2V(\tau,\bar{\mu}_n)\geq\varepsilon} I(\tau, \bar{\mu}) &= \inf_{0<p<1-\varepsilon/2:\tau(A_n)=p,\mu(A_n)=p+\varepsilon/2} D(\tau(A_n)\|\mu(A_n)), \\
&= \inf_{0<p<1-\varepsilon/2} \left(p\log\frac{p}{p+\varepsilon/2} + (1-p)\log\frac{1-p}{1-p-\varepsilon/2} \right) \\
&= g(\varepsilon),
\end{aligned}$$

and Lemma 5.3 is proved.

PROOF OF THEOREM 5.6. Let m_n and A_n be as in the proof of Theorem 5.5. Put

$$\Delta_n = \sum_{j=1}^{m_n} |\mu(A_{n,j}) - \mu_n(A_{n,j})| + |\mu(A_n) - \mu_n(A_n)|.$$

Then by Jensen's inequality

$$\begin{aligned}
\|f_n - f\| &= \sum_{j=1}^{\infty} \int_{A_{n,j}} |f_n(x) - f(x)|\lambda(dx) \\
&\geq \sum_{j=1}^{m_n} \int_{A_{n,j}} |f_n(x) - f(x)|\lambda(dx) \\
&+ |\sum_{j=m_n+1}^{\infty} \int_{A_{n,j}} f_n(x)\lambda(dx) - \sum_{j=m_n+1}^{\infty} \int_{A_{n,j}} f(x)\lambda(dx)| \\
&\geq \sum_{j=1}^{m_n} \left| \int_{A_{n,j}} f_n(x)\lambda(dx) - \int_{A_{n,j}} f(x)\lambda(dx) \right| \\
&+ \left| \int_{A_n} f_n(x)\lambda(dx) - \int_{A_n} f(x)\lambda(dx) \right| \\
&= \sum_{j=1}^{m_n} |\mu(A_{n,j}) - \mu_n(A_{n,j})| + |\mu(A_n) - \mu_n(A_n)| \\
&= \Delta_n.
\end{aligned}$$

On the other hand

$$\begin{aligned}
\|f_n - f\| &\leq \|f_n - \mathbb{E}f_n\| + \|\mathbb{E}f_n - f\| \\
&= \sum_{j=1}^{\infty} |\mu(A_{n,j}) - \mu_n(A_{n,j})| + \|\mathbb{E}f_n - f\| \\
&\leq \sum_{j=1}^{m_n} |\mu(A_{n,j}) - \mu_n(A_{n,j})| + |\mu(A_n) - \mu_n(A_n)| \\
&\quad +2\mu(A_n) + \|\mathbb{E}f_n - f\| \\
&= \Delta_n + 2\mu(A_n) + \|\mathbb{E}f_n - f\|.
\end{aligned}$$

By definition $A_n \subset S_n^c$, and by assumption $S_n \uparrow \mathcal{R}^d$, therefore $\mu(A_n) \to 0$, so because of $\|\mathbb{E}f_n - f\| \to 0$

$$2\mu(A_n) + \|\mathbb{E}f_n - f\| \to 0,$$

and consequently Theorem 5.6 is proved if

$$\lim_{n\to\infty} \frac{1}{n} \log \mathbb{P}\{\Delta_n > \varepsilon\} = -g(\varepsilon), \tag{5.11}$$

which follows from Lemma 5.3 for $J_n = \Delta_n$.

For cubic partition assume that the density f has a compact support and is continuously differentiable, then for consistent histogram f_n

$$\mathbb{E}(\|f_n - f\|) \leq \frac{c_1}{\sqrt{nh_n^d}} + c_2 h_n.$$

If $h_n = cn^{-1/(d+2)}$ then

$$\mathbb{E}(\|f_n - f\|) \leq Cn^{-1/(d+2)}.$$

(Devroye, Györfi (1985) and Beirlant, Györfi (1998)). For consistent histogram f_n

$$\frac{\|f_n - f\| - \mathbb{E}\|f_n - f\|}{\sqrt{Var(\|f_n - f\|)}} \xrightarrow{\mathcal{D}} \mathcal{N}(0,1),$$

and

$$\limsup_{n\to\infty} \; nVar(\|f_n - f\|) \leq 1 - \frac{2}{\pi}$$

(Beirlant, Györfi and Lugosi (1994), Berlinet, Devroye and Györfi (1995) and Beirlant and Györfi (1998)).

If μ and ν are absolutely continuous with respect to a σ-finite measure λ with densities f and g respectively, then the I-divergence of μ and ν becomes the so-called divergence between f and g, i.e.

$$I(\mu,\nu) = \int_{\mathcal{R}^d} f(x) \log \frac{f(x)}{g(x)} \lambda(dx) = D(f,g).$$

Theorem 5.2 implies a negative result on density estimation for a dominated class:

Theorem 5.7. (GYÖRFI AND VAN DER MEULEN (1994)) *Given any sequence of density estimators $\{f_n\}$ there always exists a density f with finite differential entropy*

$$H(f) = -\int f(x)\log f(x)dx$$

and with arbitrary many derivatives such that

$$D(f, f_n) = \infty \ a.s.$$

Barron (1988) was one of the first to consider the problem of estimating a probability density function such that the density estimate is consistent in I-divergence. His results were generalized in Barron, Györfi and van der Meulen (1992) showing that if one imposes a certain condition on the class of distributions from which we are estimating the unknown one, namely that there exists a known probability measure ν such that $I(\mu, \nu) < \infty$, then one can construct a distribution estimator which is a.s. consistent in information divergence for all distributions in the class. As it is well-known, the condition $I(\mu, \nu) < \infty$ implies that μ is absolutely continuous with respect to ν, so it is reasonable to create a distribution estimate, which has a density with respect to ν. Introduce the notation

$$f(x) = \frac{d\mu}{d\nu}(x)$$

and for a partition $\mathcal{P}_n = \{A_{n,1}, A_{n,2}, \ldots, A_{n,m_n}\}$, $0 < m_n < n$, $n \geq 2$, assume that $\nu(A_{n,i}) \geq h_n = 1/m_n$. For a given sequence $\{a_n\}$, $0 < a_n < 1$ with

$$\lim_{n\to\infty} a_n = 0$$

consider the following density estimate:

$$f_n(x) = (1 - a_n)\mu_n(A_n(x))/h_n + a_n, \tag{5.12}$$

For the choice

$$a_n = \frac{1}{nh_n + 1} \tag{5.13}$$

f_n becomes a density estimator introduced by Barron (1988).

Theorem 5.8. (BARRON, GYÖRFI AND VAN DER MEULEN (1992)) *If $I(\mu, \nu) < \infty$ and*

$$\lim_{n\to\infty} h_n = 0, \quad \lim_{n\to\infty} nh_n = \infty.$$

and

$$\limsup \frac{1}{nh_n a_n} \leq 1$$

then

$$\lim_{n\to\infty} D(f, f_n) = 0 \ a.s.$$

The asymptotic normality of I-divergence of Barron's estimate has been proved by Berlinet, Györfi and van der Meulen (1997).

Berlinet (1995) made a comprehensive review on the asymptotic normality of the various global errors for nonparametric estimates.

5.4 Choosing Between Two Densities

Consider the following simple situation: g_n and f_n are two density estimates, and we must select the best one, that is, $\arg\min(\int|f_n - f|, \int|g_n - f|)$. More precisely, given the sample $X_1, \ldots, X_n$ distributed according to density f, we are asked to construct a density estimate ϕ_n such that

$$\int|\phi_n - f| \approx \min\left(\int|f_n - f|, \int|g_n - f|\right).$$

This simple problem turns out to be surprisingly difficult, even if the estimates f_n and g_n are fixed densities, not depending on the data.

The solution we propose here is fundamental, and is at the heart of the matter. First we introduce some notation. The empirical measure $\mu_n(A)$ of $\mu(A) \stackrel{\text{def}}{=} \int_A f$ is the measure that gives mass $\mu_n(A) = (1/n)\sum_{i=1}^{n} \mathbb{I}_{[X_i \in A]}$ to set A. The set $A = A(f_n, g_n) = \{x : f_n(x) > g_n(x)\}$ will be called the *Scheffé set* for the ordered pair (f_n, g_n), as we recall Scheffé's identity,

$$\int|f_n - g_n| = 2\int(f_n - g_n)_+ = 2\int_A f_n - 2\int_A g_n,$$

valid whenever $\int f_n = \int g_n = 1$.

We define the *Scheffé estimate* f_n^* as follows:

$$f_n^* = \begin{cases} f_n & \text{if } |\int_A f_n - \mu_n(A)| < |\int_A g_n - \mu_n(A)|, \\ g_n & \text{otherwise.} \end{cases}$$

Note that this is not a symmetric definition in f_n and g_n! The situations in which we wish to carry out a selection are innumerable: we may want to pick the best of two bandwidths in kernel estimates, or the best of two summation sizes in series or wavelet estimates, or the best of a histogram and a kernel estimate, or the best of a parametric gamma density estimate and a nonparametric unimodal Grenander estimate. Scheffé's theorem was at the basis of our approach. The idea for selecting sets of the form $f > g$ when dealing with L_1 norms in density estimation was grabbed from Yatracos (1985) and developed by Devroye and Lugosi (1996, 1997) in the context of kernel density estimation.

Theorem 5.9. *Let f_n and g_n be two density estimates with $\int f_n = \int g_n = 1$. For the Scheffé* *estimate f_n^*, we have*

$$\int |f_n^* - f| \le 3\min\left(\int |f_n - f|, \int |g_n - f|\right) + 4\max_{A\in\mathcal{A}}\left|\int_A f - \mu_n(A)\right|,$$

where $\mathcal{A} = \{\{f_n > g_n\}, \{g_n > f_n\}\}$.

PROOF. Denote by ξ_n the best density estimate:

$$\xi_n = \begin{cases} f_n & \text{if } \int |f_n - f| < \int |g_n - f|, \\ g_n & \text{otherwise.} \end{cases}$$

Clearly,

$$\begin{aligned} \int |f_n^* - f| &\le \int |\xi_n - f| + \int |\xi_n - f_n^*| \\ &= \int |\xi_n - f| + \int |f_n - g_n| \mathbb{I}_{[\xi_n \equiv f_n, f_n^* \equiv g_n]} \\ &\quad + \int |f_n - g_n| \mathbb{I}_{[\xi_n \equiv g_n, f_n^* \equiv f_n]} \\ &\stackrel{\text{def}}{=} I + II + III. \end{aligned}$$

We consider II. Let $E = [\xi_n \equiv f_n, f_n^* \equiv g_n]$ and $A = A(f_n, g_n)$:

$$\begin{aligned} II &= \int |f_n - g_n| \mathbb{I}_E \\ &= 2\left|\int_A f_n - \int_A g_n\right| \mathbb{I}_E \\ &= 2\left|\int_A f_n - \mu_n(A)\right| \mathbb{I}_E + 2\left|\int_A g_n - \mu_n(A)\right| \mathbb{I}_E \\ &\le 4\left|\int_A f_n - \mu_n(A)\right| \mathbb{I}_E \\ &\le 4\int_A |\xi_n - f| \mathbb{I}_{[E]} + 4\left|\int_A f - \mu_n(A)\right| \mathbb{I}_{[E]} \\ &\le 2\int |\xi_n - f| + 4\left|\int_A f - \mu_n(A)\right|. \end{aligned}$$

For the middle step above, consider the cases $\int_A f_n > \mu_n(A) > \int_A g_n$ and $\int_A f_n > \int_A g_n > \mu_n(A)$ separately. Similarly, by switching f_n and g_n, we see that

$$III \le 2\int |\xi_n - f| + 4\left|\int_{A(g_n, f_n)} f - \mu_n(A(g_n, f_n))\right|.$$

Thus,

$$\int |f_n^* - f| \le 3\min\left(\int |f_n - f|, \int |g_n - f|\right) + 4\sup_{A\in\mathcal{A}}\left|\int_A f - \mu_n(A)\right|.$$

□

The Scheffé estimate thus has an error that is within

$$E_n \stackrel{\text{def}}{=} 4\sup_{A\in\mathcal{A}}\left|\int_A f - \mu_n(A)\right|$$

of three times the best possible error. Note that this exceedance E_n only depends upon f_n and g_n through the shape of the set A.

EXAMPLE: CHOOSING BETWEEN TWO NORMAL DENSITIES. Assume that f_n and g_n are both normal densities (perhaps determined by the data), and we must decide between them. Then A is either of the form (a,b) or $[a,b]^c$, and thus, $\mathcal{A}$ is contained in the collection $\mathcal{B}_2$ of unions of two intervals. But then,

$$\sup_{A\in\mathcal{B}_2}\left|\int_A f - \mu_n(A)\right| \le 2\sup_x |F_n(x) - F(x)|,$$

where F is the distribution function for F, and F_n is the empirical distribution function. But we know from Massart's tightening (1990) of the Dvoretzky–Kiefer–Wolfowitz inequality (1956) that

$$\mathbb{E}\left\{\sup_x |F_n(x) - F(x)|\right\} \le \frac{1}{\sqrt{n}}.$$

Therefore, if we use the empirical measure μ_n, using *the same data* that were used in obtaining f_n and g_n, regardless of the dependence of f_n and g_n upon these data, we have

$$\mathbb{E}\left\{\int |f_n^* - f| - 3\min\left(\int |f_n - f|, \int |g_n - f|\right)\right\} \le \frac{8}{\sqrt{n}}$$

for all n. While the coefficient "3" may not be acceptable to some, we should stress nevertheless the simplicity of the procedure. Indeed, the set $A = \{f_n > g_n\}$ is very easily determined by the solution of a quadratic equation, and $\mu_n(A)$ is indeed easy to compute. Furthermore, the inequality above applies to any normal density estimate, with arbitrary dependence on the data, and to all densities.

Consider a more sophisticated situation, in which a normal density estimate f_n is challenged by a histogram estimate g_n having k bins (of data-dependent widths, and data-dependent heights in the bins). Then both $A = A(f_n, g_n)$ and $A(g_n, f_n)$ consist of unions of at most $k+2$ disjoint intervals (we call the collection of such sets $\mathcal{B}_{k+2}$). Thus, by the argument presented above,

$$\sup_{A\in\mathcal{B}_{k+2}}\left|\int_A f - \mu_n(A)\right| \le 2(k+2)\sup_x |F_n(x) - F(x)|.$$

Therefore, by Massart's inequality,

$$\begin{aligned}\mathbb{E}\left\{\int |f_n^* - f| - 3\min\left(\int |f_n - f|, \int |g_n - f|\right)\right\} &\leq 8(k+2)\mathbb{E}\left\{\sup_x |F_n(x) - F(x)|\right\} \\ &\leq \frac{8(k+2)}{\sqrt{n}}.\end{aligned}$$

This inequality, while useful, is much too weak in general, when k grows with n. This may be explained by noting its generality: it remains valid if a mean-spirited adversary is allowed to select the bins and bin heights in the worst possible manner so as to produce the worst possible selection f_n^*. If $\mu_n(A)$ in the definition of f_n^* is based upon an independently drawn sample of size m, then in the two examples above, $\mathcal{A}$ would consist of two sets, and by the Cauchy–Schwarz inequality, for any fixed $A \in \mathcal{A}$:

$$\begin{aligned}\mathbb{E}\left\{\left|\mu_n(A) - \int_A f\right| \middle| X_1,\ldots,X_n\right\} &\leq \sqrt{\mathbb{E}\left\{\left(\mu_n(A) - \int_A f\right)^2 \middle| X_1,\ldots,X_n\right\}} \\ &= \mathbf{Var}\{\mu_n(A)|X_1,\ldots,X_n\} \leq \sqrt{\frac{1}{4m}}.\end{aligned}$$

Conditioning and then unconditioning, we have

$$\mathbb{E}\left\{\int |f_n^* - f| - 3\min\left(\int |f_n - f|, \int |g_n - f|\right)\right\} \leq 4\sqrt{\frac{1}{m}}.$$

We must think of m as a part of the data set aside for testing purposes, and will call these data test data. Typically, of course, m is much smaller than n. In the former example (normal versus normal), this test data approach does not help. However, in the second example (normal versus any-histogram), the obtained bound is better than the previous one as soon as $m/n \geq 1/(k+2)^2$. So we distinguish between μ_n based on the original data and μ_n based on test data. Mostly, for simple parametric density estimates, the original data approach is best.

THE FACTOR THREE. One may wonder whether the factor three is necessary. It turns out that it cannot be removed as the following result from Devroye and Lugosi (2001) shows.

Theorem 5.10. *For any $t < 1$, $\varepsilon > 0$, there exist three densities, g, h, f, such that if $f_n = h$, $g_n = g$, then the Scheffé estimate f_n^* based upon a sample of size n drawn from f has the following lower bound:*

$$\frac{\lim_{n\to\infty}\mathbb{E}\{\int |f_n^* - f|\}}{\min\left(\int |g - f|, \int |h - f|\right)} > 2 - \varepsilon.$$

Furthermore, $\lim_{n\to\infty}\mathbb{E}\int |f_n^* - f| = t$.

WHY NOT MAXIMUM LIKELIHOOD? In spite of the necessity of a constant factor larger than one, the Scheffé estimate is the only one we know for which the estimate is guaranteed to be within a constant factor of the smaller error plus a uniformly controllable term. One may wonder why we decided not to use another method of selection, such as the maximum likelihood. This is simply because the maximum likelihood criterion, even when applied under idealized circumstances, does not minimize L_1. Consider the real line, with two density estimates: $f_0 = \frac{1}{2}\mathbb{I}_{[-1,1]}$ and $g_\delta = \mathbb{I}_{[\delta,1+\delta]}$, where $\delta \in (0,1)$ is to be picked. Assume that the true density f is the uniform density on $[0,1]$. If the data are $X_1,\ldots,X_n$, the maximum likelihood choice ϕ_n would be defined by

$$\phi_n \stackrel{\text{def}}{=} \arg\max\left(\prod_{i=1}^n f_0(X_i), \prod_{i=1}^n g_\delta(X_i)\right).$$

Clearly, everything depends upon N, the number of data points in $[0,\delta)$, a binomial (n,δ) random variable. We note that g_δ is picked if and only if $N=0$. Also, $\int |f_0 - f| = 1$, $\int |g_\delta - f| = 2\delta$. Let $\delta < \frac{1}{2}$ so that g_δ is the better estimate. Clearly,

$$\begin{aligned}\mathbb{E}\left\{\int |\phi_n - f|\right\} - \int |g_\delta - f| &= (1-2\delta)\mathbb{P}\{N>0\}\\ &= (1-2\delta)(1-(1-\delta)^n)\\ &\to 1-2\delta,\end{aligned}$$

as $n \to \infty$. Thus, while for small δ the expected error for the best density is a respectable 2δ, the maximum likelihood estimate makes, with probability tending to one, a catastrophic blunder.

WHY NOT LEAST SQUARES? Others may want to minimize $\int (f_n - f)^2$, the L_2 distance. But even if we were given f, and even if that f were square integrable, we might still make catastrophic choices. Consider this situation: f is once again the uniform density on $[0,1]$, and we are indeed given this information. Let ε be a small positive number, and assume that both density estimates ignore the data: f_ε is our catastrophic candidate—it is $1-\varepsilon$ on $[0,1]$, and ε^3 on $[1, 1+1/\varepsilon^2]$ (a long skinny tail, if you wish). Now, $g_\varepsilon = 1$ on $[-\varepsilon^2, 1-\varepsilon^2]$. Verify that $\int |g_\varepsilon - f| = 2\varepsilon^2 < 2\varepsilon = \int |f_\varepsilon - f|$, so that the choices are not even close! However, $\int (f_\varepsilon - f)^2 = \varepsilon^2 + \varepsilon^4 < 2\varepsilon^2 = \int (g_\varepsilon - f)^2$, so that minimizing the L_2 error, even with f given, picks the wrong density from the set $\{f_\varepsilon, g_\varepsilon\}$! Therefore, for universal properties in density estimation, criteria that are based on the square integrated distance are doomed. In contrast, the Scheffé estimate does not make such bad mistakes.

SELECTION FROM A FINITE SET. Consider now the selection problem with k candidates f_{ni}, $1 \le i \le k$, such that $\int f_{ni} = 1$ for all i. Let A_{ij}, $i<j$, denote the Scheffé set $A(f_{ni}, f_{nj}) = \{x : f_{ni} > f_{nj}\}$.

We say that f_{ni} wins against f_{nj} when

$$\left|\int_{A_{ij}} f_{ni} - \mu_n(A_{ij})\right| < \left|\int_{A_{ij}} f_{nj} - \mu_n(A_{ij})\right|.$$

Of course, for each pair $i < j$, if f_n^* denotes the winner, we have by Theorem 5.9:

$$\begin{aligned}\int |f_n^* - f| &\\ &\leq 3\min\left(\int |f_{ni} - f|, \int |f_{nj} - f|\right) + 4 \sup_{A\in\{A_{ij},A_{ji}\}} \left|\int_A f - \mu_n(A)\right| \\ &= 3\min\left(\int |f_{ni} - f|, \int |f_{nj} - f|\right) + 4\Delta,\end{aligned}$$

where

$$\Delta = \sup_{A\in\{A_{ij},A_{ji};1\leq i<j\leq k\}} \left|\int_A f - \mu_n(A)\right|.$$

Let us thus run a little competition with $k(k-1)/2$ matches, one for each ordered pair $i < j$. For each f_{ni}, we total the number of wins, and declare the density estimate with the maximum number of wins the Scheffé tournament winner. Ties are broken by picking the smallest index.

Theorem 5.11. *For the Scheffé tournament winner f_n^*, we have*

$$\int |f_n^* - f| \leq 9 \min_i \int |f_{ni} - f| + 16\Delta.$$

PROOF. Let $m = \min_i \int |f_{ni} - f|$. Group the density estimates as follows: in group 0, we place all those with $\int |f_{ni} - f| = m$; in group 1, those with $\int |f_{ni} - f| \in (m, 3m+4\Delta]$; in group 2, those with $\int |f_{ni} - f| \in (3m+4\Delta, 9m+16\Delta]$; in group 3, those with $\int |f_{ni} - f| > 9m+16\Delta$. Let the number of density estimates in each group be denoted by n_0, n_1, n_2, and n_3, respectively. From Theorem 5.9, it is clear that any group 0 estimate must win against any group 2 or 3 estimate, and thus, their number of wins is at least $n_2 + n_3$. Any group 3 estimate must lose to any group 0 or 1 estimate, and thus, its number of wins is at most $n_2 + n_3 - 1$. Therefore, the Scheffé tournament winner cannot be from group 3. □

We may also declare another winner. We define

$$\Delta_i = \sup_{A\in\mathcal{A}} \left|\int_A f_{ni} - \mu_n(A)\right|,$$

where $\mathcal{A} = \{A_{ij}, A_{ji} : 1 \leq i < j \leq k\}$. The minimum distance estimate ψ_n is that f_{ni} of smallest index that minimizes Δ_i. It is called a minimum distance estimate because it minimizes the distance to the empirical measure in a metric that is reminiscent of the total variation or L_1

distance—it would be the total variation distance if $\mathcal{A}$ were replaced by the class of Borel sets, $\mathcal{B}$. Only, the total variation distance

$$\sup_{B\in\mathcal{B}}\left|\int_B f-\mu_n(B)\right|=1$$

for any density f and any empirical measure μ_n: just let $B=\{X_1,\ldots,X_n\}$, for example. If each computation of $\int_A f_{ni}$ requires one time unit, then the computation of the Scheffé tournament winner requires $k(k-1)/2$ time units, while that for the minimum distance estimate requires $k^2(k-1)/2$ time units—it is about k times computationally more intensive. This extra cost of course draws benefits, as is shown in the next theorem.

Theorem 5.12. *For the minimum distance estimate ψ_n, we have*

$$\int|\psi_n-f|\le 3\min_i\int|f_{ni}-f|+4\Delta.$$

PROOF. Let $\psi_n=f_{ni}$ and let f_{nj} be any density estimate minimizing $\int|f_{n\ell}-f|$ over all ℓ. Assume that $j\neq i$. Then, clearly,

$$\int|\psi_n-f|\le\int|f_{nj}-f|+\int|f_{ni}-f_{nj}|.$$

Now, assuming without loss of generality that $i<j$:

$$\begin{aligned}
\int|f_{ni}-f_{nj}| &= 2\sup_{A\in\{A_{ij},A_{ji}\}}\left|\int_A f_{ni}-\int_A f_{nj}\right|\\
&\le 2\sup_{A\in\mathcal{A}}\left|\int_A f_{ni}-\int_A f_{nj}\right|\\
&\le 2\sup_{A\in\mathcal{A}}\left|\int_A f_{ni}-\mu_n(A)\right|+2\sup_{A\in\mathcal{A}}\left|\int_A f_{nj}-\mu_n(A)\right|\\
&\le 4\sup_{A\in\mathcal{A}}\left|\int_A f_{nj}-\mu_n(A)\right| \quad (\text{by definition of } \psi_n=f_{ni})\\
&\le 4\sup_{A\in\mathcal{A}}\left|\int_A f_{nj}-\int_A f\right|+4\sup_{A\in\mathcal{A}}\left|\int_A f-\mu_n(A)\right|\\
&\le 4\sup_{B\in\mathcal{B}}\left|\int_B f_{nj}-\int_B f\right|+4\Delta\\
&= 2\int|f_{nj}-f|+4\Delta
\end{aligned}$$

which together with the first inequality is all that is needed. □

SELECTING A NORMAL ESTIMATE. If we continue our examples, we note that if we are given any number k of normal density estimates, then the class $\mathcal{A} \subseteq \mathcal{B}_2$, the class of intervals or complements of intervals. Thus, we have

$$\sup_{A\in\mathcal{B}_2}\left|\int_A f-\mu_n(A)\right| \le 2\sup_x |F_n(x)-F(x)|,$$

where F is the distribution function for F, and F_n is the empirical distribution function, we have, for $t>0$, if μ_n is the standard empirical measure,

$$\begin{aligned}
\mathbb{E}\left\{\sup_{A\in\mathcal{A}}\left|\int_A f-\mu_n(A)\right|\right\} &\le \mathbb{E}\left\{\sup_{A\in\mathcal{B}_2}\left|\int_A f-\mu_n(A)\right|\right\}\\
&\le \mathbb{E}\left\{2\sup_x |F_n(x)-F(x)|\right\}\\
&\le \frac{2}{\sqrt{n}}.
\end{aligned}$$

Thus, for the minimum distance estimate ψ_n:

$$\mathbb{E}\left\{\int|\psi_n-f|-3\min_i\int|f_{ni}-f|\right\}\le\frac{8}{\sqrt{n}}.$$

Note, in particular, that this bound does not depend upon k. In fact, we may select the best among k arbitrary normal density estimates, and regardless of the value of k, we will obtain the same performance guarantees. In fact, Theorem 5.12 applies also to the infinite selection problem, although computations become a nightmare. With a selection from a possibly uncountably infinite class of densities, we can no longer index by integers i, and we can no longer be sure that minima are attained. The set-up must therefore be slightly modified. Also, we can no longer run Scheffé tournaments. Luckily, the minimum distance method remains valid.

5.5 The Minimum Distance Estimate

Let our density estimates be parametrized by $\theta \in \Theta$, where θ could represent the mean–variance pair in a normal density estimate, or the bandwidth in a kernel estimate, or the number of terms in a wavelet estimate. Let $f_{n,\theta}$ denote the density estimate with parameter θ. Let

$$\mathcal{A}=\{\{f_{n,\theta}>f_{n,\theta'}\}:\theta\in\Theta,\theta'\in\Theta,\theta\neq\theta'\}.$$

Define

$$\Delta_\theta=\sup_{A\in\mathcal{A}}\left|\int_A f_{n,\theta}-\mu_n(A)\right|.$$

We define *the minimum distance estimate* ψ_n as any density estimate selected from among those density estimates $f_{n,\theta}$ with

$$\Delta_\theta < \inf_{\theta^* \in \Theta} \Delta_{\theta^*} + 1/n.$$

The $1/n$ here is added to ensure the existence of such a density estimate.

Theorem 5.13. *Let $\{f_{n,\theta} : \theta \in \Theta\}$ be an arbitrary class of density estimates satisfying $\int f_{n,\theta} = 1$. For the minimum distance estimate ψ_n as defined above for an infinite selection problem, we have*

$$\int |\psi_n - f| \le 3 \inf_{\theta \in \Theta} \int |f_{n,\theta} - f| + 4\Delta + \frac{3}{n},$$

where $\Delta = \sup_{A \in \mathcal{A}} |\int_A f - \mu_n(A)|$.

PROOF. Mimic the proof of Theorem 5.12. The details are left to the reader as an easy exercise. □

The theorem above makes a crucial connection between the error of the selected density estimate and the supremum error Δ for empirical probability measures over certain classes of sets. This permits us thus to use the rich theory of the uniform convergence of empirical measures initiated by Vapnik and Chervonenkis to density estimation.

Define $\mathcal{F} = \{f_n = f_\theta : \theta \in \Theta\}$. The Yatracos class $\mathcal{A}$ is the collection of all sets of the form $\{x : f_\theta(x) > f_{\theta'}(x)\}$, $\theta \ne \theta'$, and μ_n is the standard empirical measure based on the sample $X_1, \ldots, X_n$. Then Theorem 5.12 asserts that for all f:

$$\int |f_n - f| \le 3 \inf_{\theta \in \Theta} \int |f_\theta - f| + 4 \sup_{A \in \mathcal{A}} \left| \int_A f - \mu_n(A) \right| + \frac{3}{n}.$$

In particular, if $f \in \mathcal{F}$, then

$$\int |f_n - f| \le 4 \sup_{A \in \mathcal{A}} \left| \int_A f - \mu_n(A) \right| + \frac{3}{n}$$

The right-hand side of the last inequality is of the order of $n^{-1/2}$ whenever $\mathcal{A}$ has a finite Vapnik–Chervonenkis dimension. The problem from this point on is purely combinatorial. In the rest of this section we work out examples for different classes.

SERIES ESTIMATES. Let $\psi_1, \ldots, \psi_k$ be fixed basis functions from $\mathcal{R}^d$ to $\mathcal{R}$ such that $\int \psi_i = r_i$ for all $1 \le i \le k$. We define the class $\mathcal{F}_k$ as the class of all linear combinations of the basis functions $f_\theta = \sum_{i=1}^k a_i \psi_i$ with coefficient vector $\theta = (a_1, \ldots, a_k)$ satisfying $\sum_{i=1}^k a_i r_i = 1$. The latter condition assures that all candidates f_θ have integral equal to one, a necessary condition to apply Theorem 5.12 for the performance of the minimum distance estimate. In this case, all sets $A_{\theta,\theta'} = \{x : f_\theta(x) > f_{\theta'}(x)\}$ are of the form

$$\left\{x : \sum_{i=1}^{k} b_i \psi_i(x) > 0\right\},$$

where $\theta' = (a'_1, \ldots, a'_k)$, and $b_i = a_i - a'_i$. Lemma 1.6 shows that the Vapnik–Chervonenkis dimension of $\mathcal{A}$ is at most k, so invoking section 1.4.5, we obtain, for the minimum distance estimate f_n:

$$\mathbb{E}\int |f_n - f| \leq 3 \inf_{g \in \mathcal{F}_k} \int |g - f| + c\sqrt{\frac{k}{n}} + \frac{3}{n},$$

where c is a universal constant. For a survey on orthogonal series estimates and their L_1 aspects, we refer to Chapter 12 of Devroye and Györfi (1985). Surveys of approximation properties of orthogonal series may be found in Butzer and Nessel (1971).

PARAMETRIC ESTIMATES: EXPONENTIAL FAMILIES. Assume that the data are believed to be drawn from a normal density on the real line. This leads us to the construction of the class

$$\mathcal{F} = \left\{ f_{m,\sigma}(x) = \frac{1}{\sigma\sqrt{2\pi}} e^{-(x-m)^2/2\sigma^2} : m \in \mathcal{R}, \sigma > 0 \right\}.$$

To determine the performance of the minimum distance estimate, we must estimate the Vapnik–Chervonenkis dimension of the corresponding Yatracos class $\mathcal{A}$. Note that every set in $\mathcal{A}$ may be written in the form of $\{x : ax^2 + bx + c \geq 0\}$ for some coefficients $a, b, c \in \mathcal{R}$, that is, $\mathcal{A}$ contains either closed intervals, or the union of two closed half-infinite intervals. The Vapnik–Chervonenkis dimension of this class is easily seen to be three.

The argument above may be significantly generalized as follows. A family $\mathcal{F}$ of densities on $\mathcal{R}^d$ is called an *exponential family* if each density in $\mathcal{F}$ may be written in the form

$$f_\theta(x) = c\alpha(\theta)\beta(x)e^{\sum_{i=1}^{k} \pi_i(\theta)\psi_i(x)},$$

where θ belongs to some parameter set Θ, $\psi_1, \ldots, \psi_k : \mathcal{R}^d \to \mathcal{R}$, $\beta : \mathcal{R}^d \to [0, \infty)$, $\alpha, \pi_1, \ldots, \pi_k : \Theta \to \mathcal{R}$ are fixed functions, and c is a normalizing constant. Examples of exponential families include classes of Gaussian, gamma, beta, Rayleigh, and Maxwell densities.

Theorem 5.14. *Let $\mathcal{F}$ be an exponential family of densities defined as above, and let f_n be the minimum distance estimate based on this class. Then for any $f \in \mathcal{F}$:*

$$\mathbb{E}\int |f_n - f| \leq c\sqrt{\frac{k+1}{n}} + \frac{3}{n},$$

where c is a universal constant. Moreover, for any density f:

$$\mathbb{E}\int |f_n - f| \leq 3 \inf_{\theta \in \Theta} \int |f_\theta - f| + c\sqrt{\frac{k+1}{n}} + \frac{3}{n}.$$

PROOF. Note that $f_\theta(x) > f_{\theta'}(x)$ if and only if

$$\sum_{i=1}^{k}(\pi_i(\theta) - \pi_i(\theta'))\psi_i(x) + \log\frac{\alpha(\theta)}{\alpha(\theta')} > 0.$$

The class of functions appearing on the left-hand side spans a $(k+1)$-dimensional vector space, and so Lemma 1.6 implies that the Vapnik–Chervonenkis dimension of $\mathcal{A}$ is at most $k+1$. □

NEURAL NETWORK ESTIMATES. A feed-forward one-hidden-layer *neural network* (or simply neural network) with k hidden nodes is a function $\phi : \mathcal{R}^d \to \mathcal{R}$ of the form

$$\phi(x) = \sum_{i=1}^{k} a_i\sigma(b_i^T x + c_i),$$

where $\sigma : \mathcal{R} \to \mathcal{R}$ is a fixed *activation function*, $a_i, c_i \in \mathcal{R}$, and $b_i \in \mathcal{R}^d$, $i = 1,\ldots,k$, are the parameters of the network, and $b_i^T x$ denotes the usual inner product of the vectors b_i and $x \in \mathcal{R}^d$. In order to generate a class eligible for the minimum distance method, we must require that $\int\sigma < \infty$. Define $\mathcal{F}$ as the set of neural networks ϕ. Once again, to obtain a performance bound for the minimum distance density estimate, we need upper bounds for the Vapnik–Chervonenkis dimension of the associated Yatracos class. Each set in the class is of the form

$$\left\{x : \sum_{i=1}^{k}\left(a_i\sigma(b_i^T x + c_i) - a_i'\sigma(b_i'^T x + c_i')\right) > 0\right\}.$$

Observe that $\mathcal{A}$ is a subclass of the class of all sets of the form $\{x : \psi(x) > 0\}$, where ψ is a neural network of $2k$ hidden nodes. Computation of the Vapnik–Chervonenkis dimension of such classes of sets has been the subject of intensive research in pattern recognition and learning theory. It is known that the Vapnik–Chervonenkis dimension may be finite or infinite, depending on the activation function σ. The monograph of Anthony and Bartlett (1999) contains, among other things, the following useful result.

Theorem 5.15. (ANTHONY AND BARTLETT, 1999) *Let $\Theta \subset \mathcal{R}^m$ be a parameter set, and consider a class of functions $g_\theta : \mathcal{R}^d \to \mathcal{R}$ parametrized by Θ. Assume that for each $\theta \in \Theta$ and $x \in \mathcal{R}^d$, $g_\theta(x)$ may be computed by at most t steps of an algorithm which in each step executes one of the following operations: the arithmetic operations $+, -, \cdot$, and $/$ on real numbers; the exponential function e^x on real numbers; an indicator function of the form $\mathbb{I}_{[x>a]}$, $\mathbb{I}_{[x\geq a]}$, $\mathbb{I}_{[x<a]}$, $\mathbb{I}_{[x\leq a]}$, $\mathbb{I}_{[x=a]}$, $\mathbb{I}_{[x\neq a]}$, for $a \in \mathcal{R}$. Then the Vapnik–Chervonenkis dimension V of the class of sets $\{\{x : g_\theta > 0\} : \theta \in \Theta\}$ is at most*

$$V \leq t^2 m(m + 19\log_2(9m)).$$

If the exponential function is evaluated at most q times for each pair (x,θ), then

$$V \leq m^2(q+1)^2 + 11m(q+1)(t + \log_2(9m(q+1))).$$

The above theorem guarantees the finiteness of the Vapnik–Chervonenkis dimension of the neural network class for a whole host of all activation functions, though in many cases it does not provide the best possible bound. Theorem 5.15 may be used in many other situations, not just for neural network classes. Related results appear in Anthony and Bartlett (1999), who build on work of Goldberg and Jerrum (1995), Khovanskii (1991), Karpinsky and Macintyre (1997), Macintyre and Sontag (1993), and Koiran and Sontag (1997). Several related results are also surveyed in Devroye, Györfi, and Lugosi (1996).

MIXTURE CLASSES, RADIAL BASIS FUNCTION NETWORKS. Consider first the class of all mixtures of k normal densities in $\mathcal{R}$, that is, the class of all densities of form

$$\phi(x) = \frac{1}{\sqrt{2\pi}} \sum_{i=1}^{k} \frac{p_i}{\sigma_i} e^{-(x-m_i)^2/2\sigma_i^2},$$

where $(p_1,\ldots,p_k)$ is a probability vector, $\sigma_1,\ldots,\sigma_k$ are positive numbers, and $m_1,\ldots,m_k$ are arbitrary real numbers. The corresponding Yatracos class contains all sets of the form

$$\begin{aligned}\{x : \phi(x) > \phi'(x)\} \\ = \left\{ x : \sum_{i=1}^{k} \left(\frac{p_i}{\sigma_i} e^{-(x-m_i)^2/2\sigma_i^2} - \frac{p_i'}{\sigma_i'} e^{-(x-m_i')^2/2\sigma_i'^2} \right) > 0 \right\}.\end{aligned}$$

This class of sets may be written in the form given in Theorem 5.15 with $m = 6k$, where the parameter set Θ is a subset of the $6k$-dimensional vector space of the parameters $p_i, p_i', m_i, m_i', \sigma_i, \sigma_i'$, $i = 1,\ldots,k$. Simple counting reveals that given x and the vector of parameters, each function may be computed with no more than $t = 17k$ operations admitted by Theorem 5.15. Hence, we obtain that the Vapnik–Chervonenkis dimension of the Yatracos class based on all mixtures of k univariate normal densities is bounded by a constant times k^4. We suspect that this estimate is loose. Nevertheless, the finiteness of the Vapnik–Chervonenkis dimension implies that the error of the minimum distance estimate is not more than a constant times $k^2/\sqrt{n}$, whenever the true density f is, in fact, a mixture of k normals.

The same argument may be generalized to the multivariate setting. Consider now the class $\mathcal{F}$ of all mixtures of k normal densities over $\mathcal{R}^d$:

$$\phi(x) = \sum_{i=1}^{k} \frac{p_i}{\sqrt{(2\pi)^d \det(\Sigma_i)}} e^{-\frac{1}{2}(x-m_i)^T \Sigma_i^{-1}(x-m)},$$

where $(p_1,\ldots,p_k)$ is a probability vector, $\Sigma_1,\ldots,\Sigma_k$ are positive definite $d \times d$ matrices, and $m_1,\ldots,m_k$ are arbitrary elements of $\mathcal{R}^d$. Then it is clear from Theorem 5.15 that the Vapnik–Chervonenkis dimension of the corresponding Yatracos class is finite, and it is a matter of straightforward counting to obtain an explicit upper bound. In any case, if f is a mixture of

k normal densities, and is estimated by the minimum distance estimate based on the class defined here, then

$$\mathbb{E}\left\{\int |f_n - f|\right\} \leq \sqrt{\frac{c_{k,d}}{n}},$$

where $c_{k,d}$ is a constant depending on k and d. Radial basis function networks originated in the work of Poggio and Girosi (1990) and were analyzed in a series of subsequent papers, such as in Krzyżak and Linder (1998).

MIXTURES OF EXPONENTIALS. Often one may obtain much sharper estimates by direct methods. As an illustration, consider the class $\mathcal{F}$ of all mixtures of k exponential densities (i.e., translations and scales of $e^{-x}, x \geq 0$). Again, it suffices to bound the Vapnik–Chervonenkis dimension of

$$\mathcal{A} = \{x : f(x) > g(x); f, g \in \mathcal{F}\}.$$

A member set in this class is thus of the form

$$\left\{x : \sum_{i=1}^{2k} a_i e^{-b_i x} I_{[x > c_i]} > 0\right\},$$

where $a_i, c_i \in \mathcal{R}$ and $b_i > 0$ are free parameters. Now clearly, we have at most $k+1$ intervals defined by the thresholds c_i. On the other hand, on each of these intervals, a set of the form $\{x : \sum_{i=1}^{2k} a_i e^{-b_i x} > 0\}$ defines at most $2k+1$ intervals (try showing this!). Therefore, each set in the class $\mathcal{A}$ is the union of at most $(2k+1)(k+1)$ intervals, and the Vapnik–Chervonenkis dimension of $\mathcal{A}$ is not more than $2(2k+1)(k+1) = O(k^2)$. Contrast this with the $O(k^4)$ bound obtainable by Theorem 5.15.

5.6 The Kernel Density Estimate

In this section, we get our first taste of real analysis, starting with some results on the approximations of functions in L_1. The problem is that f cannot be approximated in L_1 by μ_n, the empirical measure, as the total variation distance between any density f and any atomic measure (like μ_n) is 1. Thus, the approximation itself must have a density. The kernel estimate provides this: it smooths the empirical measure μ_n.

We define the convolution density $f * K$ as the density of $X + Y$, where X has density f and Y has density K on $\mathcal{R}^d$. Think of this as a perturbation Y applied to X. We note that

$$f * K(x) = \int f(z) K(x - z)\, dz.$$

The definition above also holds when f and/or K are absolutely integrable functions, and in that case we have

$$\int |f * K| \leq \int |f| \times \int |K|,$$

a fact that is easy to check by change of integration, and will be referred to as Young's inequality. The convolution operation too lowers the total variation distance: for any densities f,g and for any integrable function K:

$$\int |f*K-g*K| \le \int |K| \int |f-g|.$$

To see this, we merely apply Young's inequality:

$$\int |f*K-g*K| = \int |(f-g)*K| \le \int |f-g|*|K| = \int |f-g| \int |K|.$$

It is clear that if Y (with density K) is concentrated near 0, then $f*K$ should be close to f. Indeed, we have the following fundamental approximation theorem from real analysis.

Theorem 5.16. *Let K be an arbitrary integrable function on $\mathcal{R}^d$ (i.e., $\int |K| < \infty$), and let f be a density on $\mathcal{R}^d$. Denoting $K_h(x) = (1/h^d)K(x/h)$, $x \in \mathcal{R}^d, h > 0$, we have*

$$\lim_{h \downarrow 0} \int \left| f*K_h - f\int K \right| = 0.$$

PROOF. We may assume without loss of generality that $\int K \in \{0,1\}$. We will prove the statement when $\int K = 1$, leaving the $\int K = 0$ case as an easy exercise. Assume first that the statement is true for a dense subspace of functions g. Then

$$\begin{aligned}\int |f*K_h - f| &\le \int |f-g|*|K_h| + \int |f-g| + \int |g*K_h - g| \\ &\le \left(\int |K|+1\right) \int |f-g| + o(1)\end{aligned}$$

as $h \to 0$. Here we made use of Young's inequality:

$$\int |f*g| \le \int |f| \cdot \int |g|$$

valid for any integrable functions f,g. The first term on the right-hand side can be made as small as desired by the choice of g and the finiteness of $\int |K|$. So, we only need to prove the theorem for a dense subclass, such as the class of Lipschitz densities of compact support (cf. Lemma 5.1). Thus, let f be Lipschitz with constant C (i.e., $|f(x)-f(y)| \le C\|x-y\|$, $x,y \in \mathcal{R}^d$), and supported on $[-M,M]^d$ for finite M. Let $L = KI_A$ where I is the indicator function, and $A = [-r,r]^d$ is a large cube of our choice. Then, by Young's inequality,

$$\begin{aligned}
&\int |f*K_h - f| \\
&\leq \int \left| f*L_h - f\int L_h \right| + \int |f| \int |K_h - L_h| + \int |f*(K_h - L_h)| \\
&\leq \int\int |f(x-y) - f(x)||L_h(y)|\,dy\,dx + 2\int |K-L| \\
&\leq \int_{[-M-r,M+r]^d} \left(\int C\|y\||L_h(y)|dy \right) dx + 2\int |K-L| \\
&\leq (2M+2r)^d Crh\sqrt{d} \int |L| + 2\int |K-L| \\
&\leq (2M+2r)^d Crh\sqrt{d} \int |K| + 2\int |K-L| \\
&= o(1) + 2\int |K-L|,
\end{aligned}$$

which is as small as desired by choice of r (and thus L). □

If our goal is to find a density estimate for which $\int |f_n - f|$ is small, we might make use of the fact that $\int |f*K_h - f|$ is small when h is small and $\int K = 1$. Functions K with $\int |K| < \infty$ and $\int K = 1$ will be called kernels. Indeed, $f*K_h$ may in turn be approximated by $\mu_n * K_h$, where μ_n is the empirical measure. More explicitly, $\mu_n * K_h$ is

$$f_n(x) = \frac{1}{n}\sum_{i=1}^{n} K_h(x - X_i) = \frac{1}{nh^d}\sum_{i=1}^{n} K\left(\frac{x - X_i}{h}\right)$$

which is nothing but the kernel estimate first proposed and developed by Akaike (1954), Parzen (1962), and Rosenblatt (1956). We may pick h and K as a function of n and/or the data and write $h = h(X_1, \ldots, X_n)$, for example.

Theorem 5.17. (CONSISTENCY) *Let K be a fixed kernel, and let h depend on n only. If $h \to 0$ and $nh^d \to \infty$ as $n \to \infty$, then $\mathbb{E}\{\int |f_n - f|\} \to 0$.*

PROOF. Note that

$$\int |f_n - f| \leq \int |f*K_h - f| + \int |\mu_n * K_h - f*K_h|.$$

The first term on the right-hand side, also called the bias term, tends to 0 by Theorem 5.16. There are two tools that we will use repreatedly in the text. First of all, if ν is any probability measure, and K and L are kernels, then, by Young's inequality, $\int |\nu * K_h - \nu * L_h| \leq \int |K - L|$. This inequality applies in particular when ν is f and when it is μ_n. Thus,

$$\int |\mu_n * K_h - f*K_h| \leq \int |\mu_n * L_h - f*L_h| + 2\int |K-L|.$$

The last term can be made as small as desired by choice of L from among bounded kernels with support on a compact hypercube. The second trick uses the fact that $\int \nu * K_h = 1$ for any probability measure ν and kernel K. Thus,

$$\int |\mu_n * L_h - f * L_h| = 2\int (f * L_h - \mu_n * L_h)_+$$

and therefore

$$\begin{aligned}
&\mathbb{E}\left\{\int |\mu_n * L_h - f * L_h|\right\} \\
&= \int \mathbb{E}\{|\mu_n * L_h - f * L_h|\} \\
&= 2\int \mathbb{E}\{(f * L_h - \mu_n * L_h)_+\} \\
&= 2\int \min\left(f * L_h, \mathbb{E}\{(f * L_h - \mu_n * L_h)_+\}\right) \\
&\leq 2\int \min\left(f * L_h, \sqrt{\mathbb{E}\{(f * L_h - \mu_n * L_h)^2\}}\right) \\
&\leq 2\int \min\left(f, \sqrt{\mathbb{E}\{(f * L_h - \mu_n * L_h)^2\}}\right) + 2\int |f * L_h - f| \\
&= o(1).
\end{aligned}$$

The last step follows from $nh^d \to \infty$, the dominated convergence theorem, and Theorem 5.16. Here we used the fact that $\mu_n * L_h - f * L_h = \sum_{i=1}^n Z_i$, where the Z_i's are i.i.d. zero mean random variables that have variance bounded by $(L_h)^2 * f/n^2$. Thus, $\mathbb{E}\{(\mu_n * L_h - f * L_h)^2\}$ is bounded by $(L^2)_h * f/(nh^d)$, which is $(f\int L^2 + o(1))/(nh^d)$ at all Lebesgue points for x. □

The remarkable thing about Theorem 5.17 is that it is valid whenever the kernel $\int K = 1$. For example, if we take K to be the density that is uniformly distributed on the ball of radius 1 centered at $(1000, 0, \ldots, 0)$, then the average $f * K_h$ calculates an integral that does not even include the origin. Furthermore, it is not even necessary that K be a density. Also, there is no possiblity of improving the conditions $h \to 0$ and $nh^d \to \infty$ (Devroye, 1983; Devroye and Györfi, 1985). In some cases, we have bandwidths that depend upon the data. The proof of the following theorem is left as an exercise.

Theorem 5.18. *Let K be a fixed kernel. Let the bandwidth H be an arbitrary function of the data such that $H \to 0$ and $nH^d \to \infty$ in probability as $n \to \infty$ (i.e., for every $\varepsilon > 0$, $\lim_{n\to\infty} \mathbb{P}\{H + 1/(nH^d) > \varepsilon\} = 0$). Then $\mathbb{E}\{\int |f_n - f|\} \to 0$.*

Consider the quantity $\int |f_n - g|$ where g is an arbitrary function of x and may be substituted by f or $f * K_h$. This quantity is very stable in the sense that if one X_i changes value and the other

$n-1$ data points remain fixed (and denoting the new kernel estimate by f_n^*), then $\int |f_n - f_n^*| \leq 2\int |K|/n$. By the bounded difference inequality (Theorem 1.8), we thus have

$$\mathbb{P}\left\{\left|\int |f_n - g| - \mathbb{E}\int |f_n - g|\right| \geq t\right\} \leq 2e^{-nt^2/2(\int |K|)^2}, \; t > 0.$$

In particular, we have

Theorem 5.19. (DEVROYE, 1987, 1988A, 1991) *If* $\sqrt{n}\mathbb{E}\int |f_n - f| \to \infty$, *then*

$$\frac{\int |f_n - f|}{\mathbb{E}\int |f_n - f|} \to 1 \text{ in probability.}$$

The concentration inequality in Theorem 5.19 shows that $\int |f_n - f| - \mathbb{E}\int |f_n - f|$ is of the order of $O(1/\sqrt{n})$. In fact, $\sqrt{n}(\int |f_n - f| - \mathbb{E}\int |f_n - f|)$ tends to a normal limit (Csörgő and Horváth, 1988; Beirlant and Mason, 1995).

Theorem 5.19 thus establishes the relative stability of the L_1 error for kernel estimates. It implies that the expected L_1 error is a good measuring stick because the actual error never deviates substantially from it. We will see further on that for all nonnegative kernels, the $\sqrt{n}$ condition is satisfied.

CHOOSING THE BANDWIDTH. In bandwidth selection, one is interested in functions $H = H(X_1, \ldots, X_n)$ such that $\int |f_{n,H} - f|$ comes close to $\inf_h \int |f_{n,h} - f|$, where $f_{n,h}$ makes the dependence upon h explicit. A bandwidth H with the property that

$$\limsup_{n\to\infty} \frac{\mathbb{E}\{\int |f_{n,H} - f|\}}{\inf_h \mathbb{E}\{\int |f_{n,h} - f|\}} \leq C(f)$$

for some finite $C(f)$ is called a universal bandwidth. It is uniformly universal with constant C if we may replace $C(f)$ by C for all f. This problem has been attacked by many. However, many attempts can be discarded from the outset. Some authors suggested minimizing $\int (f_{n,h} - f)^2$ with respect to h. Even if f were known, and even if f were square integrable, the bandwidth thus selected would not be universally useful. For example, let us denote by H a data-dependent bandwidth for which $\int (f_{n,H} - f)^2 \sim \inf_h \int (f_{n,h} - f)^2$ (so that H is the L_2-optimal choice; for bounded densities, Stone (1984) shows how one can do this; Wegkamp (2000) provides a newer proof). Assume that K is the uniform density on $[-1,1]$. For the Cauchy density $1/(\pi(1+x^2))$, $H \sim c/n^{1/5}$ in probability, yet one can check that $\mathbb{E}\{\int |f_{n,H} - f|\}/\inf_h \mathbb{E}\{\int |f_{n,h} - f|\} \to \infty$ as the optimal bandwidth for L_1 is larger than $n^{-1/5}$. In other words, even within the class of ultra-smooth densities (such as the Cauchy density), minimizing L_2 is just the wrong thing to do—intuitively, squaring tends to squash errors in the tails and make them unimportant.

Others have attempted to pick h by maximum likelihood, for example by maximization of $\prod_{i=1}^n f_{n-1,i,h}(X_i)$, where $f_{n-1,i,h}$ is the kernel estimate based on the $n-1$ data points obtained after deleting X_i. While for densities with compact support on $\mathcal{R}$, the maximizing H is indeed consistent, i.e., $H + 1/(nH^d) \to 0$ in probability (Devroye and Györfi, 1985), there is again no relationship with the total variation criterion. If the density f has heavier than exponential tails, the method is not even consistent. The maximum likelihood cross-validation method was studied in detail by Broniatowski, Deheuvels and Devroye (1989).

Hall and Wand (1988) looked at the asymptotic expansion of $\mathbb{E}\{\int |f_{n,h} - f|\}$ and minimized the main asymptotic terms to obtain a recipe for h as a function of n, f and K. They then estimate the unknown quantity involving f from the data, and propose this as a plug-in bandwidth estimate. For sufficiently smooth and small-tailed densities, and for positive kernels, they were able to show that $\mathbb{E}\{\int |f_{n,H} - f|\} \sim \inf_h \mathbb{E}\{\int |f_{n,h} - f|\}$. However, their method, and all other plug-in methods we are aware of, are not universal bandwidths.

The double kernel method uses a pair of kernels, K and L, and picks $H = \arg\min_h \int |f_{n,h} - g_{n,h}|$, where $f_{n,h}$ and $g_{n,h}$ are the kernel estimates with kernels K and L, respectively. Assume that $d = 1$. If the characteristic functions of K and L do not coincide on an open interval about the origin, then the choice H is consistent (Devroye, 1989a). Furthermore, if K and L are symmetric, bounded kernels of compact support, $K \geq 0$, $\int x^2 L(x)\,dx = 0$, if both K and L are L_1-Lipschitz (i.e., $\int |K_1 - K_h| \leq C(h-1)$ for some $C < \infty$ and all $h > 1$, and similarly for L), and if f is absolutely continuous, and f' is absolutely continuous, $\int |f''| < \infty$ and $\int \sqrt{\sup_{|y|\leq 1} f(x+y)}\,dx < \infty$, then

$$\limsup_{n\to\infty} \frac{\mathbb{E}\{\int |f_{n,H} - f|\}}{\inf_h \mathbb{E}\{\int |f_{n,h} - f|\}} \leq \frac{1+\rho}{1-\rho}$$

where $\rho = 4\sqrt{\int L^2 / \int K^2}$ is a constant that can be made as small as desired by choice of L. The behavior outside the smoothness class described above is unknown: for example, we do not know whether the double kernel bandwidth is universal, let alone uniformly universal. The methodology studied in this lecture series and in Devroye and Lugosi (2001) is meant to fill the gap. It allows us to obtain a uniformly universal bandwidth with constant 3, and in fact, to obtain even stronger nonasymptotic uniformity:

$$\sup_f \frac{\mathbb{E}\{\int |f_{n,H} - f|\}}{\inf_h \mathbb{E}\{\int |f_{n,h} - f|\}} \leq 3 + o(1).$$

Note in particular that this result does not depend on the dimension, and that the bound is uniform over all f. Bandwidth selection methods are surveyed by Berlinet and Devroye (1994), Cao, Cuevas and González-Manteiga (1994), and Devroye (1997). A detailed study of the L_1 behavior of the kernel estimate in $\mathcal{R}^d$ is provided by Holmström and Klemelä (1992).

CHOICE OF THE KERNEL. The selection of the pair (h, K) may be tackled as one problem, and indeed, h may be absorbed into K as a scale factor. Asymptotic evidence for smooth densities shows that for large sample sizes, the shape of the optimal kernel is unique. For example, for

$\mathcal{R}^1$, classical L_2 theory (Watson and Leadbetter, 1963) shows that for L_2 errors, among all positive kernels, the Epanechnikov kernel (Epanechnikov, 1969) $K(x) = \max(3/4(1-x^2), 0)$ is best possible. For $\mathcal{R}^d$, Deheuvels (1977) showed the L_2 optimality of $c\max((1-\|x\|^2)^d, 1)$. For the L_1 error, there is evidence that the Epanechnikov kernel is also best among all positive kernels (see the discussion in the next section). For these reasons, authors have typically decoupled the choices of K and h, and picked K fixed as one of these asymptotically optimal kernels.

RATES OF CONVERGENCE. The rate of convergence of the L_1 error of the standard kernel estimate is well understood in $\mathcal{R}$ but much less so in $\mathcal{R}^d$. For simplicity, we stick to $\mathcal{R}$ in this section. In that case, we note the following:

A. $\inf_f \liminf_n \inf_h n^{2/5}\mathbb{E}\{\int |f_{nh} - f|\} \geq A(K)$ where $A(K) \geq 0.86$ is a function of K only, and is minimized for the Epanechnikov kernel $K(x) = 3/4(1-x^2)_+$ (Devroye and Penrod, 1984; see also Devroye and Györfi, 1985). The $n^{-2/5}$ rate is thus a universal lower bound beyond which we cannot go within the class of standard kernel densities.

B. There is no uniform rate of convergence: $\sup_f \inf_h \mathbb{E}\{\int |f_{nh} - f|\} = 2$.

C. There is no universal rate of convergence for individual densities: for any sequence $a_n \downarrow 0$, there exists a density f such that for all n large enough,

$$\inf_h \mathbb{E}\int |f_{nh} - f| > a_n.$$

See Birgé (1986), Devroye (1983, 1995).

5.7 Additive Estimates and Data Splitting

Assume that we are given a class of density estimates parametrized by $\theta \in \Theta$, such that $f_{n,\theta}$ denotes the density estimate with parameter θ. Our goal is to construct a density estimate f_n whose L_1 error is (almost) as small as that of the best estimate among the $f_{n,\theta}$, $\theta \in \Theta$. Applying the minimum distance estimate directly to this class is often problematic because of the dependence of each estimate in the class and the empirical measure μ_n. Consider, as a basic example, the class of kernel estimates

$$f_{n,h}(x) = \frac{1}{nh^d}\sum_{i=1}^{n} K\left(\frac{x - X_i}{h}\right)$$

parametrized by the smoothing factor $h \in (0, \infty)$, where K is a fixed nonnegative function with $\int K = 1$. It is easy to see that if

$$\mathcal{A}_\Theta = \mathcal{A} = \{\{f_{n,h} > f_{n,h'}\} : h, h' \in (0,\infty), h \neq h'\},$$

then

$$\sup_{A\in\mathcal{A}}\left|\int_A f_{n,h}-\mu_n(A)\right|=1$$

for all h, so the minimum distance method selects a degenarate estimate.

We remedy this problem by introducing artificial independence between the estimates and the empirical measure. This may be achieved by holding out m samples from the design of the density estimates, and using the empirical measure based on the held-out samples to construct the minimum distance estimate. More precisely, let $m<n$, and define $\mathcal{A}_\Theta$ as the *Yatracos class* of subsets of $\mathcal{R}^d$ (corresponding to the family of density estimates $f_{n,\theta}$, $\theta\in\Theta$) as the class of all sets of the form

$$A_{\theta_1,\theta_2}=\left\{x: f_{n-m,\theta_1}(x)>f_{n-m,\theta_2}(x)\right\},\ \theta_1,\theta_2\in\Theta.$$

We select a parameter θ_n from Θ by minimizing the distance

$$\Delta_\theta=\sup_{A\in\mathcal{A}_\Theta}\left|\int_A f_{n-m,\theta}-\mu_m(A)\right|.$$

over all $\theta\in\Theta$, where μ_m denotes the empirical measure defined by the subsample $X_{n-m+1},\ldots,X_n$. If the minimum does not exist, we select θ_n such that $\Delta_{\theta_n}<\inf_{\theta^*\in\Theta}\Delta_{\theta^*}+1/n$. Define $f_n=f_{n-m,\theta_n}$. Using Theorem 5.13 conditionally, we immediately obtain:

Theorem 5.20. *If* $\int f_{n-m,\theta}=1$ *for all* $\theta\in\Theta$, *then for the minimum distance estimate* f_n *as defined above, we have*

$$\int|f_n-f|\le 3\inf_{\theta\in\Theta}\int|f_{n-m,\theta}-f|+4\Delta+\frac{3}{n},$$

where

$$\Delta=\sup_{A\in\mathcal{A}_\Theta}\left|\int_A f-\mu_m(A)\right|.$$

In order to use Theorem 5.20, first we have to make sure that $\inf_{\theta\in\Theta}\int|f_{n-m,\theta}-f|$ is not much larger than $\inf_{\theta\in\Theta}\int|f_{n,\theta}-f|$, that is, holding out m samples does not hurt. In the next section we will see that for many important families of estimates, this is indeed the case. The second part of the analysis is then purely combinatorial, as upper bounds for the value of Δ may be obtained by bounding the shatter coefficients of the class $\mathcal{A}_\Theta$.

Many classical nonparametric density estimates are *additive* estimates, that is, they can be written in the form

$$g_n(x)=\frac{1}{n}\sum_{i=1}^n K(x,X_i),$$

where $K:\mathcal{R}^d\times\mathcal{R}^d\to\mathcal{R}$ is a measurable function, and $\int K(x,y)dx=1$ for all y. We say that the additive estimate g_n is *regular* if for each x, $\mathbb{E}\{|K(x,X)|\}<\infty$. Examples of additive estimates include the kernel, histogram, series, and wavelet estimates. Theorem 5.21 below is a straightforward extension of a slightly less general inequality in Devroye and Lugosi (1996).

Theorem 5.21. *Let Θ be a class of parameters, and assume that each density estimate $f_{n,\theta}(x) = (1/n)\sum_{i=1}^{n} K_\theta(x, X_i)$ is additive and regular. Denote $J_{n,\theta} = \int |f_{n,\theta} - f|$. If $m > 0$ is a positive integer such that $2m \leq n$, then*

$$\frac{\inf_{\theta\in\Theta} \mathbb{E}\{J_{n-m,\theta}\}}{\inf_{\theta\in\Theta} \mathbb{E}\{J_{n,\theta}\}} \leq 1 + \frac{2m}{n-m} + 8\sqrt{\frac{m}{n}}.$$

This means that by decreasing the sample size to $n-m$, the performance of the best estimate in the class cannot deteriorate by more than a constant factor. If m is small relative to n, the loss in the L_1 error is negligible. The proof uses some simple results about sums of independent random variables developed in Lemmas 5.4 through 5.6.

Lemma 5.4. *Let X and Y be independent random variables, and let $\mathbb{E}\{Y\} = 0$. Then $\mathbb{E}\{|X + Y|\} \geq \mathbb{E}\{|X|\}$.*

PROOF. We write $X = \mathbb{E}\{X + Y|X\}$, and use Jensen's inequality:

$$\mathbb{E}\{|X|\} = \mathbb{E}\{|\mathbb{E}\{X + Y|X\}|\} \leq \mathbb{E}\{|X + Y|\}.$$

□

Lemma 5.5. (KHINCHINE'S INEQUALITY) *Let $a_1, \ldots, a_n$ be real numbers, and let $\sigma_1, \ldots, \sigma_n$ be i.i.d. sign variables with $\mathbb{P}\{\sigma_1 = 1\} = \mathbb{P}\{\sigma_1 = -1\} = 1/2$. Then*

$$\mathbb{E}\left\{\left|\sum_{i=1}^{n} a_i\sigma_i\right|\right\} \geq \frac{1}{\sqrt{2}}\sqrt{\sum_{i=1}^{n} a_i^2}.$$

The best constant in Khintchine's inequality is due to Szarek (1976) (see also Haagerup, 1978). The basis of the proof given here is Lemma 7.1 in Devroye and Györfi (1985).

Compare this with the closely related upper bound

$$\mathbb{E}\left\{\left|\sum_{i=1}^{n} a_i\sigma_i\right|\right\} \leq \sqrt{\mathbb{E}\left\{\left(\sum_{i=1}^{n} a_i\sigma_i\right)^2\right\}} = \sqrt{\sum_{i=1}^{n} a_i^2}.$$

PROOF. The proof of the inequality may be found in Szarek (1976). Here we give a short proof with a suboptimal constant ($1/\sqrt{3}$ instead of $1/\sqrt{2}$). First note that for any random variable X with finite fourth moment,

$$\mathbb{E}\{|X|\} \geq \frac{\left(\mathbb{E}\{X^2\}\right)^{3/2}}{\left(\mathbb{E}\{X^4\}\right)^{1/2}}.$$

Indeed, for any $a > 0$, the function $1/x + ax^2$ is minimal on $(0,\infty)$ when $x^3 = 1/(2a)$. Thus,

$$\frac{x+ax^4}{x^2} \geq (2a)^{1/3} + \frac{a}{(2a)^{2/3}} = 3/2(2a)^{1/3}.$$

Replace x by $|X|$ and take expectations:

$$\mathbb{E}\{|X|\} \geq 3/2(2a)^{1/3}\mathbb{E}\{X^2\} - a\mathbb{E}\{X^4\}.$$

The lower bound, considered as a function of a, is maximized if we take $a = 1/2\left(\mathbb{E}\{X^2\}/\mathbb{E}\{X^4\}\right)^{3/2}$ Resubstitution yields the claim. Applying the above inequality for $X = \sum_{i=1}^n a_i\sigma_i$ gives

$$\mathbb{E}\left\{\left|\sum_{i=1}^n a_i\sigma_i\right|\right\} \geq \frac{\left(\sum_{i=1}^n a_i^2\right)^{3/2}}{\sqrt{\sum_{i=1}^n a_i^4 + 3\sum_{i\neq j} a_i^2 a_j^2}} \geq \frac{1}{\sqrt{3}}\sqrt{\sum_{i=1}^n a_i^2},$$

where we used $\sum_{i=1}^n a_i^4 + 3\sum_{i\neq j} a_i^2 a_j^2 \leq 3(\sum_{i=1}^n a_i^2)^2$. □

Finally, we generalize the above inequality for general sums of independent random variables:

Lemma 5.6. *Let $Y_1,\ldots,Y_n$ be i.i.d. zero mean random variables. Then*

$$\mathbb{E}\left\{\left|\sum_{i=1}^n Y_i\right|\right\} \geq \sqrt{\frac{n}{8}}\mathbb{E}\{|Y_1|\}.$$

PROOF. The proof uses symmetrization. We introduce the i.i.d. random variables $Y_1',\ldots,Y_n'$, distributed as the Y_i's, and independent of them. Let $\sigma_1,\ldots,\sigma_n$ be i.i.d. sign variables with $\mathbb{P}\{\sigma_1 = -1\} = \mathbb{P}\{\sigma_1 = 1\} = 1/2$, independent of $Y_1, Y_1', \ldots, Y_n, Y_n'$. Then

$$\begin{aligned}
\mathbb{E}\left\{\left|\sum_{i=1}^n Y_i\right|\right\} &= 1/2\left(\mathbb{E}\left\{\left|\sum_{i=1}^n Y_i\right|\right\} + \mathbb{E}\left\{\left|\sum_{i=1}^n Y_i'\right|\right\}\right) \\
&\geq 1/2\mathbb{E}\left\{\left|\sum_{i=1}^n (Y_i - Y_i')\right|\right\} \\
&= 1/2\mathbb{E}\left\{\left|\sum_{i=1}^n \sigma_i(Y_i - Y_i')\right|\right\} \\
&= 1/2\mathbb{E}\mathbb{E}\left\{\left|\sum_{i=1}^n \sigma_i(Y_i - Y_i')\right| \,\middle|\, Y_1, Y_1', \ldots, Y_n, Y_n'\right\} \\
&\geq \frac{1}{2\sqrt{2}}\mathbb{E}\sqrt{\sum_{i=1}^n (Y_i - Y_i')^2} \quad \text{(by Lemma 5.5)} \\
&\geq \frac{\sqrt{n}}{2\sqrt{2}}\mathbb{E}\{|Y_i - Y_i'|\} \\
&\geq \frac{\sqrt{n}}{2\sqrt{2}}\mathbb{E}\{|Y_i|\} \quad \text{(by Lemma 5.4)}.
\end{aligned}$$

□

Lemma 5.7. *For any density estimate g_n,*

$$\mathbb{E}\int |f - g_n| \geq \tfrac{1}{2}\mathbb{E}\int |g_n - \mathbb{E}g_n|.$$

PROOF. Sum the two inequalities

$$\mathbb{E}\int |f - g_n| \geq \int |f - \mathbb{E}g_n| \quad \text{(Jensen's inequality)},$$

and

$$\mathbb{E}\int |f - g_n| \geq \mathbb{E}\int |g_n - \mathbb{E}g_n| - \int |f - \mathbb{E}g_n| \quad \text{(triangle inequality)}.$$

□

PROOF OF THEOREM 5.21. Note the following:

$$\begin{aligned}
\inf_{\theta\in\Theta} \mathbb{E}\{J_{n-m,\theta}\} &\leq \inf_{\theta\in\Theta} \mathbb{E}\{J_{n,\theta}\} \times \sup_{\theta\in\Theta}\left(\frac{\mathbb{E}\{J_{n-m,\theta}\}}{\mathbb{E}\{J_{n,\theta}}\}\right) \\
&= \inf_{\theta\in\Theta} \mathbb{E}\{J_{n,\theta}\} \times \left(1 + \sup_{\theta\in\Theta} \frac{\mathbb{E}\{J_{n-m,\theta} - \mathbb{E}\{J_{n,\theta}\}\}}{\mathbb{E}\{J_{n,\theta}\}}\right).
\end{aligned}$$

The supremum is rewritten as follows:

$$\begin{aligned}
\sup_{\theta\in\Theta} \frac{\mathbb{E}\{J_{n-m,\theta} - \mathbb{E}\{J_{n,\theta}\}\}}{\mathbb{E}\{J_{n,\theta}\}} &\leq \sup_{\theta\in\Theta} \frac{\mathbb{E}\{\int |f_{n-m,\theta} - f_{n,\theta}|\,dx\}}{\mathbb{E}\{J_{n,\theta}\}} \\
&\leq 2\sup_{\theta\in\Theta} \frac{\mathbb{E}\{\int |f_{n-m,\theta} - f_{n,\theta}|\,dx\}}{\mathbb{E}\{\int |f_{n,\theta} - \mathbb{E}f_{n,\theta}|\,dx\}},
\end{aligned}$$

where we used Lemma 5.7. Fix x and θ for now. Introduce

$$Y_i = K_\theta(x, X_i) - \mathbb{E}\{K_\theta(x, X)\},$$

and denote the partial sums of Y_i's by $S_j = Y_1 + \cdots + Y_j$. By assumption, for fixed x and θ, the first absolute moment of Y_1 is finite. Then observe the following:

$$n|f_{n-m,\theta} - f_{n,\theta}| = \left|\frac{m}{n-m}(Y_1 + \cdots + Y_{n-m}) - (Y_{n-m+1} + \cdots + Y_n)\right|$$

so that

$$\mathbb{E}\{n|f_{n-m,\theta} - f_{n,\theta}|\} \leq \frac{m}{n-m}\mathbb{E}\{|S_{n-m}|\} + \mathbb{E}\{|S_m|\}.$$

Also, $n|f_{n,\theta}-\mathbb{E}f_{n,\theta}|=|S_n|$, which implies $\mathbb{E}\{n|f_{n,\theta}-\mathbb{E}f_{n,\theta}|\}=\mathbb{E}\{|S_n|\}$. Still holding x and θ fixed, we bound the following ratio:

$$\begin{aligned}
\frac{\mathbb{E}\{|f_{n-m,\theta}-f_{n,\theta}|\}}{\mathbb{E}\{|f_{n,\theta}-\mathbb{E}f_{n,\theta}|\}} &\le \frac{(m/(n-m)\mathbb{E}\{|S_{n-m}|\}+\mathbb{E}\{|S_m|\}}{\mathbb{E}\{|S_n|\}} \\
&\le \frac{m}{n-m}+\frac{\mathbb{E}\{|S_m|\}}{\mathbb{E}\{|S_n|\}} \\
&\quad \text{(because } \mathbb{E}\{|S_n|\}\ge \mathbb{E}\{|S_{n-m}|\}) \\
&\le \frac{m}{n-m}+\frac{\mathbb{E}\{|S_m|\}}{\sqrt{\lfloor n/m\rfloor/8}\,\mathbb{E}\{|S_m|\}} \quad \text{(by Lemmas 5.4 and 5.6)} \\
&\le \frac{m}{n-m}+4\sqrt{\frac{m}{n}} \quad \text{(if } 2m\le n).
\end{aligned}$$

This implies that for any fixed θ:

$$\mathbb{E}\int |f_{n-m,\theta}-f_{n,\theta}|\,dx \le \left(\frac{m}{n-m}+4\sqrt{\frac{m}{n}}\right)\mathbb{E}\int |f_{n,\theta}-\mathbb{E}f_{n,\theta}|\,dx.$$

The result now follows without work. □

Combining Theorems 5.20, 5.21 and 1.9, we readily obtain:

Theorem 5.22. *Let the set Θ determine a class of regular additive density estimates with $\int f_{n-m,\theta}=1$ for all $\theta\in\Theta$. Then for all n, $m\le n/2$, Θ, and f:*

$$\begin{aligned}
\mathbb{E}\int |f_n-f| &\le 3\inf_{\theta\in\Theta}\mathbb{E}\int |f_{n,\theta}-f|\left(1+\frac{2m}{n-m}+8\sqrt{\frac{m}{n}}\right) \\
&\quad +8\mathbb{E}\left\{\sqrt{\frac{\log 2\mathbb{S}_{\mathcal{A}_\Theta}(m)}{m}}\right\}+\frac{3}{n}.
\end{aligned}$$

We note that $\mathbb{S}_{\mathcal{A}_\Theta}(m)$ is a random variable that depends upon $X_1,\ldots,X_{n-m}$, since the definition of $\mathcal{A}_\Theta$ involves these data points. In most applications, $\mathcal{A}_\Theta$ can be bounded uniformly over all values of $X_1,\ldots,X_{n-m}$.

EXAMPLE: HISTOGRAM ESTIMATE. The simplest additive estimates are histograms. A histogram density estimate based on a partition P of $\mathcal{R}^d$ is defined by

$$f_{n,P}(x)=\frac{\mu_n(A_P(x))}{\lambda(A_P(x))},$$

where μ_n is the empirical measure based on the sample $X_1,\ldots,X_n$, λ is the Lebesgue measure on $\mathcal{R}^d$, $A_P(x)$ denotes the cell of the partition P into which x falls, and a/∞ is defined to be zero.

As a first application of Theorem 5.22, we consider the selection of a partition for a histogram. Formally, let $\mathcal{P}$ be a family of partitions of $\mathcal{R}^d$, and to each partition $P \in \mathcal{P}$, assign the corresponding histogram estimate $f_{n,P}(x)$. We use the minimum distance estimate based on data splitting to select a partition from the class $\mathcal{P}$, thus obtaining the density estimate f_n. First observe that any histogram estimate is regular and additive. However, if P contains a cell with infinite Lebesgue measure, then $f_{n,P}$ does not necessarily integrate to one, which makes the use of Theorem 5.22 illegitimate. To fix this problem, we map $\mathcal{R}^d$ to $[0,1]^d$, and consider only partitions of the cube.

To apply Theorem 5.22, we merely need to obtain upper bounds for the shatter coefficients $\mathbb{S}_{\mathcal{A}_\mathcal{P}}(m)$, where $\mathcal{A}_\mathcal{P}$ is the Yatracos class of all sets of the form

$$\{x : f_{n-m,P_1}(x) > f_{n-m,P_2}(x)\}, \qquad P_1, P_2 \in \mathcal{P}.$$

As a prototype example, consider the class $\mathcal{P}$ of partitions of $\mathcal{R}$ into intervals of length h, where $h = 2^k$ for some integer k. The estimate $f_{n,P}$ based on such a partition is called a regular histogram estimate. To be specific, assume that each partition is anchored at the origin, that is, 0 lies on the boundary of two cells. The parameter k may take infinitely many possible values, and our goal is to select a close-to-optimal value. The following combinatorial argument guarantees, via Theorem 5.22, that the minimum distance estimate may be successfully used for the selection of h from a dyadic collection of interval lengths.

Lemma 5.8.

$$\mathbb{S}_{\mathcal{A}_\mathcal{P}}(m) \le (m+1)n^2.$$

Also, for all n, $m \le n/2$, and f, if f_n is the regular histogram estimate picked by the minimum distance method for $\mathcal{P}$, we have

$$\begin{aligned}\mathbb{E}\int |f_n - f| &\le 3 \inf_{\theta\in\mathcal{P}} \mathbb{E}\int |f_{n,\theta} - f| \left(1 + \frac{2m}{n-m} + 8\sqrt{\frac{m}{n}}\right) \\ &\quad + 8\sqrt{\frac{\log(2(m+1)n^2)}{m}} + \frac{3}{n}.\end{aligned}$$

PROOF. Each partition P_k in $\mathcal{P}$ is indexed by a (possibly negative-valued) integer k. Let $A_k(x)$ be the cell of P_k containing x. Observe that the value of the vector

$$z_k = \left(\sum_{i=1}^{n-m} \mathbb{I}_{[y_1 \in A_k(X_i)]}, \ldots, \sum_{i=1}^{n-m} \mathbb{I}_{[y_m \in A_k(X_i)]}\right)$$

can take at most n different values when we vary k but keep the y_i's and X_i's fixed. To see this, observe that there exists a large integer k_0 such that all data points fall in the same cell of P_{k_0}. Then for all $k \ge k_0$ the value of z_k is the same. As k decreases in steps of size one, z_k can only change when there exists a cell $[a,b)$ of P_k such that $[a,(a+b)/2)$ and $[(a+b)/2,b)$ both contain at least one data point among $\{X_1,\ldots,X_{n-m},y_1,\ldots,y_m\}$. But this can happen for at most $n-1$ different values of k. Thus,

$$|\{(z_k,z_\ell):k,\ell\in\{0,\pm 1,\pm 2,\ldots\}\}|\le n^2.$$

If w and w' are two possible values of the vector z_k, let $U_{w,w'}$ be the set of all pairs (k,ℓ) such that $(z_k,z_\ell)=(w,w')$. Fix $w=(w_1,\ldots,w_m)$ and $w'=(w'_1,\ldots,w'_m)$. If $(k,\ell)\in U_{w,w'}$, then $y_i\in A_{k,\ell}=\{x: f_{n-m,k}(x)>f_{n-m,\ell}(x)\}$ if and only if

$$\frac{w_i}{2^k}>\frac{w'_i}{2^\ell}.$$

Therefore,

$$\begin{aligned}&|\{\{y_1,\ldots,y_m\}\cap A_{k,\ell}:(k,\ell)\in U_{(w,w')}\}|\\ &\quad\le|\{(\mathbb{I}_{[w_1\ge cw'_1]},\ldots,\mathbb{I}_{[w_m\ge cw'_m]}):c>0\}|\le m+1.\end{aligned}$$

Thus, $\mathbb{S}_{\mathcal{A}_P}(m)\le(m+1)n^2$. □

For various results on the regular histogram estimate we refer to Devroye and Györfi (1985). Consistency of the histogram estimate based on data-dependent partitions is investigated in Lugosi and Nobel (1996). Selection of the partition based on penalized maximum likelihood methods is studied by Barron, Birgé, and Massart (1999) and Castellan (2000).

5.8 Bandwidth Selection for Kernel Estimates

This section is about the choice of the bandwidth (or smoothing factor) $h\in(0,\infty)$ of the standard kernel estimate

$$f_{n,h}(x)=\frac{1}{nh^d}\sum_{i=1}^n K\left(\frac{x-X_i}{h}\right).$$

We assume that $K:\mathcal{R}^d\to\mathcal{R}$ is a fixed kernel with $\int K=1$. We study the minimum distance estimate based on data-splitting as described in the previous chapter, that is, we fix $m\le n/2$, define a class of densities $\mathcal{F}=\{f_{n-m,h}:h>0\}$, and project the empirical measure μ_m (defined on the held-out data $X_{m+1},\ldots,X_n$) on $\mathcal{F}$ by the minimum distance method. Clearly, every estimate in $\mathcal{F}$ is additive, so if, in addition, K is a bounded function, then Theorem 5.22 applies. To apply the bound of this theorem, we need to compute useful upper bounds for the shatter coefficient $\mathbb{S}_{\mathcal{A}}(m)$ of the class

$$\mathcal{A}\stackrel{\text{def}}{=}\{\{x:f_{n-m,h}(x)>f_{n-m,h'}(x)\}:\ h,h'>0\}.$$

Of course, this depends on n, m, and the kernel function K. Obtaining meaningful (i.e., polynomial in m and n) upper bounds may be difficult, or even impossible for general kernels. Thus, we begin our study with estimates based on simple kernels. The material of this section is based on Devroye and Lugosi (1997). We will not discuss implementation issues, for which we refer to Devroye (1997).

We consider kernels of the form

$$K(x) = \sum_{i=1}^{k} \alpha_i \mathbb{I}_{A_i}(x),$$

where $k < \infty$, $\alpha_1, \ldots, \alpha_k \in \mathcal{R}$, and $A_1, \ldots, A_k$ are Borel sets in $\mathcal{R}^d$ with the following property: the intersection of an infinite ray $\{x : x = tx_0, t \geq 0\}$, anchored at the origin, with any A_i is an interval.

Observe that all convex sets and all star-shaped sets satisfy the above requirement. (A set A is called star-shaped if $x \in A$ implies $\lambda x \in A$ for all $\lambda \in [0,1]$.) The A_i's need not be disjoint. Kernels of this type are called *Riemann kernels* of parameter k. Important special cases include the uniform densities on ellipsoids, balls, and hypercubes. The key combinatorial result is the following:

Lemma 5.9. *Let $K = \sum_{i=1}^{k} \alpha_i \mathbb{I}_{A_i}$ be a Riemann kernel of parameter k. For all $m \geq 1$ we have*

$$\mathbb{S}_{\mathcal{A}}(m) \leq (m+1)(1+2km(n-m))^2.$$

PROOF. Let $r = n - m$. Define the vector

$$z_u = \left(\sum_{i=1}^{r} K\left(\frac{y_1 - X_i}{u}\right), \ldots, \sum_{i=1}^{r} K\left(\frac{y_m - X_i}{u}\right) \right) \in \mathcal{R}^m.$$

As we increase u from zero to infinity, each component of z_u changes every time $(y_j - X_i)/u$ enters or leaves a set A_q, $1 \leq q \leq k$ for some X_i, $1 \leq i \leq r$. Note that for fixed $(y_j - X_i)$, the evolution is along an infinite ray anchored at the origin. By our assumption on the possible form of the sets A_q, the number of different values a component can take in its history (as $u \uparrow \infty$) is bounded by $2kr$. As there are m components, the cardinality of the set of different values of z_u is bounded as

$$|\{z_u : u > 0\}| \leq 1 + 2kmr.$$

Thus,

$$|\{(z_u, z_v) : u, v > 0\}| \leq (1 + 2kmr)^2.$$

Let $\mathcal{W} = \{(w, w') : (w, w') = (z_u, z_v) \text{ for some } u, v > 0\}$. For fixed $(w, w') \in \mathcal{W}$, let $U_{(w,w')}$ denote the collection of all (u, v) such that $(z_u, z_v) = (w, w')$. For $(u, v) \in U_{(w,w')}$, we have

$$y_i \in A_{u,v} \quad \text{if and only if} \quad w_i \geq \left(\frac{u}{v}\right)^d w'_i,$$

where w, w' have components w_i, w'_i, respectively, $1 \le i \le m$. Thus,

$$|\{\{y_1,\ldots,y_m\}\cap A_{u,v} : (u,v) \in U_{(w,w')}\}|$$
$$\le \left|\left\{\left(\mathbb{I}_{[w_1 \ge cw'_1]},\ldots,\mathbb{I}_{[w_m \ge cw'_m]}\right) : c \ge 0\right\}\right| \le m+1.$$

But then

$$|\{\{y_1,\ldots,y_m\}\cap A_{u,v} : (u,v) > 0\}|$$
$$\le (m+1)|U_{(w,w')}| \le (m+1)(1+2kmr)^2.$$

□

For Riemann kernels, Lemma 5.9 is the last missing link. Plugging the upper bound of the lemma into Theorem 5.22, we obtain a nonasymptotic, density-free inequality.

Most kernel functions used in practice are not Riemann kernels, and bounding the complexity of the Yatracos class based on such kernels is difficult. Luckily, most kernels can be well approximated by Riemann kernels, and this suggests the following solution for the bandwidth selection problem: first select a positive integer k and a Riemann kernel $K' = \sum_{i=1}^k \alpha_i \mathbb{I}_{A_i}$ such that

$$\int |K - K'| \le \frac{1}{n}.$$

Note that this is always possible if K is Riemann integrable. Now define the kernel estimates using the approximating kernel

$$f'_{n-m,h}(x) = \frac{1}{n-m} \sum_{i=1}^{n-m} K'_h(x - X_i)$$

for all $h > 0$. Finally, select the smoothing factor H by the minimum distance estimate over this class, and define the density estimate

$$f_n(x) = \frac{1}{n-m} \sum_{i=1}^{n-m} K_H(x - X_i).$$

This H is called the Riemann approximation bandwidth and f_n is the Riemannian kernel estimate. The size of the smallest k for which this is possible depends on the kernel K. We call this the *kernel complexity* κ_n of K:

$$\kappa_n = \min\left\{k : \exists K' = \sum_{i=1}^k \alpha_i \mathbb{I}_{A_i} \text{ such that } \int |K - K'| \le \frac{1}{n}\right\}.$$

The main result of this section is the following performance bound for f_n:

Theorem 5.23. *Let K be a bounded kernel with kernel complexity κ_n, and let $m \le n/2$. Then for all densities f, the Riemannian kernel estimate f_n satisfies*

$$\mathbb{E}\int |f_n - f| \le 3\left(1+\frac{2m}{n-m}+8\sqrt{\frac{m}{n}}\right)\inf_h \mathbb{E}\int |f_{n,h}-f| + 8\sqrt{\frac{\log\left(2(m+1)(1+2\kappa_n m(n-m))^2\right)}{m}}+\frac{31}{n}.$$

PROOF. Observe that for each h, $\int |f_{n-m,h} - f'_{n-m,h}| \le \int |K-K'|$. Thus,

$$\begin{aligned}
&\mathbb{E}\int |f_n - f| \\
&\le \mathbb{E}\int |f'_{n-m,H} - f| + \int |K-K'| \\
&\le 3\left(1+\frac{2m}{n-m}+8\sqrt{\frac{m}{n}}\right)\inf_h \mathbb{E}\int |f'_{n,h}-f| \\
&\quad +8\sqrt{\frac{\log\left(2(m+1)(1+2\kappa_n m(n-m))^2\right)}{m}}+\int |K-K'|+\frac{3}{n} \\
&\quad \text{(by the results of the previous section)} \\
&\le 3\left(1+\frac{2m}{n-m}+8\sqrt{\frac{m}{n}}\right)\inf_h \mathbb{E}\int |f_{n,h}-f| \\
&\quad +8\sqrt{\frac{\log\left(2(m+1)(1+2\kappa_n m(n-m))^2\right)}{m}} \\
&\quad +\left(1+3\left(1+\frac{2m}{n-m}+8\sqrt{\frac{m}{n}}\right)\right)\int |K-K'|+\frac{3}{n}.
\end{aligned}$$

□

To understand the implications of Theorem 5.23, consider the simplest, though suboptimal choice $m = \lfloor n/2 \rfloor$. Then we obtain

$$\mathbb{E}\int |f_n - f| \le 43 \inf_h \mathbb{E}\int |f_{nh} - f| + c\sqrt{\frac{\log(n\kappa_n)}{n}}+\frac{31}{n},$$

where c is a universal constant, independent of f and K. Taking m as a smaller fraction of n, we may decrease the factor 43 to close to 3, at the expense of increasing the factor c in front of the second term. As we will see later, for most kernels the first term converges to zero much slower than $\sqrt{\log n/n}$ for *all* densities. Therefore, the first term asymptotically dominates the second one if the kernel complexity κ_n is a polynomial function of n. In the next section we show that this is the case for all important kernels.

KERNEL COMPLEXITY: UNIVARIATE EXAMPLES. In this section we provide a list of examples of kernels on $\mathcal{R}$ whose complexity κ_n is bounded by a polynomial of n. Let us warm up with two toy examples:

UNIFORM KERNELS. If $K(x) = \mathbb{I}_A(x)$ for a star-shaped set A, then obviously $\kappa_n = 1$ for all $n > 1$.

ISOSCELES TRIANGULAR DENSITY. If $K(x) = (1 - |x|)_+$, then elementary calculation shows that for all n, $\kappa_n \leq n+1$.

Most important kernels are nonnegative, unimodal, and symmetric about the origin. For such kernels, the following bound is useful.

SYMMETRIC UNIMODAL KERNELS. Consider a symmetric unimodal density (i.e., $K \geq 0$ and $\int K = 1$) on the real line. Let β be the last positive value for which $\int_\beta^\infty K \leq 1/(4n)$. Partition $[0,\beta]$ and $[-\beta,0]$ into $N = \lceil 4nK(0)\beta \rceil$ equal intervals. On each interval, let K' be constant with value equal to the average of K over that interval. Let $\gamma = \int_\beta^\infty K/K(\beta)$, and set $K'(x) = K(\beta)$ on $[\beta, \beta+\gamma]$ and $[-\beta-\gamma, -\beta]$. Note that $\int K' = 1$, $\int |K - K'| \leq 1/n$, and that K' is Riemann with parameter $k \leq 2N+2 \leq 8nK(0)\beta + 10$. Thus, $\kappa_n \leq 8nK(0)\beta + 10$.

EXAMPLE 1 (BOUNDED COMPACT SUPPORT KERNELS). If $K(x) \leq a\mathbb{I}_{[-b,b]}(x)$ and K is symmetric, nonnegative, and unimodal (such as the Epanechnikov kernel), then $\kappa_n \leq 8nab + 10$.

EXAMPLE 2 (THE NORMAL KERNEL). When $K(x) = e^{-x^2/2}/\sqrt{2\pi}$, we have $K(0) = 1/\sqrt{2\pi}$. Since for $\beta \geq 1$,

$$\int_\beta^\infty \frac{1}{\sqrt{2\pi}} e^{-x^2/2}\,dx \leq \frac{1}{\sqrt{2\pi}} \frac{1}{\beta} e^{-\beta^2/2} \leq \frac{1}{\sqrt{2\pi}} e^{-\beta^2/2},$$

we may take $\beta = \sqrt{2\log(4n/\sqrt{2\pi})}$. Thus, for all $n > 1$:

$$\kappa_n \leq \frac{8n\sqrt{\log n}}{\sqrt{\pi}} + 10.$$

EXAMPLE 3 (THE CAUCHY KERNEL). Take $K(x) = 1/(\pi(1+x^2))$. Note that $K(0) = 1/\pi$, and that $\beta = \pi/(4n)$ will do. Therefore,

$$\kappa_n \leq \frac{32n^2}{\pi^2} + 10.$$

EXAMPLE 4 (KERNELS WITH POLYNOMIAL TAILS). Note that if K is a symmetric unimodal density, and $|K(x)| \leq c/(1+|x|^{\gamma+1})$ for some $c < \infty$, $\gamma > 0$, then $\kappa_n = O(n^{1+1/\gamma})$. In fact, for most cases of interest, $\kappa_n = O(n^\alpha)$ for some finite constant $\alpha > 0$.

Sometimes it may be beneficial to use kernels which may take negative values. The next bound will be useful even for kernels that oscillate infinitely many times.

LIPSCHITZ KERNELS WITH A POSSIBLY HEAVY TAIL. Let K be a univariate kernel that is Lipschitz with Lipschitz constant C, and assume that $|K(x)| \le D/x^2$ for another constant D. Then

$$\kappa_n \le 1 + 32CD^2n^3.$$

PROOF. Take $r = 4Dn$ and note that $\int_{|x|>r} |K| \le 2D/r = 1/(2n)$. Partition $[-r,r]$ into q equal intervals of length $2r/q$ each. Define a Riemann kernel K' of order q taking a constant value on each of these intervals, equal to the average of K over the intervals. By the Lipschitz condition, on any such interval A, $\int_A |K-K'| \le C(2r/q)^2/2 = 2Cr^2/q^2$. Let $\gamma^+ = (1/K(r))\int_r^\infty K$ and $\gamma^- = (1/K(-r))\int_{-\infty}^{-r} K$, and set $K'(x) = K(r)$ on $[r, r+\gamma^+]$ and $K'(x) = K(-r)$ on $[-r-\gamma^-, r]$. Thus, $\int K' = 1$ and

$$\int |K-K'| \le \frac{1}{2n} + \frac{2Cr^2}{q} \le \frac{1}{n}$$

provided that $q \ge 2Cr^2n = 32CD^2n^3$. □

Finally, we mention a huge class of kernels, containing nearly every one-dimensional kernel.

KERNELS OF BOUNDED VARIATION. If K is symmetric and a difference of two monotone functions, that is, $K = K_1 - K_2$, $K_1 \downarrow 0$, $K_2 \downarrow 0$ on $[0,\infty)$, then each K_1, K_2 may be approximated as above. Thus, in particular, if K is of *bounded variation*, and $|K(x)| \le c/(1+|x|^{\gamma+1})$ for some $c < \infty$, $\gamma > 0$, then we may approximate with $\kappa_n = O(n^{1+1/\gamma})$.

PRODUCT KERNELS. If $K = K_1 \times \cdots \times K_d$ is a product of d univariate kernels, and if we approximate K_i with K_i' with parameter $\kappa_{nd}^{(i)}$ for all i (where $\kappa_{nd}^{(i)}$ is the kernel complexity of K_i of precision $1/(nd)$), and form $K' = K_1' \times \cdots \times K_d'$, then K' is a weighted sum of indicators of product sets, and it is Riemann with parameter not exceeding $\prod_{i=1}^d \kappa_{nd}^{(i)}$. Furthermore,

$$\begin{aligned}
\int |K-K'| &\le \int |K_1 \times \cdots \times K_{d-1} \times K_d - K_1 \times \cdots \times K_{d-1} \times K_d'| \\
&\quad + \ldots \\
&\quad + \int |K_1 \times K_2' \cdots \times K_d' - K_1' \times K_2' \cdots \times K_d'| \\
&\le d\left(\frac{1}{nd}\right) \\
&= \frac{1}{n}.
\end{aligned}$$

Thus, κ_n is bounded by $\prod_{i=1}^d \kappa_{nd}^{(i)}$.

KERNELS THAT ARE FUNCTIONS OF $\|x\|$. Assume that $K(x) = M(\|x\|)$, where M is a bounded nonnegative monotone decreasing function on $[0,\infty)$. Then we may approximate M by a stepwise constant function M', and use the Riemann kernel $K'(x) = M'(\|x\|)$ in the estimate as an approximation of K. Clearly,

$$\int |K(x) - K'(x)|\,dx = \int_0^\infty c_d u^{d-1} |M(u) - M'(u)|\,du,$$

where c_d is d times the volume of the unit ball in $\mathcal{R}^d$. We may define M' as follows. Let β be the largest positive number for which $\int_\beta^\infty c_d u^{d-1} M(u)\,du \le 1/(2n)$. Partition $[0,\beta]$ into $N = \lceil 2nc_d M(0)\beta^d \rceil$ equal intervals. On each interval, let M' equal to the average of M over that interval. Let $\gamma = \int_\beta^\infty c_d u^{d-1} M(u)\,du/M(\beta)$, and set $M'(u) = M(\beta)$ on $u \in [\beta, \beta+\gamma]$, and let $M'(u) = 0$ for $u > \gamma$. Clearly $\int K' = 1$, and that K' is Riemann with parameter $k = N + 1 \le 2nc_d K(0)\beta^d + 2$. Moreover,

$$\begin{aligned}
&\int |K(x) - K'(x)|\,dx \\
&= \int_0^\beta c_d u^{d-1} |M(u) - M'(u)|\,du + \int_\beta^\infty c_d u^{d-1} |M(u) - M'(u)|\,du \\
&\le \frac{1}{2n} + c_d \beta^{d-1} \int_0^\beta |M(u) - M'(u)|\,du \\
&\le \frac{1}{2n} + c_d \beta^{d-1} \frac{M(0)\beta}{N} \\
&\le \frac{1}{n}.
\end{aligned}$$

Thus,

$$\kappa_n \le 2nc_d M(0)\beta^d + 2.$$

THE MULTIVARIATE STANDARD NORMAL KERNEL. We may apply the bound of the previous paragraph to the multivariate normal density. First note that it suffices to take $\beta = 2\sqrt{2\log n}$. From this, we deduce that the kernel complexity is

$$\kappa_n = O(n \log^{d/2} n).$$

ASYMPTOTIC OPTIMALITY. One important corollary of Theorem 5.23 is that asymptotically the error of the estimate stays within a factor of three of that of the kernel estimate with the best possible smoothing factor.

Theorem 5.24. (DEVROYE AND LUGOSI, 1996, 1997) *Let K be a bounded nonnegative kernel on the real line with complexity κ_n bounded by some polynomial of n. If $m/n \to 0$ and*

$m/(n^{4/5}\log n) \to \infty$ as $n \to \infty$, then for all densities f on $\mathcal{R}$:

$$\sup_f \limsup_{n\to\infty} \frac{\mathbb{E}\{\int |f_n - f|\}}{\inf_h \mathbb{E}\{\int |f_{n,h} - f|\}} \le 3.$$

The statement is an easy consequence of the fact that by Theorem 5.23, with the given choice of m,

$$\mathbb{E}\int |f_n - f| \le (3 + o(1)) \inf_h \mathbb{E}\int |f_{nh} - f| + o(n^{-2/5}),$$

and the following lower bound due to Devroye and Penrod (1984).

Lemma 5.10. *Let K be a nonnegative kernel on the real line. Then for any density f:*

$$\liminf_{n\to\infty} n^{2/5} \inf_h \mathbb{E}\int |f_{nh} - f| \ge 0.86.$$

5.9 References

Abou-Jaoude, S. (1976). Conditions nécessaires et suffisantes de convergence L_1 en probabilité de l'histogramme pour une densité. *Annales de l'Institut Henri Poincaré* , XII, 213-231.

Akaike, H. (1954). "An approximation to the density function," *Annals of the Institute of Statistical Mathematics,* vol. 6, pp. 127–132.

Anthony, M. and Bartlett, P. L. (1999). *Neural Network Learning: Theoretical Foundations,* Cambridge University Press, Cambridge.

Azuma, K. (1967). "Weighted sums of certain dependent random variables," *Tohoku Mathematical Journal,* vol. 37, pp. 357–367.

Barron, A. R. (1988). The convergence in information of probability density estimates. *IEEE ISIT, Kobe, Japan.*

Barron, A. R. (1989). Uniformly powerful goodness of fit tests. *Ann. Statist.* 17, 107-124.

Barron, A. Birgé, L. and Massart, P. (1999). "Risk bounds for model selection via penalization," *Probability Theory and Related Fields,* vol. 113, pp. 301–415.

Barron, A. R., Györfi, L. and van der Meulen, E. C. (1992). Distribution estimates consistent in total variation and in two types of information divergence. *IEEE Trans. on Information Theory,* 38, pp. 1437-1454.

Beirlant, J., Berlinet, A. and Györfi, L. (1999). On piecewise linear density estimation. *Statistica Neerlandica*, 53, pp. 287-308.

Beirlant, J., Devroye, L., Györfi, L. and Vajda I. (2001). Large deviations of divergence measures on partitions. *J. Statistical Planning and Inference*, 93, pp. 1-16.

Beirlant, J. and Györfi, L. (1998). On the L_1-error in histogram density estimation: the multidimensional case. *J. Nonparametric Statistics*, 9, pp. 197-216.

Beirlant, J., Györfi, L. and Lugosi, G. (1994). On the asymptotic normality of the L_1- and L_2- errors in the histogram density estimation. *Canadian J. Statistics*, 22, pp. 309-318.

Beirlant, J. and Mason, D. M. (1995). "On the asymptotic normality of L_p-norms of empirical functionals," *Mathematical Methods of Statistics*, vol. 4, pp. 1–19.

Berlinet, A. (1995). Central limit theorems in functional estimation. *Bulletin of the International Statistical Institute*, 56, pp. 531-548.

Berlinet, A. and Devroye, L. (1994). "A comparison of kernel density estimates," *Publications de l'Institut de Statistique de l'Université de Paris*, vol. 38, pp. 3–59.

Berlinet, A., Devroye, L. and Györfi, L. (1995). Asymptotic normality of L_1 error in density estimation. *Statistics*, 26, pp. 329-343.

Berlinet, A., Györfi, L. and van der Meulen, E. (1995). The asymptotic normality of relative entropy in multivariate density estimation. *Publications de l'Institut de Statistique de l'Université de Paris*, 41, pp. 3-27.

Birgé, L. (1986). On estimating a density using Hellinger distance and some other strange facts. *Probability Theory and Related Fields*, 71, pp. 271-291.

Bretagnolle, J. and Huber, C. (1979). "Estimation des densités: Risque minimax," *Zeitschrift für Wahrscheinlichkeitstheorie und verwandte Gebiete*, vol. 47, pp. 119–137.

Broniatowski, M., Deheuvels, P. and Devroye, L. (1989). "On the relationship between stability of extreme order statistics and convergence of the maximum likelihood kernel density estimate," *Annals of Statistics*, vol. 17, pp. 1070–1086.

Butzer, P. L. and Nessel, R. J. (1971). *Fourier Analysis and Approximation*, Birkhäuser-Verlag, Basel, 1971.

Cao, R., Cuevas, A. and González–Manteiga, W. (1994). "A comparative study of several smoothing methods in density estimation," *Computational Statistics and Data Analysis*, vol. 17, pp. 153–176.

Castellan, G. (2000). "Sélection d'histogrammes ou de modèles exponentiels de polynomes par morceaux à l'aide d'un critère de type Akaike," Thèse, Mathématiques, Université de Paris-Sud.

Cline, D. B. H. (1988). "Admissible kernel estimators of a multivariate density," *Annals of Statistics*, vol. 16, pp. 1421–1427.

Cline, D. B. H. (1990) "Optimal kernel estimation of densities," *Annals of the Institute of Statistical Mathematics*, vol. 42, pp. 287–303.

Csiszár, I. (1967). Information-type measures of divergence of probability distributions and indirect observations. *Studia Sci. Math. Hungar.*, 2, pp. 299-318.

Csiszár, I. and Körner, J. (1981). *Information Theory: Coding Theorems for Memoryless Systems*. Academic Press, New York.

Csörgő, M. and Horváth, L. (1988). "Central limit theorems for L_p-norms of density estimators," *Probability Theory and Related Fields*, vol. 80, pp. 269–291.

Deheuvels, P. (1977). "Estimation nonparamétrique de la densité par histogrammes generalisés," *Publications de l'Institut de Statistique de l'Université de Paris*, vol. 22, pp. 1–23.

Dembo, A. and Zeitouni, O. (1992). *Large Deviations Techniques and Applications*. Jones and Bartlett Publishers.

Devroye, L. (1983a). The equivalence of weak, strong and complete convergence in L1 for kernel density estimates. *Annals of Statistics*, 11, pp. 896–904.

Devroye, L. (1983b). On arbitrary slow rates of global convergence in density estimation. *Zeitschrift für Wahscheinlichkeitstheorie und verwandte Gebiete*, 62, pp. 475-483.

Devroye, L. (1987). *A Course in Density Estimation*, Birkhäuser-Verlag, Boston.

Devroye, L. (1988a). The kernel estimate is relatively stable. *Probability Theory and Related Fields*, 77, pp. 521–536.

Devroye, L. (1988b) "Asymptotic performance bounds for the kernel estimate," *Annals of Statistics*, vol. 16, pp. 1162–1179.

Devroye, L. (1989a) "The double kernel method in density estimation," *Annales de l'Institut Henri Poincaré*, vol. 25, pp. 533–580.

Devroye, L. (1989b). "Nonparametric density estimates with improved performance on given sets of densities," *Statistics (Mathematische Operationsforschung und Statistik)*, vol. 20, pp. 357–376.

Devroye, L. (1991). "Exponential inequalities in nonparametric estimation," in: *Nonparametric Functional Estimation and Related Topics*, (edited by G. Roussas), pp. 31–44, NATO ASI Series, Kluwer Academic, Dordrecht.

Devroye, L. (1997). "Universal smoothing factor selection in density estimation: Theory and practice (with discussion)," *Test*, vol. 6, pp. 223–320.

Devroye, L. (1995). Another proof of a slow convergence result of Birgé. *Statistics and Probability Letters*, 23, pp. 63-67.

Devroye, L. and Györfi, L. (1985). *Nonparametric Density Estimation: the L_1 View*. Wiley.

Devroye, L. and Györfi, L. (1990). No empirical measure can converge in total variation sense for all distributions. *Annals of Statistics*, 18, pp. 1496-1499.

Devroye, L., Györfi, L. and Lugosi, G. (1996). *A Probabilistic Theory of Pattern Recognition.* Springer Verlag.

Devroye, L. and Lugosi, G. (1996). "A universally acceptable smoothing factor for kernel density estimation," *Annals of Statistics*, vol. 24, pp. 2499–2512.

Devroye, L. and Lugosi, G. (1997). "Nonasymptotic universal smoothing factors, kernel complexity and Yatracos classes," *Annals of Statistics*, vol. 25, pp. 2626–2637.

Devroye. L. and Lugosi, G. (2001). *Combinatorial Methods in Density Estimation*, Springer-Verlag, New York.

Devroye, L. and Penrod, C. S. (1984). "Distribution-free lower bounds in density estimation," *Annals of Statistics*, vol. 12, pp. 1250–1262.

Devroye L. and Wand, M. P. (1993). "On the effect of density shape on the performance of its kernel estimate," *Statistics*, vol. 24, pp. 215–233.

de Guzmán, M. (1975). *Differentiation of Integrals in R^n*, Lecture Notes in Mathematics #481, Springer-Verlag, Berlin.

de Guzmán, M. (1981). *Real Variable Methods in Fourier Analysis*, North-Holland, Amsterdam.

Dvoretzky, A. Kiefer, J. and Wolfowitz, J. (1956). "Asymptotic minimax character of a sample distribution function and of the classical multinomial estimator," *Annals of Mathematical Statistics*, vol. 33, pp. 642–669.

Epanechnikov, V. A. (1969). "Nonparametric estimation of a multivariate probability density," *Theory of Probability and its Applications*, vol. 14, pp. 153–158.

Goldberg P. and Jerrum, M. (1995). "Bounding the Vapnik-Chervonenkis dimension of concept classes parametrized by real numbers," *Machine Learning*, vol. 18, pp. 131–148.

Györfi, L., Páli, I. and van der Meulen, E. C. (1994). There is no universal source code for infinite alphabet. *IEEE Trans. on Information Theory*, 40, pp. 267-271.

Györfi, L. and van der Meulen, E. C. (1994). There is no density estimate consistent in information divergence for all densities. *Transactions of the Twelfth Prague Conference on Information Theory, Statistical Decision Functions, Random Processes*, pp. 88-90.

Haagerup, U. (1978). "Les meilleures constantes de l'inégalité de Khintchine," *Comptes Rendus des Séances de l'Académie des Sciences de Paris. Séries A*, vol. 286, pp. 259–262.

Hall, P. and Wand, M. P. (1988). "Minimizing L_1 distance in nonparametric density estimation," *Journal of Multivariate Analysis*, vol. 26, pp. 59–88.

Hoeffding, W. (1963). "Probability inequalities for sums of bounded random variables," *Journal of the American Statistical Association*, vol. 58, pp. 13–30.

Holmström, L. and Klemelä, J. (1992). "Asymptotic bounds for the expected L_1 error of a multivariate kernel density estimator," *Journal of Multivariate Analysis*, vol. 40, pp. 245–255.

Kemperman, J. H. B. (1969). An optimum rate of transmitting information. *Ann. Math. Statist.*, 40, pp. 2156-2177.

Karpinski, M. and Macintyre, A. (1997). "Polynomial bounds for VC dimension of sigmoidal and general pfaffian neural networks," *Journal of Computer and System Science*, vol. 54, pp. 169–176.

Khovanskii, A. G. (1991). "Fewnomials," in: *Translations of Mathematical Monographs*, vol. 88, American Mathematical Society, Providence, RI.

Koiran, P. and Sontag, E. D. (1997). "Neural networks with quadratic VC dimension," *Journal of Computer and System Science*, vol. 54, pp. 190–198.

Krzyżak, A. and Linder, T. (1998). "Radial basis function networks and complexity regularization in function learning," *IEEE Transactions on Neural Networks*, vol. 9, pp. 247–256.

Kullback, S. (1967). A lower bound for discrimination in terms of variation. *IEEE Trans. Information Theory*, 13, pp. 126-127.

LeCam, L. (1973). "Convergence of estimates under dimensionality restrictions," *Annals of Statistics*, vol. 1, pp. 38–53.

Ledoux, M. (1996). "On Talagrand's deviation inequalities for product measures," *ESAIM: Probability and Statistics*, vol. 1, pp. 63–87.

Liese, F. and Vajda, I. (1987). *Convex statistical distances*. Teubner, Leipzig.

Lugosi, G. and Nobel, A. (1996). "Consistency of data-driven histogram methods for density estimation and classification," *Annals of Statistics*, vol. 24, pp. 687–706.

Macintyre, A. and Sontag, E. D. (1993). "Finiteness results for sigmoidal neural networks," in: *Proceedings of the 25th Annual ACM Symposium on the Theory of Computing*, pp. 325–334, Association of Computing Machinery, New York.

Massart, P. (1990). "The tight constant in the Dvoretzky–Kiefer–Wolfowitz inequality," *Annals of Probability*, vol. 18, pp. 1269–1283.

McDiarmid, C. (1989). "On the method of bounded differences," in: *Surveys in Combinatorics 1989*, vol. 141, pp. 148–188, London Mathematical Society Lecture Notes Series, Cambridge University Press, Cambridge.

Nadaraya, E. A. (1974). "On the integral mean square error of some nonparametric estimates for the density function," *Theory of Probability and its Applications*, vol. 19, pp. 133–141.

Parzen, E. (1962). On the estimation of a probability density function and the mode. *Annals of Mathematical Statistics*, 33, pp. 1065-1076.

Poggio, T. and Girosi, F. (1990). "A theory of networks for approximation and learning," *Proceedings of the IEEE*, vol. 78, pp. 1481–1497.

Rényi, A. (1959). On the dimension and entropy of probability distributions. *Acta Math. Acad. Sci. Hungar.*, 10, pp. 193-215.

Rosenblatt, M. (1956). Remarks on some nonparametric estimates of a density function. *Annals of Mathematical Statistics*, 27, pp. 832-837.

Rosenblatt, M. (1971). "Curve estimates," *Annals of Mathematical Statistics*, vol. 42, pp. 1815–1842.

Sanov, I. N. (1957). On the probability of large deviations of random variables. *Mat. Sb.*, 42, pp. 11-44 (English translation in *Sel. Transl. Math. Statist. Prob.*, 1, (1961), pp.213-244).

Scheffé, H. (1947). A useful convergence theorem for probability distributions. *Ann. Math. Statist.* 18, pp. 434-458.

Stein, E. M. (1970). *Singular Integrals and Differentiability Properties of Functions*, Princeton University Press, Princeton, NJ.

Stone, C. J. (1984). "An asymptotically optimal window selection rule for kernel density estimates," *Annals of Statistics*, vol. 12, pp. 1285–1297.

Szarek, S. J. (1976). "On the best constants in the Khintchine inequality," *Studia Mathematica*, vol. 63, pp. 197–208.

Talagrand, M. (1995). "Concentration of measure and isoperimetric inequalities in product spaces," *Institut des Hautes Etudes Scientifiques. Publications Mathématiques*, vol. 81, pp. 73–205.

Talagrand, M. (1996). "A new look at independence," *Annals of Probability*, vol. 24, pp. 1–34.

Vajda, I. (1970). Note on discrimination information and variation. *IEEE Trans. Information Theory*, IT-16, 771-773.

Vajda, I. (1989). *Theory of Statistical Inference and Information*. Kluwer Academic Publishers.

Vapnik, V. N. and Chervonenkis, A. Ya. (1971). "On the uniform convergence of relative frequencies of events to their probabilities," *Theory of Probability and its Applications*, vol. 16, pp. 264–280.

Watson, G. S. and Leadbetter, M. R. (1963). "On the estimation of the probability density, I," *Annals of Mathematical Statistics*, vol. 34, pp. 480–491.

Wegkamp, M. (2000). "Quasi universal bandwidth selection for kernel density estimators," *Canadian Journal of Statistics*. To appear.

Wheeden, R. L. and Zygmund, A. (1977). *Measure and Integral*, Marcel Dekker, New York.

Yatracos, Y. G. (1985). "Rates of convergence of minimum distance estimators and Kolmogorov's entropy," *Annals of Statistics*, vol. 13, pp. 768–774.

Yatracos, Y. G. (1988). "A note on L_1 consistent estimation," *Canadian Journal of Statistics*, vol. 16, pp. 283–292.

6

GENETIC PROGRAMMING APPLIED TO MODEL IDENTIFICATION

M. Sebag
Ecole Polytechnique, Palaiseau, France

6.1 Summary

This chapter is interested in model identification from empirical data in the context of experimental sciences.

The approach investigated here is based on Genetic Programming, which pertains to the field of evolutionary computation. Evolutionary algorithms are stochastic population-based optimization methods, known for tackling ill-posed and large-size optimization problems, and renown for their flexibility. The distinctive characteristic of genetic programming (GP) compared to other evolutionary algorithms such as genetic algorithms, is that GP allows for optimization in tree-structured search spaces (e.g. non parametric function spaces; computer programs).

Evolution-based learning should be viewed as complementary to statistical learning, in the light of the bounded rationality advocated by H.A. Simon (1982):

> *In complex real-world situations, optimization becomes approximate optimization since the description of the real world is radically simplified until reduced to a degree of complication that the decision maker can handle. Satisficing seeks simplification in a somewhat different direction, retaining more of the detail of the real-world situation, but settling for a satisfactory, rather than approximate best, decision.*

Indeed, statistical learning beautifully works at setting the data fitting objectives as a well-posed optimization problem, bringing new results about the complexity, convergence and risk of learning at a rapid pace (Vapnik (1998), Cristianini & Shawe-Taylor (2000)).

However, optimal solutions from a data fitting perspective are not necessarily optimal from an engineering perspective. The vocation of the discovered model is to be used, either through an integrated process or through human-machine interaction. In both cases, additional criteria might be relevant to model identification.

To illustrate the first case, the efficiency of a model might be significantly perturbed through

some noise in the integrated process. The search for an optimal model must anticipate these perturbations, e.g. noise must be provisioned for during model identification.
To illustrate the second case, a model might fail to be used only because it violates the expert's prior domain knowledge. Therefore, it might be desirable to restrict the search to syntactically and/or semantically acceptable solutions.

In this perspective, model identification is viewed as an ill-posed optimization problem, involving various objectives besides data fitting. It is claimed that the search space and the learning objectives should be described in a way best reflecting the expert prior knowledge. The rationale behind this claim is that prior knowledge simultaneously allows the learning algorithms to consider very large and complex hypothesis spaces, and keeps the effective complexity within affordable limits since only a small part of the search space needs be effectively explored.

From this perspective, the main advantage of genetic programming is its ability to i) accommodate ill-behaved optimization problems; ii) explore complex search spaces; iii) accommodate constraints (syntactical or other constraints) on the solutions.

The flexibility of the approach is illustrated in the domain of model identification for Mechanics of Solids. Non parametric identification is applied to the discovery of materials behavioral laws. The first application reported considers the space of rheological models (tree-structured assemblies of elementary components). The second application reported discusses the combination of genetic programming and context free grammars, to enforce the discovery of dimensionally consistent expressions of physical variables.

6.2 Introduction

Forty years ago, three places in the world simultaneously and independently gave rise to what is now called artificial evolution (AE), or evolutionary computation (EC); both terms will be used interchangeably throughout the chapter. On the east coast of US emerged Genetic Algorithms (Holland (1975), Goldberg (1989)). On the west coast emerged Evolutionary Programming (Fogel, Owens, & Walsh (1966), Fogel (1995)), while Evolution Strategies emerged in Germany (Rechenberg (1973), Schwefel (1981)).

Artificial evolution was historically inspired by natural evolution, particularly the Darwinian model for the evolution of biological populations. Fascinated by the outstanding performances of biological evolution, computer scientists and engineers independently dreamed of reproducing these performances in an artificial milieu, that of computers. Of course, practical realizations had to wait until computers had gain enough power; only then could one think of emulating the many generations needed to obtain competent biological species and individuals, whichever the definition of competence is.

Needless to say that artificial evolution crudely apes Darwinian evolution. Factually, AE proceeds by stochastically and iteratively recombining and mutating *individuals*, i.e. elements in the search space devised for the problem domain. The milieu is encoded as a function $\mathcal{F}$ on this search space, termed *fitness* function. The milieu induces a pressure on individuals; after the Darwinian motto — *the survival of the fittest* — only the fittest individuals will enter the circuit of recombination and mutation. Ultimately, i.e. after a sufficient number of operations have been performed, fit individuals are obtained, which correspond to the optima or nearly optima of function $\mathcal{F}$.

Concretely, AE constitutes a 0th order stochastic optimization process. The only requirement on the function $\mathcal{F}$ to be optimized is that its value must be computable for all points in the search space; but $\mathcal{F}$ does not have to be convex, differentiable, or even continuous everywhere.
Like other stochastic optimization methods such as simulated annealing (Kirkpatrick, Gelatt, & Vecchi (1983)) or taboo search (Glover & Kochenberger (1995)), AE should be reserved to ill-posed optimization problems (optimization of non-convex or highly multimodal or non differentiable fitness functions). The distinctive feature of AE is that it evolves a set of individuals, termed a *population*, whereas simulated annealing or taboo search evolve a single individual. Being a population-based method considerably increases the range of action of AE, regarding the simultaneous discovery of multiple optima (multimodal optimization (Sareni & Krähenbühl (1998))), multi-criteria optimization (Deb (2001)), constrained optimization (Michalewicz & Schoenauer (1996)), and inverse problems (see for instance Richards, Moriarty, & Miikkulainen (1998).

These many possibilities are based on exploiting the interactions between the individuals in the population. As a counterpart, due to the complexity of these interactions, the theoretical foundations of AE lag far behind its practical achievements. AE has been studied from a theoretical point of view only for its simplest variations (Cerf (1996), Rabinovich & Wigderson (1999)) or for most simple problems (Schwefel (1981), Bäck (1995)). Although such studies shed useful light on the artificial evolution machinery, and greatly increase its scientific respectability, it must be noted that they did not lead insofar to practical progresses, either because they rely on unrealistic assumptions, or because the schedules founding convergence results are slower by orders of magnitude than those used in practice.

This chapter is concerned with a particular variation of artificial evolution, named Genetic Programming. The term Genetic Programming (GP) was coined by John Koza to intend the extension of genetic algorithms to tree-structured search spaces (Koza (1992)). A most interesting instance of tree-structured space is that of computer programs. Koza's pioneering first book, which founded GP as a discipline on its own, is entitled *Genetic Programming: On the Programming of Computers by Natural Evolution* (Koza (1992)). Though the theory of GP is

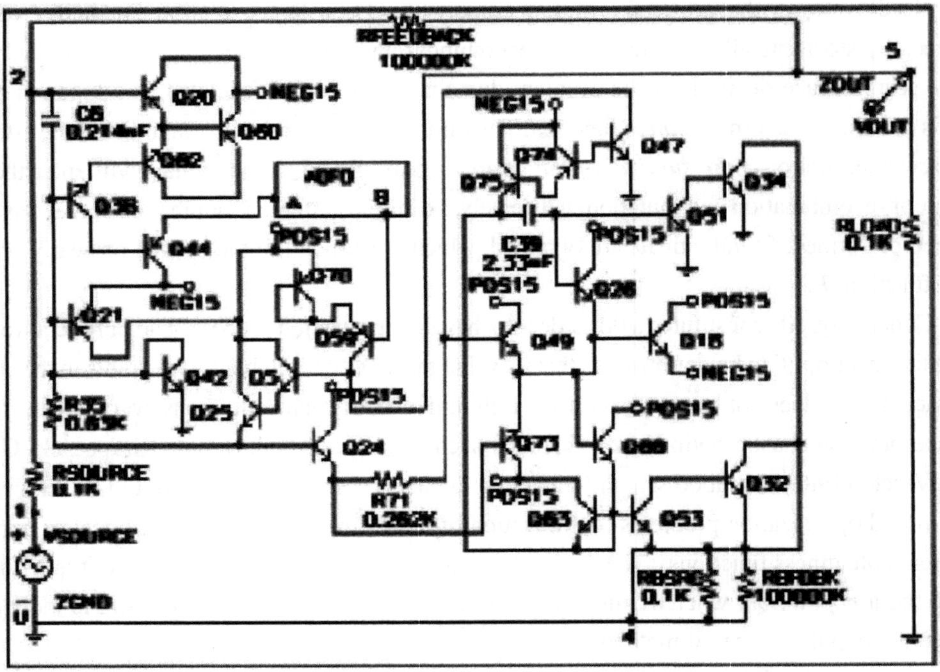

FIGURE 6.1: Non Parametric Optimization of Circuits with Genetic Programming

in its infancy compared to that of other evolutionary algorithms, its practical achievements are among the most impressive of the whole AE field (Banzhaf *et al.* (1998), Koza (1999)). As an example (Fig. 6.1[1]), GP was shown to successfully tackle the design of analogic circuits meeting prescribed specifications (Bennett-III *et al.* (1999)). This example is particularly illustrative of the current possibilities and restrictions of GP.

- First and foremost, to our best knowledge GP is the only optimization technique that could be applied in the space of analogical circuits (graphs connecting three types of elementary devices, transistors, capacities and resistances). Analogical circuits are commonly designed by human experts, having more than ten years of expertise and practice; and this expertise remains largely implicit.

- Second, this application was made possible by the existence of a strong and operational expertise, the 217,000 line SPICE circuit simulator (Simulation Program with Integrated Circuit Emphasis), developed at Berkeley University (Quarles, Pederson, & Sangiovanni-

[1]Reproduced from (Bennett-III *et al.* (1999)) with kind permission of J. Koza.

Vincentelli (1994)). SPICE was used to simulate the behavior of (almost) all genetic individuals; from this behavior was computed the individual fitness; non simulatable individuals were associated a deterrent fitness value.

- Last, the total computational cost is considerable; the success of the application required huge populations (over 600,000 individuals) to be evolved for a few hundred generations. By the way, the huge population sizes commonly used in GP raise a natural objection, that GP might be just another word for Monte-Carlo methods in structured search spaces. Still, all comparative experiments show that GP significantly outperforms random sampling based on a uniform or pre-defined distribution.

The topic addressed in this chapter is the identification of scientific laws from empirical data. Indeed, a wealth of parametric and non parametric approaches allows regression from numerical or mixed (numerical and categorical) data, ranging from data analysis (Breiman *et al.* (1984)) to neural nets (Cun (1987), Hecht-Nielsen (1989)), symbolic machine learning (Mitchell (1997)), support vector machines (Vapnik (1995), Cristianini & Shawe-Taylor (2000)) and ensemble learners (Breiman (1998), Schapire (1999)).

GP distinctive feature is that it allows for learning laws or models expressed in the very language of the expert. In opposition, all abovementioned approaches will provide the expert with black-box models. The only exception is that of symbolic machine learning; the *machine discovery* area (Langley, Simon, & Bradshaw (1983), Falkenhainer & Michalski (1986), Dzeroski & Todorovski (1993), Washio & Motoda (1997), Todorovski & Dzeroski (2001), Washio, Motoda, & Niwa (2001)) tackles the discovery of intelligible models from experimental data. However, machine discovery puts severe requirements on either the data properties (varying only two variables at a time) or the model search space, which makes it hardly practical for scientists and engineers.

To summarize, GP features several unique characteristics in the field of machine learning and modeling:

- Syntactic restrictions — optimization in a structured search space.
  GP can deal with a structured search space, matching the scientific standards for the application domain. This feature is illustrated on the non-parametric identification of rheological models in mechanics (Sebag, Schoenauer, & Maitournam (1997)) (Section 6.4.6). Rheological models, graphically similar to electric circuits (Fig. 6.7(a)), encode the behavioral law of materials. They convey deep knowledge about the material to be modeled, through an arrangement of pure elastic, viscous and plastic components.

- Functional restrictions — constrained optimization.
  GP offers the expert the opportunity to take into account his/her prior guesses regarding

the shape or properties of the target solution (e.g. existence of bifurcations, stability at infinity). This way, functional criteria can be taken into account even though they cannot be expressed at a syntactic level.

The main advantage compared to non-parametric learning is that GP and evolutionary computation can accommodate non continuous, non convex or non differentiable constraints.

- Implicit preferences — multi-modal optimization.
  As mentioned earlier, evolutionary algorithms can handle multi-modal optimization; specific heuristics allow evolution to simultaneously search for all optima of the fitness function. In such cases, the final population will offer the expert many equally fit, though different, models. Therefore, the choice of the expert can be based on criteria other than those encoded through the fitness function (e.g. manufacturing commodities, aesthetics, ...).

  It must be noted that there is no such thing as several equally good solutions in most non parametric learning approaches. Most often, the learner computes the optimal solution of a well posed, regularized problem (Vapnik (1998)), or greedily finds some locally optimal solution (Breiman *et al.* (1984), Rumelhart & McClelland (1986)). This way, the learner avoids overfitting the data, and/or its computational complexity is made tractable.

  In artificial evolution, tractability is manually ensured by limiting the computational resources allotted to learning, granted that the computational cost is the major weakness of EC.
  Overfitting avoidance is one among all learning objectives. The expert has the opportunity i) to declare all of the learning criteria (maximizing the empirical accuracy, minimizing the risk of overfitting, accommodating the environment constraints), ii) to launch the search for finding all non-dominated solutions (Pareto front) (Deb (2001)), iii) and to decide from those solutions the good trade-off between the learning objectives.

This chapter is organized as follows. Next section briefly summarizes the main ideas of artificial evolution and evolutionary computation. Section 6.4 describes canonical genetic programming (Koza (1992), Koza (1994), Banzhaf *et al.* (1998)) and presents an example application in the field of Mechanical Modeling, devoted to the identification of rheological models (1D dynamic behavioral laws) from experimental data (Schoenauer *et al.* (1996)).

Section 6.5 discusses how to take into account prior domain knowledge within the GP frame. To set the ideas, a common yet restrictive domain knowledge for modeling in experimental sciences is dimensional consistency; the target models must be dimensionally consistent with respect to the units of the application domain.

The extension of GP to address syntactic constraints is achieved by combining artificial evolution and grammars (Gruau (1996)). How Grammar-guided GP can address non parametric dimensionally consistent identification is illustrated on another problem taken from Mechanical Modeling, behavioral law identification from indentation tests (Ratle & Sebag (2000)).

6.3 Artificial Evolution

This section reviews the principles of population-based stochastic optimization methods at work in the three main evolutionary computation trends, namely genetic algorithms (GA) (Holland (1975), Goldberg (1989), Mitchell (1996)), evolution strategies (ES) (Rechenberg (1973), Schwefel (1981), Bäck (1995)) and evolutionary programming (EP) (Fogel, Owens, & Walsh (1966), Fogel (1995)). These principles are explained within binary search spaces, historically considered by GAs, and real valued spaces, historically considered by ESs. The tree-structured setting considered by genetic programming will be developed in section 6.4.

6.3.1 Sketch of the algorithm

Evolutionary algorithms are concerned with optimizing a function $\mathcal{F}$, termed *fitness function*, defined on some search space Ω.

$$\mathcal{F} : \Omega \mapsto \mathbf{R} \qquad \text{Find } Argmax(\mathcal{F})$$

For historical reasons, Artificial Evolution terminology is much inspired from biological evolution. An element in search space Ω is termed *individual*. The value $\mathcal{F}(X)$ at point X is the *fitness* of X. A *population* Π is a set of P individuals (not necessarily all distinct). Evolution performs a sequence of *generation* steps, elaborating the next population Π_{t+1} from the current population Π_t.

Table 6.1 presents a glossary of the terms commonly used, with their notation and equivalent in the standard optimization framework.

Initialization.

The first step in an evolutionary algorithm is to create the initial population Π_0. Π_0 is usually obtained as a uniform random sample of the search space Ω. Many works suggest that this step, which is often underestimated and poorly documented in reported applications, is key to the success of artificial evolution (Iba (1996), Kallel & Schoenauer (1997), Daida (1999)). The fitness of all individuals in the first population is computed.

Generation : population update rule.

At each generation, until the stop criterion is satisfied, population Π_t is replaced by population

TABLE 6.1: Artificial Evolution glossary

Individual space	Ω	Search space
Fitness	$\mathcal{F}$	Function to optimize
Individual	X	Element of Ω
Population	Π	Set of P individuals
Current population	Π_t	Set of P individuals at time t
Genetic material		All subsets of Ω overlapping Π_t
Generation		Transformation of Π_t into Π_{t+1}
Selection		Application from Ω^P onto Ω^P
Crossover		Application from Ω^2 onto Ω^2
("sexual" reproduction)		*Variation operator*
Mutation		Application from Ω onto Ω
("asexual" reproduction)		*Variation operator*

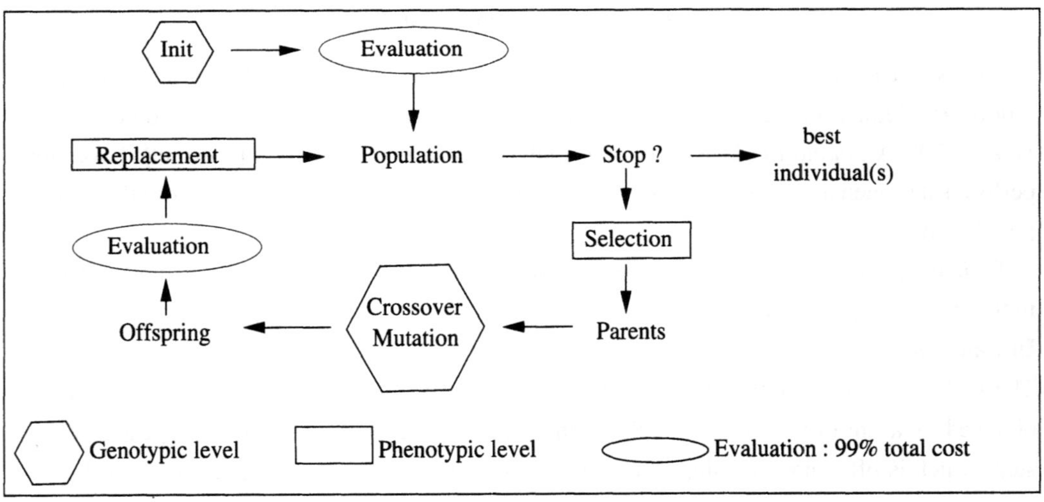

FIGURE 6.2: The Cycle of Artificial Evolution

Π_{t+1}. A generation involves five steps:

Selection determines the individuals in the current population Π_t that will be used to generate the next population. Individuals that are not selected do not transmit any information to the rest of evolution. Selected individuals, termed *parents*, undergo variation operators, crossover and mutation.

Mutation produces an individual termed *offspring* from a individual parent. The offspring is obtained by stochastic perturbation of the parent.

Crossover, or recombination, produces two offspring individuals from two parents. The offspring are obtained by perturbing the parents; the difference compared to mutation is that the perturbation applied onto one parent is conditioned by the other parent.

Evaluation computes the fitness of all offspring individuals obtained by crossover and mutation.

Replacement constructs the next population Π_{t+1} from the current offspring and possibly the parents in population Π_t. Though selection and replacement are distinguished for historical reasons, they clearly play similar roles in evolution.

Stopping criterion.
No results relative to the convergence speed of evolution, compatible with realistic computation costs, are currently available (Cerf (1996)). The system has to be stopped in a manual way: the expert usually prescribes the total number of fitness evaluations allotted to the optimization process, or the maximal computation cost bringing no improvement in the optimization results – or sets some target fitness value.

6.3.2 *Precautions of use*

The main caveat is that artificial evolution should be employed for ill-posed optimization problems only. Whenever gradient-based optimization methods (in continuous spaces) or deterministic combinatorial optimization methods (in discrete spaces) are applicable and give satisfactory results, no one should even think of artificial evolution. Experimental comparisons show that EC is consistently slower by two or three orders of magnitude than deterministic methods, when the latter apply. (Note that the combination of evolution and local optimization can be a judicious option, since any local optimizer can be cast as a mutation operator.)

But EC can successfully be used to optimize functions that are non convex, non differentiable, or not continuous everywhere, and that are possibly defined on exotic search spaces (see Section 6.4.6 for an example). The only assumption done, known as *strong causality* principle (Rechenberg (1973)), is that points that are "sufficiently close" in the search space should almost surely have close fitness values; strong causality can indeed be seen as a relaxed continuity condition.

Incidentally, should the strong causality principle not hold, only Monte-Carlo methods could be considered appropriate; but random exploration is but one particular, extreme, mode of EC.

Along this line, EC can be used as an exploratory learning and optimization approach, allowing preliminary investigations about the hypothesis search space, the most suited learning criteria, and the difficulty of the problem. These investigations might of course lead to recast the problem at hand into a more standard optimization problem, which can thereafter be efficiently solved by existing techniques.

Two other recommendations are relevant to artificial evolution. These are empirical recommendations, emerged over 20 years through experimental trials and errors on artificial and real-world fitness landscapes.

A first recommendation is that **genetic diversity must be preserved**, meaning that the population must contain individuals which are distinct and sufficiently different. At one extreme, the population is formed of P copies of a given individual. Evolving such a population is germane to simulated annealing (Kirkpatrick, Gelatt, & Vecchi (1983)) (actually, it would entail significant overhead compared to simulated annealing).

At the other extreme, all individuals in the population are independent. This entails that the population is exponentially less sensitive to hazards than every single individual, (e.g. less prone to being trapped into a local optimum). The robustness of evolution is thus directly related to the population diversity.

Moreover, the population diversity might also be encouraged to address specific goals, such as multi-modal or multi-objective optimization. Multi-modal optimization aims at finding all global (and possibly local) optima of the fitness landscape. Multi-modal AE (Mahfoud (1995), Sareni & Krähenbühl (1998)) evolves the genetic population in order to sample these optima. The population diversity is ensured at the selection level through penalizing gregarious individuals, or clustering the population and keeping only a few representative individuals per cluster. Multi-objective optimization deals with several, usually antagonistic, optimization criteria (e.g. minimize empirical accuracy AND structural complexity). Multi-objective AE (Deb (2001)) evolves the genetic population in order to sample the Pareto front of non-dominated individuals (i.e. such that there exists no other individual better or equal for all criteria). The population diversity is again ensured at the selection level, as individuals are ranked according to their being dominated.

A second recommendation regards the balance between **Exploration and Exploitation**. The Exploration versus Exploitation dilemma (Goldberg (1989)) refers to the equilibrium evolution should maintain, in order to ignore no region in the search space (exploration), while concentrating the efforts in the promising regions (exploitation). Incidentally, the same dilemma arises in most contexts dealing with incomplete information, ranging from game theory to management studies.

The exploration strength commands the convergence of evolution towards the global optima of the fitness function (Bäck (1995)). Clearly, if evolution does not have the ability to visit the whole search space, no global convergence result may hold. The ergodic exploration of the search space must be enforced through the variation operators, mostly mutation (Radcliffe (1991), Rudolph (1994)).

The exploitation strength commands the rapidity of convergence of evolution toward global or local optima of the fitness landscape. From the strong causality principle, optimal individuals should be close to good individuals. Evolution should therefore intensively visit the neighborhood of the good individuals in the current population, i.e. exploit the good individuals previously discovered. Otherwise, i.e. if good individuals are not exploited, artificial evolution boils down to a random walk in the search space. The conditions for exploitation are realized through selection, pruning the less fit individuals in the current population. Exploitation itself is achieved mainly through crossover, and partly through mutation.

As noted above, achieving an optimal trade-off between exploration and exploitation is a general concern in game theory (Gittins (1989)). Artificial evolution has in particular been analyzed in the view of the one-arm bandit problem (Holland (1975)). However, the main results regarding the optimal balance between exploration and exploitation assume a fixed reward distribution (Lai & Robbins (1985)) ; and this assumption clearly does not hold for AE. Indeed, as the evolutionary population gradually moves toward fitter regions, the reward distributions associated to exploration and exploitation actions are modified too. This point shall be returned to in the perspectives for further research (section 6.6).

The trade-off between exploration and exploitation is thus adjusted manually, depending on the problem at hand. If exploration dominates, evolution more or less behaves like a random walk. Inversely, if exploitation dominates, artificial evolution gets trapped in the first basins of attraction it meets, and premature convergence occurs.

6.3.3 Representation

Artificial evolution inherits from biological evolution the distinction between the description of an individual (e.g. DNA code) and its manifestation (shape or behavior), which might be conditioned by the milieu. The description of an individual, element of the search space Ω, is referred to as genotypic information. The manifestation of an individual is referred to as phenotypic information.

Although the aims of biological evolution, if any, are unclear, the aim of artificial evolution clearly is to find individuals with optimal manifestations. In other words, the goal of artificial evolution is defined at the phenotypic level.

The distinction might appear redundant and unnecessary as phenotype-genotype are in one-to-one correspondence in most learning settings (e.g. a logistic function, a SVM are canonically represented as a fixed size real-valued vector). But in more complex spaces, the representation might induce classes of equivalences on the search space. For instance, due to permutations, there are several possible codings for the architecture + weights of a neural net. The phenotype would correspond to the neural net up to permutations; the genotype would correspond to a particular coding of this neural net.

Although a natural concern would be to design the problem representation in order to ensure a one-to-one genotype-phenotype mapping, this might conflict with the expert "natural" representation. Ideally, the problem representation is jointly designed by the evolution expert and the problem domain expert. The phenotype-genotype mapping concretizes the trade-off between the understanding of the problem domain and the particular fitness landscape on one hand, and the understanding of artificial evolution, its exploration and exploitation abilities, on the other hand.

All in all, this distinction leads to decompose the evolution components into three categories:

- at the genotypic level are initialization, crossover and mutation; the former produces genotypes by sampling the search space; the latters manipulate parent genotypes and produce offspring genotypes.
- at the phenotypic level are selection and replacement, which do not care about genotypes and only depend on the individual fitness.
- the milieu is modeled by the fitness function, which transforms a genotype into a phenotype and allows the evaluation of the phenotype.

The genotypic representation of the problem domain must be designed in adequacy with the crossover and mutation operators. The goal is to increase the probability of finding a "good path" in the fitness landscape. This in turn depends on the average number of crossover and mutation moves needed to transform a population into another one (traveled distance). The reader interested in a probabilistic formulation of the combined effects of representation, crossover, mutation and selection in terms of Markov chains on the populations space is referred to (Rabinovich & Wigderson (1999)).

Early EC works have advocated the choice of binary representations (minimal alphabet principle); some pros and cons of the minimal alphabet dogma are respectively discussed by (Goldberg (1989)) and (Antonisse (1989)).

The current state of the art recommends to tailor the representation of the problem domain to account for any prior knowledge about the application domain (Radcliffe (1991), Janikow (1993), Daida (1999)).

Same recommendations are done regarding the variation operators and the genotype-phenotype mapping.

One-to-one mapping have been recommended in early studies, based on the fact that exploiting distinct equivalent genotypes entails a useless computational overhead, at best, and causes evolution to significantly slow down in the worst case (Radcliffe (1991)).

This recommendation was later modulated, for the following reason. Empirically, many-to-one genotype-phenotype mappings tend to create fitness plateaus, all individuals in a genotypic region having same fitness. Evolution admittedly slows down when wandering on a plateau. Nonetheless, the chances of escaping from a plateau might be better than those of escaping from a single local optimum. This is because the exploitation of a plateau is less harmful to the genetic diversity of the population, everything else being equal; and the better the genetic diversity, the shorter the traveled distance to other populations.

Most generally, any change in the problem representation will prompt to reconsider the variation operators and the initialization procedure as well.

The components of an evolutionary algorithm will now be detailed, distinguishing the genotypic, or representation-dependent components — initialization and variation operators — and the phenotypic, or fitness-dependent components — selection, replacement and stopping criterion.

The description of the genotypic components (initialization, crossover and mutation) will be illustrated on the two most popular genotypic spaces, the binary ($\Omega = \{0,1\}^n$) and the real-valued case ($\Omega \subset \mathbf{R}^n$).

6.3.4 *Initialization*

Following the genetic diversity principle, the initial population must be as diversified as possible. In the meanwhile, the optimization task is obviously facilitated if good individuals appear in the initial population.

The key point is to define the diversity criterion and the corresponding sampling mechanism, suited to the optimization problem at hand.

Binary case

The standard binary initialization procedure constructs P individuals, the i-th bit of each individual X being uniformly set to 0 or 1. This procedure was unquestioned for years; it does ensure that all individuals in the search space have same probability to appear in the initial population.

However, applications in shape optimization led to reconsider the above procedure (Kallel & Schoenauer (1997)). A standard representation for shape optimization is binary genotypes.

Assuming that the problem domain is divided into finite elements, a given shape X is coded as a binary string, the i-th bit being 1 or 0 depending on whether the i-th element belongs to shape X.

In this frame, the above random initialization procedure samples individuals/shapes that fill about half the problem domain, after the Central Limit theorem. This bias might, or might not, be suited to the shape optimization problem at hand.

A "more uniform" initialization procedure, in this case, was designed as follows (Kallel & Schoenauer (1997)): for each individual X in the initial population
– draw p_X uniformly in $(0,1)$; p_X stands for the material density in shape X;
– the i-th bit in X is set to 1 with probability p_X.

Real-valued case

In the real-valued setting, initialization is straightforward when considering a bounded domain (e.g. $\Omega = [0,1]^n$). The i-th component of each individual is then uniformly drawn in $[0,1]$. Clearly, no such straightforward initialization procedure exists in the case of unbounded domains. Usually, bounds are decided on the basis of preliminary experiments; preliminary experiments might use a uniform sampling of both the mantissa and the exponent of floating point numbers.

Discussion

The initialization step might also be guided from prior knowledge about the problem domain. For instance, it is possible to introduce in the initial population some solutions which have been manually worked out. However, since these solutions are much better than random ones, they might soon crowd the population due to the selection effects. Some care must be exercised to prevent the loss of genetic diversity.

Last, the expert might guide evolution by setting intermediate goals, through specifying a sequence of fitness functions $\mathcal{F}_1, \mathcal{F}_2, \ldots, \mathcal{F}_K = \mathcal{F}$. This powerful heuristics was first proposed to design an automaton (a pair of articulated legs) performing human-like walk (de Garis (1990)). As this goal proved too difficult to reach from scratch, a sequence of increasingly difficult goals was considered instead. In the first stage, the goal was being able to stand up without falling down ($\mathcal{F}_1$ measuring the number of time steps before the automaton falls down, starting from a random initial position). In a second stage, the goal was being able to move the legs as much as possible without falling down ($\mathcal{F}_2$ measures the number and amplitude of moves before the automaton falls down). The third and last stage was finally concerned with optimizing the distance $\mathcal{F}_3$ walked by the automaton, and the optimization proved successful.
Goals were successively considered by evolution, the initial population chosen for tackling goal $\mathcal{F}_i$ being the final population evolved with fitness function $\mathcal{F}_{i-1}$. This allows for gradually im-

proving the competence of individuals; the current population incorporates some information about the fitness landscapes explored in the chain of evolution. These cumulative fitness effects are referred to as Behavioral Memory of evolution (de Garis (1990), Schoenauer & Xanthakis (1993)).

Behavioral memory can in particular be employed to achieve evolutionary constrained optimization (Schoenauer & Xanthakis (1993), Michalewicz & Schoenauer (1996)). The series of fitness landscapes corresponds to the satisfaction of an increasing number of constraints. The last population so obtained samples the admissible region, i.e. individuals in the search space that satisfy all constraints. In the last stage, this set of admissible individuals is taken as initial population and it undergoes evolution to optimize the actual objective function. It is emphasized that preserving the population diversity all along the process is even more important than for standard optimization. The reader interested in evolutionary constrained optimization is referred to (Deb (2001)) for a comprehensive presentation.

6.3.5 *Crossover*

Canonical crossover produces with probability p_c a pair $(X'\ Y')$ of offspring from a pair (X,Y) of parents; with probability $1-p_c$, the offspring are copies of the parents.

Binary case

In a binary setting, crossover crudely mimics biological sexual reproduction; offspring are obtained by patching subparts of the parents according to randomly chosen cut-points.
In a real-valued setting, the offspring are randomly selected in the hyper-rectangle defined from the two parents.

$$\begin{array}{rcl}
\text{Crossover}: \quad \Omega\times\Omega & \mapsto & \Omega\times\Omega \\
\begin{array}{r} X=(X_1,\ldots,X_n) \\ Y=(Y_1,\ldots Y_n) \end{array} & \mapsto & \begin{array}{l} X'=(X'_1,\ldots,X'_n) \\ Y'=(Y'_1,\ldots Y'_n) \end{array} \\
X,Y\in\{0,1\}^n & \mapsto & \left\{\begin{array}{l} X'_i=X_i \\ Y'_i=Y_i \end{array}\right. OR \left\{\begin{array}{l} X'_i=Y_i \\ Y'_i=X_i \end{array}\right. \\
X,Y\in\mathbf{R}^n & \mapsto & \left\{\begin{array}{l} X'_i=(1-\alpha_i)X_i+\alpha_i Y_i \\ Y'_i=(1-\alpha_i)Y_i+\alpha_i X_i \end{array}\right.
\end{array}$$

Practically, binary crossover most often uses one or two cut points. The general formulation given above corresponds to uniform crossover (Syswerda (1989)).

Real-valued case

Real-valued crossover has been less studied than binary crossover. Some recommendations have been formulated regarding barycentric weights α_i, which should allow the offspring to fall outside the hyper rectangle defined from the parents (α_i uniformly selected in $[-.5, 1.5]$ rather than $[0, 1]$, as done in early works) (Michalewicz (1996)). This prevents crossover from automatically contracting the convex hull of the population.

Discussion

During crossover, offspring are produced by sampling a given distribution (e.g. uniform distribution of all individuals that share the common bits of both parents in uniform binary crossover, uniform distribution on the convex hull defined from the parents in the real-valued case). The difference compared to mutation (see section 6.3.6 below) is that the offspring distribution with respect to one parent (amplitude and direction of perturbations) is conditioned by the other parent.

In other words, the crossover distribution is biased from the current population. In particular, its amplitude (the average distance between an offspring and a parent) generally depends on the diversity of the population. In the earlier generations, crossover effects large perturbations (e.g. mutating half of the bits in binary case, or generating offspring half-way from both parents in real-valued case); it behaves as an exploration operator. However, as the population climbs toward more fit regions, its diversity is reduced, and the crossover amplitude decreases accordingly; crossover thereafter behaves more as an exploitation operator.

The fact that the crossover effects are biased from the current population is both a strength and a weakness. On one hand, crossover alone cannot counteract the loss of population diversity. For instance in a binary setting, if a coordinate has same value over the whole population (genetic drift), the diversity cannot be recovered through crossover and evolution has to wait for a lucky mutation.

On the other hand, the crossover distribution is dictated from the parents; it encapsulates an estimation of the optima distribution, as parents are more fit than random individuals. Due to this additional information, it is expected that the individual neighborhoods visited by crossover will be more fit than those explored by mutation. Accordingly, crossover might make larger and more profitable steps, thereby speeding up the evolution progress toward fit and fitter regions.

The good case performance of GAs compared to single-individual stochastic methods (hill-climbers, simulated annealing, taboo search) are mostly attributed to crossover efficiency (Goldberg (1989)). The idea is that crossover allows the offspring to cumulate the "good building blocks" in the parents, while preserving the parent common assets (*your brain and my beauty*).

In the evolutionary programming framework, it is considered instead that crossover merely compensates for insufficient mutation. Well adjusted mutation can be shown to outperform mutation + crossover on part of the existing benchmark problems (Fogel & Stayton (1994)).

The debate about the comparative efficiency of mutation and crossover has been going on for long. Experiments have been designed to check whether crossover intelligently exploits the population information, or automatically allows for more exploration in the earlier stages of evolution. The hypothesis was tested by comparing the results obtained with binary crossover, against those obtained with *macro-mutation*, defined as crossing over the parent with a random individual, everything else being equal (Jones (1995)). Experimental results were again inconclusive; macro-mutation happens to outperform standard crossover on some problems, and vice versa. An interesting line of research aims at defining prior experiments or criteria in order to evaluate the relative efficiency of variation operators (Naudts & Kallel (2000)).

6.3.6 *Mutation*

Mutation constructs the offspring individual by stochastic perturbation of the parent. Like for crossover, the mutation operator is applied to all individuals of the current population with a given probability p_{mut}.

Binary case

In a binary setting, mutation flips with a low mutation probability p_m each bit in the mutated parent. In historical GAs, the probability p_{mut} to apply mutation always equals 1, and tuning the importance of mutation is left to the internal parameter p_m of the mutation operator.

$$\begin{array}{rcl} \text{Mutation}: & \Omega & \mapsto & \Omega \\ & X=(X_1,..X_n) & \mapsto & Y=(Y_1,..Y_n),\ Pr(Y_i \neq X_i)=p_m \end{array}$$

Real-valued case

In a real-valued setting, a Gaussian perturbation $\mathcal{N}(0,\sigma)$ is added to each parent component.

$$\begin{array}{rcl} \text{Mutation}: & \Omega & \mapsto & \Omega \\ & X=(X_1,..X_n) & \mapsto & Y=(Y_1,..Y_n),\ Y_i=X_i+\mathcal{N}(0,\sigma) \end{array}$$

Discussion

The mutation operator instrumentalizes a given distribution centered on the parent at hand; mutation produces offspring by sampling this implicit distribution. Depending on the shape of this distribution, mutation will perform more exploitation (the distribution is peaked around the parent, e.g. for low values of p_m or σ), or more exploration (long tail distributions).

Practically, mutation is intended to perform exploitation and visit the close neighborhood of the parent, most of the times. But mutation is also responsible for the ergodic visitation of the search space (Radcliffe (1991)) ; each point in Ω must be reachable in a finite number of mutation steps from any other point. Accordingly, mutation should with non zero probability produce an offspring arbitrarily far from the parent, i.e. it achieves exploration too. The exploration role of mutation is predominant in GAs, as the population diversity can only be restored by mutation; the crossover operator (see 6.3.5) might become ineffective when the population diversity is too poor. In the last resort, only a lucky mutation can drive evolution away from a local optimum.

The distribution of the mutation steps (genotypic distance between the offspring and the parent) commands the convergence and speed of evolution, as it controls the exploitation/exploration trade-off (together with crossover and selection).

In a binary setting, mutation distribution is a binomial law with parameter p_m; p_m must be low enough to avoid random walk, and large enough for mutation to be sufficiently effective *in each generation*. Diverse schedules have been investigated for the adjustment of p_m: the simplest one sets p_m to a user-supplied value (Goldberg (1989)) ; dynamic schedules, reducing the value of p_m from an initial value according e.g. to a hyperbolic schedule (Bäck & Schütz (1996)) have also been considered; another possibility is to adapt p_m depending on the frequency of successful mutations (when the offspring is more fit than the parent) (Rechenberg (1973), Davis (1989)). A wide variety of parameter dynamic control has been proposed in the literature; a comprehensive presentation might be found in (Hinterding, Michalewicz, & Eiben (1997)).

The mutation distribution induces a metric among individuals (traveled distance, average number of mutations needed to reach a point from another one). This metric should be considered bearing in mind the "natural" metric of the problem domain. For instance, when standard binary genotypes are used to represent real-valued parameters (a common though questionable representation for real variables), reconsidering the mutation design or the coding proves beneficial (Caruana & Schaffer (1988), Goldberg (1989)). A first option is to adjust the mutation rates depending on the bit weight, perturbing more frequently the low weight bits. Another possibility is to rather use Gray than standard binary coding; this avoids the presence of so-called Hamming cliffs (large Hamming distance) between the two consecutive integers 2^n and $2^n - 1$ and the shortcuts between 1 and $2^n + 1$) .

To the possible extent, the representation and the mutation designs must be geared to each other, and reflect the phenotypic neighborhoods.

Real-valued case revisited: self-adaptive mutation

In a real-valued setting, mutation distribution is most often a Gaussian law centered on the parent. It has been claimed that Cauchy distributions and other long tail distributions are more appropriate than Gaussian ones to evolutionary mutation because of their better exploration

Mutation ($X = (X_1, \ldots X_n, \sigma_1, \ldots, \sigma_n)$)	
$\varepsilon := \tau \times \mathcal{N}(0,1)$	Recommended : $\tau \propto \frac{1}{\sqrt{2 \times \sqrt{N}}}$
Foreach $i = 1 \ldots N$	
$\varepsilon_i := \tau' \times \mathcal{N}(0,1)$	Recommended : $\tau' \propto \frac{1}{\sqrt{2 \times N}}$
$\sigma'_i := \sigma_i \times e^{\varepsilon + \varepsilon_i}$	
$X'_i := X_i + \mathcal{N}(0, \sigma'_i)$;	
Return $X' = (X'_1, \ldots X'_n, \sigma'_1, \ldots, \sigma'_n)$	

FIGURE 6.3: Self adaptive mutation

ability (Yao, Liu, & Lin (1999)) – some experimental results support that claim. On the other hand, a theoretical analysis of evolution dynamics in the case of the sphere problem proved that a whole set of distributions (from Gaussian to Student with 5 degrees of freedom, but not including Cauchy) have the same local convergence speed (Rudolph (1997)).

The main practical point is to adjust the Gaussian law parameters. Standard deviation σ might be a scalar (isotropic mutation); it might also be a vector (there is no reason why the standard deviation should be the same for all coordinates of the search space). It might last be a matrix (as the space coordinates might not be independent) (Hansen & Ostermeier (1996)). The most widely used model is the vectorial one: mutating individual $X = (X_1, ..X_n) \in \mathbf{R}^n$ relies on the standard deviation vector $(\sigma_1, \ldots \sigma_n)$; to each coordinate X_i is added a Gaussian perturbation $\mathcal{N}(0, \sigma_i)$.

A most effective and elegant way of adjusting parameters σ_i, referred to a self-adaptive mutation, has been proposed by (Schwefel (1981)).

Self-adaptive mutation is based on extending the representation of the individuals; an individual genotype X includes besides the fitness related components X_i, the σ_i used to evolve the X_i. Parameters σ_i undergo the evolution effects, i.e. they are mutated and recombined as the X_i. This way, the range of the mutation distribution specific to the individual, is adjusted through evolution itself.

It was considered at first that, through self-adaptive mutation, evolution "optimized for free" the mutation parameters σ_i, even though genetic individuals are selected on the only basis of the X_i values.

The efficiency of self adaptive mutation in order to adjust the σ_i was later explained, as bad σ_i values adversely affect the cumulated transmission of the genetic material in an individual (the expected number of its descendants after some generations).

Indeed, a parent having (too) large σ_i values will tend to produce far away offspring — also bearing large σ_i values. The descendants of this parent might be harmed as they tend to follow a broad distribution on the search space.
Inversely, a parent having (too) small σ_i values will be discarded for its descendants can only make small steps and small improvements, in case some other individual with a larger σ_i happens to make a fitness leap.
In summary, relevant individuals, i.e. with viable descendants, bear appropriate values for the X_i (otherwise the individual could not have resisted the current selection step) and for the σ_i too (otherwise, the descendants of this individual will be prone to errand or stagnate, compromising the survival of the genetic material).
Interestingly, σ_i empirically reflects the derivative of the fitness function wrt the i-th coordinate; the σ_is can be interpreted as sensitivity estimators, implicitly constructed along evolution.

Special care about real-valued mutation must be exercised when the search space is bounded (i.e. $\Omega = [0,1]^n$). The offspring must be projected onto Ω when it has been mutated too far away. The projection must be such that it ensures a "fair" distribution of the offspring. Truncating the offspring (e.g. replacing 1.23 by 1.) implies that the boundary of the space will be more intensively visited — which again, might be appropriate or not depending on the problem. Toric projection (e.g. replacing 1.23 by .23) is not continuous. Another widely used heuristic is based on symmetrical projection (e.g. replacing 1.23 by .77).

6.3.7 *Selection and replacement*

This section details the *evolution engine* implementing artificial Darwinism – *the most adapted individuals survive and reproduce*. It includes both selection (choice of those who will reproduce) and replacement (choice of those who will survive onto the next generation) as represented on Figure 6.3.1.

Selection and replacement both choose some individuals in the current population based on their fitnesses. Their common features are that both favor individuals with better fitness, and both can be either deterministic or stochastic. The main difference is that selection associates an integer to each individual in the population (the number of times the individual will participate into the reproduction), while replacement associates a binary value to each parent and offspring (whether the individual will survive, i.e. belong to the next population).

Historically, most evolutionary algorithms (GAs, ES and EP) used a single one of selection and replacement to enforce the selection pressure on the populations.

Selection

Selection chooses the individuals in the current population that will become parents. Only parents might transmit their genetic material to the next population. Since only selected individuals will be exploited, selection is the main operator for controlling the strength of exploitation.

At one extreme (maximal selection pressure), selection only retains the best individual in the population. The next population is thus formed of clones of this best individual, genetic diversity is lost, and artificial evolution boils down to stochastic hill-climbing.

At the other extreme (no selection), all individuals might become parents, artificial evolution only performs exploration and boils down to a random walk in the search space.

Any intermediate strategy can be obtained through adjusting the selection effects. But how to best adjust the strategy depending on the problem at hand remains so far a skill more than a science.

Historical GAs involve a stochastic selection of the parents; the probability of selecting an individual is proportional to its fitness. The sampling mechanism is a roulette wheel, based on the normalized fitness in the population. Algorithmically, roulette wheel operates as follows. A number r is uniformly selected in $[0, \sum_{j=1}^{P} \mathcal{F}(X_j)]$; the i-individual in the population is selected for r belonging to $[\sum_{j<i} \mathcal{F}(X_j), \sum_{j\leq i} \mathcal{F}(X_j)]$.

The selection distribution probability can be rescaled in different ways, in order to control either its maximal value, or its regularity. Fitness scaling allows for controlling the maximal value, also termed *fitness pressure*, i.e. the expected number of times the best individual is selected (Goldberg (1989)). Fitness ranking allows for linearizing the distribution, the expected number of times an individual is selected being proportional to its rank in the population (Goldberg (1989)).

Another very popular selection method, that has been designed to handle some deficiencies of proportional selection is the tournament selection: to select one individual, T individuals are uniformly drawn from the population and the best one is returned. The size T of the tournament determines the selective pressure.

Historical EP and ES do not use any selection: all individuals of the current population have the same probability of becoming parents. Artificial Darwinism is ensured through the replacement procedure. In EP, each parent gives birth to one offspring (using mutation only) while in ES, each parent gives birth to a number of offspring (using mutation only in early ES; mutation and crossover are both used in modern ES variants).

Replacement

Replacement forms the next population from the current parents and offspring, deciding the ones that will survive.

Historical EP uses a stochastic replacement (Fogel (1995), Blickle & Thiele (1995)) based on a global tournament: each individual "encounters" a number T of opponents (T is the tournament size) and increases its score by 1 if it has a better fitness; the best scores survive onto the next generation. In this frame, the selective pressure is controlled from the tournament size T.

Historical ES uses a deterministic replacement to construct the next parent population from the current μ parents and their λ offspring. In the (μ, λ)-ES setting, the next μ parents are the best individuals among the offspring; in the $(\mu + \lambda)$-ES setting, the next μ parents are the best individuals among the parents *and* the offspring (Schwefel (1981), Bäck (1995)). The analysis of ES dynamics on the sphere problem provides theoretical insights into the role of parameters μ and λ, as well as optimal values for this problem (Bäck (1995)). Typical values for μ and λ in the literature are (3,20), (7,50), (10,100).

Historical GAs uses a *generational replacement*; the next population simply consists of the offspring generated from the current parents.

Elitism.
In most settings except $(\mu + \lambda)$-ES, it might happen that the best so far individual disappears from the population (that is to say, the population best fitness decreases). This phenomenon, termed fleeting discoveries (Mitchell (1996)), slows down evolution as a promising individual must be discovered several times before it can be exploited.

The elitism heuristics is meant to avoid fleeting discoveries; the best so far individual is arbitrarily put in the next population in case no better offspring have been discovered. Theoretical results show that elitism favors the premature convergence of evolution, and adversely affects the global convergence results. Nevertheless, elitism is used in most evolutionary real-world applications.

Steady-State engines

An alternative to the generation of a whole population of offspring is known as steady-state evolution. In this variation, a single parent is selected in each time step, according to one selection mechanism above. A single offspring is generated by applying the variation operators, which is then incorporated to the current population according to some replacement procedure. The tournament procedure is often used to select the parent to be replaced by the new offspring (with inverted order: the "winner" is the one with worst fitness among the T uniformly selected individuals).

Steady state evolution using tournament selection presents the advantage that the whole population needs not be available simultaneously. This is most appreciated in a parallelization perspective (asynchronous parallelization). Parallelized evolution receives much attention, given the huge computational efforts needed in real-world applications. Incidentally, parallel implementations of evolution suggest diverse heuristics in order to control and delay the diffusion of genetic material (island models (Starkweather, Whitley, & Mathias (1991)), spatial mating using 2D or 3D toric neighborhoods (Gorges-Schleuter (1998)), distributed learning (Anglano *et al.* (1998))), which are found to be useful in sequential implementations too, in order to improve the population diversity.

6.3.8 Fitness Evaluation

The fitness evaluation amounts to 99,9% of the total computational cost of evolution in real-world problems. Evolution commonly asks for some 10^5 up to 10^9 fitness evaluations. Though the fitness evaluation step is easily and/or often parallelized (Anglano *et al.* (1998), Koza *et al.* (1999)), any refinement regarding the fitness evaluation will be rewarded, and it makes it worth spending some time on the particularities of the fitness function.

First of all, it is most often implemented as a computer program. (An interesting exception is that of *subjective optimization*, where the human expert takes in charge the fitness evaluation and selection steps inside the evolution loop (Herdy (1996), Venturini *et al.* (1997)). As fatal error cases will cause evolution to stop, evolution will tend to detect many such errors previously unnoticed; even more so, as evolution explores regions which are beyond the normal range of utilization of the fitness program (e.g. using Finite Element Methods (Dasgupta & Michalewicz (1997))).

Second, the fitness program implements the domain oracle in the sense that it virtually allows the evaluation of any individual in the search space. Any available and usable knowledge about the problem domain and the expert preferences — the milieu — might be incorporated in the fitness function.

Third, the fitness function must guide evolution to the desired optima. Any rescaling or other modifications of the fitness function will do, provided they do not change the fitness order. Along the same lines, the fitness function can be modified through adding heuristic terms, in order to preserve population diversity and obtain special evolution effects (multi-modal optimization (Sareni & Krähenbühl (1998)), constrained optimization (Michalewicz & Schoenauer (1996)), constraint satisfaction (Eiben & Ruttkay (1996))).

In summary, besides a deep expertise of the problem the fitness function should reflect some understanding of the evolution machinery as well. Typically, bugs in the fitness program only harm evolution if these bugs deceptively lead evolution toward some actually poor optima. When this is the case, the only thing to do is fixing the fitness program; questioning the relevance of the fitness function is also an option (Mansanne *et al.* (1999)).

But evolution might successfully resist many bugs in the fitness program. As an example, if an unfit individual is mistaken for a very unfit individual, or conversely, evolution will not be affected for all such individuals are filtered out by selection. The evolution robustness wrt such inaccuracies in the fitness evaluation can be used on purpose to simplify the fitness function design, and achieve significant computational savings. An elegant such simplification in the domain of evolutionary robotics (the fitness function involving a robot simulator) has been proposed by (Jacobi (1998)).

Specific heuristics have been designed to handle computationally heavy fitness functions (e.g., involved in Mechanics, Hydrodynamics, and more generally Numerical Engineering optimization applications). The most interesting *Response surface method* builds a surrogate model of the fitness function, which is used in alternance with the true fitness. Among the most popular surrogate models are Neural Networks (Poloni & Pediroda (1997), Jin, Olhofer, & Sendhoff (2000)) and krieging (Ratle (1998), El-Beltagy, Nair, & Keane (1999)). Machine Learning approaches, based on Case-Based Reasoning (Khaleed & Hirsh (1997)) or Rule learning (Cervone, Kaufman, & Michalski (2000)) have also been proposed.
The difficulty is to determine when and how one should swap from the fitness estimate to the true fitness and back. If the same surrogate model is used for too long, the population might become too specific, and never recover its diversity enough to accommodate any further fitness estimate; instead, the results are strongly biased toward the optima of the surrogate model (El-Beltagy, Nair, & Keane (1999), Jin, Olhofer, & Sendhoff (2000)).

Some fitness landscapes adversely affecting evolution have been studied in binary framework. One of them, referred to as Needle in the Haystack, involves a large plateau with possibly many local optima besides a narrow peak (Goldberg (1989), Radcliffe (1991)). An other situation is that of wide fit regions, driving the population away from the optimum (deceptive fitness landscapes (Das & Whitley (1991), DeJong (1992))). The Royal Road problem showed unexpected difficulties, though designed to best reward the good schemas in the search space, as the first best discovered schema tend to erase the progress done in other parts of the search space (Forrest & Mitchell (1993)). Other difficult problems (Long Path problems) involve a narrow exponentially long path gradually leading to the optimum (Höhn & Reeves (1996), Garnier, Kallel, & Schoenauer (1999)).

Not surprisingly, the ideal fitness landscape is the sphere, the individual fitness being inversely correlated to its distance to the optimum (Jones & Forrest (1995)). But of course evolution is never (and never should be) presented with such simple landscapes.

6.3.9 Stop criterion

The last but not least important aspect of artificial evolution concerns the stop criterion. As can be expected from a stochastic dynamic process, the trajectory of evolution might stagnate for generations in a given basin of attraction, before it can continue its progression (Eiben & Ruttkay (1996)).

Admittedly, stopping too early will preclude the discovery of good optima. On the other hand, evolution usually spends half its time in gaining the very last fitness points.

Some remedies have been proposed, based on coupling evolution with a local, e.g. gradient-based, optimizer. A loose coupling is when the local optimizer only operates on the best individual discovered by evolution. A tight coupling uses the local optimizer as one among all variation operators, likewise mutation. This way, evolution actually explores the space of local optima instead of the whole search space (Radcliffe & Surry (1994)). The crossover and mutation operators should preferably be redesigned along with this redefinition of the search space.

Practically, the stop criterion most often restricts the total number of fitness evaluations allowed, and/or the number of consecutive evaluations bringing no fitness improvement.

6.3.10 Success criterion

Artificial evolution produces an optimum (model and associated fitness) which is viewed as a random variable. As standard for all randomized algorithms, the validation of an evolution algorithm is based on the histogram of the optima obtained over a few dozens of independent runs (everything being equal except the random seed). Two evolution engines are compared using statistical tests, most often t-tests.

From an application viewpoint, two problem categories are distinguished. The first category includes design problems (Schnier & Gero (1998)). In such problems, it makes no sense to consider the average solution obtained; the goal is to find at least one very good solution. The success of the application measures the best model ever discovered – favoring large standard deviation is indeed an option.

The second category includes process problems (control, planning); the goal is to find a model with good average fitness, and to do so with good confidence. The success of the application measures the average fitness of the models, together with small standard deviations.

6.4 Genetic Programming

The search space considered in most if not all optimization settings (including GAs, ES and EP), involves a finite fixed number of coordinates (discrete or real-valued attributes). Along these coordinates, any individual is completely defined by a vector of values.

But a recurrent dream in computer science and in artificial intelligence is known as "the program that writes the right program". Finding the "right program" is amenable to optimization and/or learning in the computer program space. However, the description of a program is hardly amenable to an attribute-value representation.

This limitation of attribute-value spaces has motivated specific extensions to handle more complex, structured search spaces. On the learning side, in the domain of Inductive Logic Programming (Muggleton & De Raedt (1994)) and Relational Learning (Quinlan (1990)) have been developed efficient systems (Quinlan (1990), Muggleton (1995), Blockeel & Raedt (1998), Sebag & Rouveirol (2000)), and successful program identification has been achieved in application domains such as chemistry (Srinivasan *et al.* (1996)) and natural language processing (Cussens (1997)), among others.

On the evolutionary side, the extension of artificial evolution to structured search spaces, among which computer languages, was addressed by Genetic Programming (Koza (1992), Koza (1994), Banzhaf *et al.* (1998)).

The GP scheme parallels standard artificial evolution: after the first population is initialized, a sequence of generations iteratively transforms the current population Π_t into Π_{t+1}, using phenotypic operators (selection and replacement) and genotypic variation operators (crossover and mutation), until the stop criterion is satisfied.

The main difference between GP and GAs concerns the structure of the genotypes. All genotypic operators (initialization, crossover and mutation) therefore have to be redefined.

6.4.1 Representation

Historically, the first GP search space was a subset of LISP language (Koza (1992)). Rapidly, GP was extended to accommodate any tree-structured search space, described from a set of leaves or terminals $\mathcal{T} = \{x, y, \ldots\}$ and a set of nodes or operators of a given arity $\mathcal{N} = \{p_k, q_l, \ldots\}$ (e.g. $+_2$ stands for the binary addition).

The search space Ω includes all well-formed expressions e, recursively defined as being either a terminal ($e = x$) or the application of a k-ary operator op_k to a list of k well-formed expressions ($e = op_k(e_1, \ldots, e_k)$). Space Ω can conveniently thought of as all sentences recognized by a grammar; this point will be discussed further on.

Example 1: Polynoms

A simple example of tree-structured space is that of polynoms of any degree of variables $x_1, \ldots, x_n$. Terminal set $\mathcal{T}$ includes all variables x_i and a particular $\mathcal{R}$ terminal, standing for any real-valued

$x+1+2\times x^2$

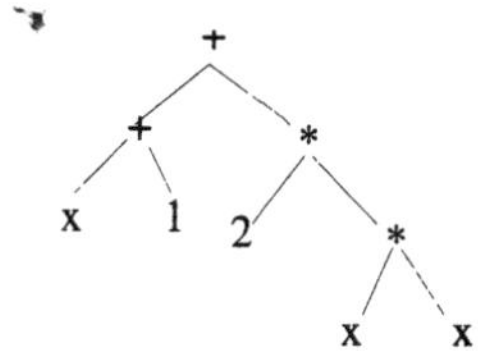

FIGURE 6.4: Polynoms and tree-structured expressions

constant:

$$\mathcal{T} = \{x_1, \dots, x_n, \mathcal{R}\}$$

Node set $\mathcal{N}$ includes binary addition, subtraction and multiplication:

$$\mathcal{N} = \{+_2, -_2, \times_2\}$$

An objection frequently raised about tree-structured representations of polynoms is that, as far as only polynoms with finite degree will be considered, the non-parametric tree-structured and the parametric (coefficient vector) representations are equivalent.

However, it must be noted that parametric (vectorial) representations for n-degree polynoms of several variables have exponential size in n. In opposition, non-parametric (tree-structured) representations allow for a compact description of *polynoms with few terms.*

In other words, tree-structured representations allow for a compact description of "simple" regions (polynoms with few terms) in large search spaces. This way, artificial evolution manages to explore large search spaces with affordable complexity, because it mostly investigates simple regions in the search space. This distinction is germane to that done between VC dimension and effective VC dimension of a hypothesis class (Vapnik, Levin, & Cun (1994)). But the underlying simplicity criterion (the 0-norm given as the number of non-0 coefficients) is not easily amenable to standard parametric learning and optimization techniques.

As a counterpart, tree-structured representations suffer from their redundancy. Many different genotypes code for a same phenotype; a polynom admit many equivalent expressions. Useless subtrees (multiplying 0 with any subtree) can be added to the tree without changing the polynom. Such useless expressions are referred to as non-coding tree segment, or *introns* in analogy with the non-coding segments in DNA code (Levenick (1991), Banzhaf *et al.* (1998)). The proliferation of introns along evolution is one main concern faced by Genetic Programming (this point shall be returned to in section 6.4.5).

In summary, the parametric (real-valued vector) and non-parametric (tree-structured) representations of polynoms confer distinct properties to the fitness landscape. Choosing the latter can only be recommended if the target model is a polynom with high degree. In any case, the coefficients of the ultimate model should be fine tuned in a further parametric optimization step.

Example 2: Computer programs

Another tree-structured space is that of computer programs. An example corresponding to a subset of the C language is presented below. The main restriction is that only functions (returning a value) are considered.

As in the general tree case, a program or function is recursively defined as a terminal, or as the result of a k-ary operator applied to k functions.

The node set $\mathcal{N}$ includes all structured instructions and numerical operations.

- Branching instructions ($if - then - else$, $switch$) are k-ary operators, calling one of the $2..k$ arguments depending on the value of the first argument.
- Loop instructions ($do - while$, $repeat - until$) are binary operators, iteratively calling both their arguments until a given argument fails, and returning the last argument called.
- An additional binary operator, seq, is introduced, sequentially calling both its arguments and returning its last argument.
- Affectation is a binary operator $=$, setting its first argument to its second argument, if possible (see below), and returning its last argument.
- Operations ($+$, $-$, $\times$, $\div$, $==$, $\&, \ldots$) are binary operators, calling sequentially both arguments and returning the expression value.

The terminal set $\mathcal{T}$ includes the input variables of the program, some local variables, and the constant $\mathcal{R}$.

This formalism can be extended to accommodate as well subroutines, or Automatically Defined Functions (Koza (1994), Rosca & Ballard (1994)).

A noteworthy difference between the two example cases, is that all trees in the polynom space correspond to actual polynoms, while many trees in the program space (e.g. affectation of a value to an non terminal) do not correspond to syntactically admissible programs. This remark will be discussed further (section 6.5).

Generative Models

Among all possibilities offered by tree structured spaces, a most interesting one is based on generative models and cellular automata (Gruau (1994)). In this framework, the phenotype is a model evolved from an embryo (elementary neural net or circuit design), through a generative, tree-structured model, describing the growth instructions applied to the embryo (Gruau (1994), Bennett-III *et al.* (1999)). Complex patterns might thus be obtained by duplicating in series or in parallel the current cells according to rather simple rules, and evolution can be tamed so as to explore the sets of simple rules, rather than the set of complex patterns.

6.4.2 Initialization

The initialization procedure parallels the definition of the search space. It recursively calls the *Tree_Generation* function, which randomly selects some terminal in $\mathcal{T}$ or operator in $\mathcal{N}$. In the former case, *Tree_Generation* returns the selected terminal. In the latter case, the arguments of the selected k-ary operator are obtained by recursively calling k times the *Tree_Generation* function; the well formed expression given as the operator with its k arguments, is returned.

This function needs be adjusted to meet practical requirements, such as enforcing some maximal tree depth D. A simple way is by using different selection distributions depending on the tree level, e.g. forcing the selection of a terminal when called at level $D-1$. More generally, the selection distribution might be tuned to reflect any prior knowledge about the operators relevance. Though most adjustments of this distribution are done manually, a promising adaptive scheme has been proposed by (Salustowicz & Schmidhuber (1997)), in the line of distribution-based evolution (Baluja & Caruana (1995)).

As mentioned earlier on, the importance of initialization cannot be overestimated. Any effort done to improve the initial population diversity is significantly rewarded, up to some orders of magnitude w.r.t. the final performances of evolution (Daida (1999)).

The most used initialization procedure, "ramped half and half" initialization (Banzhaf *et al.* (1998)) constructs trees with diversified depths and shapes. For each individual, its target depth D is uniformly selected in a given range; with probability 1/2, the individual is a full tree (terminals only appear at depth D); otherwise it is a sparse tree, with a few long branches (Soule & Foster (1997)).

6.4.3 Crossover (and Mutation)

Canonical GP crossover selects with uniform probability one node (cut point) in each parent, and constructs the offspring by swapping the subtrees attached to these nodes (Fig 6.5).

GP mutation comes with different modes. Structural mutation proceeds by crossing over the parent with a random tree. Node mutation replaces a single operator node in the tree by another operator, without modifying its arguments. "Promoting" mutation replaces a node by one of its descendant nodes while its inverse, "Inserting" mutation, actually inserts a node, eventually adding random subtrees as required by its arity. Parametric mutation, usually based on stochastic hill climbing, is devoted to the adjustment of the constants in the tree; it can be viewed as an internal parametric optimizer.

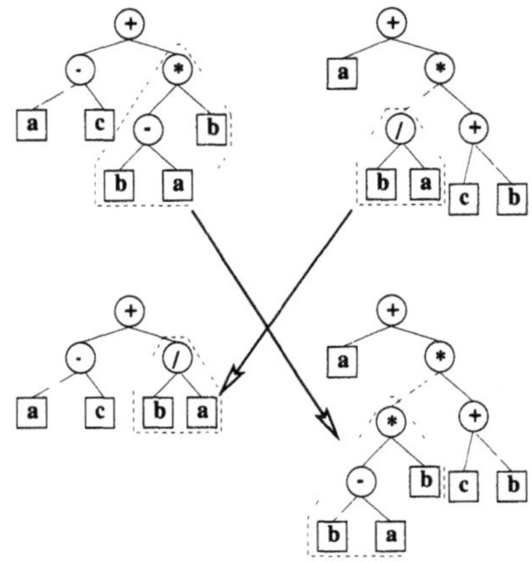

FIGURE 6.5: GP Crossover

Following the GA trend, canonical GP mostly relies on crossover and selection. Mutation plays an even lesser role in canonical GP than in GAs, for several reasons. The main motivation behind GA mutation is to resist the loss of genetic diversity. Indeed, when the population has converged to a subspace of the search space closed under crossover (genetic drift: when a given bit takes the same value over all current individuals, or when an operator or a terminal has disappeared from the population), only mutation can reintroduce diversity. But genetic drift is not experienced in GP as it might be in GAs. Almost no terminal or operator ever disappears from the current population; actually, most terminals and operators appear many times in most individuals.

Moreover, GP commonly uses much larger populations than GAs, and these populations are evolved during less generations (see section 6.4.5); all together, this explains why genetic drift is unlikely in GP.
It is therefore conjectured that, given the amount of genetic material available in the current population, crossover can emulate all intended effects of mutation (Koza (1992)).

6.4.4 Exploration (and Exploitation)

Canonical GP crossover and mutation perturb the parents in a more wild and unpredictable way than was the case for GAs. Note that crossing over an individual with itself, which would have unnoticed effects in a binary GA, will in general produce offspring trees quite different from the parent tree in GP.

Clearly, any single symbol modified in the individual tree (except perhaps constant values) will generally have a considerable impact on the individual phenotype; it might modify its fitness in an unpredictable way. From this remark, tree-structured search spaces violate the strong causality principle (Rechenberg (1973)). Henceforth, artificial evolution is hugely biased toward exploration, and should rather emulate a random walk in the GP search space. The reason why it is not the case, related to both the space structure and the dynamics of selection, is discussed in section 6.4.5.

In the meanwhile, under the pressure of computationally heavy applications (Dasgupta & Michalewicz (1997)), it was recognized that parametric optimization was more efficient in adjusting the constant values in the individual trees than GP. The coupling of parametric and non-parametric optimization is most easily done, as any parametric optimization method (gradient-based method, hill-climber, estimate optimization) can be cast as a parametric mutation operator.

The effects of parametric mutation are twofold. On one hand, parametric mutation tends to drive the search toward the local optima of the search space (see also (Radcliffe & Surry (1994), Merz & Freisleben (1999))). On the other hand, only parametric mutation can allow offspring with new tree-structures to survive after the early generations; otherwise, such offspring cannot compete against their parents, whose coefficients have been adjusted for some generations.

Globally, the use of parametric mutation allows GP to achieve some trade-off between exploration and exploitation. The evergreen problem is to allocate the computational credits between parametric and non-parametric optimization, and adjust the amount of parametric mutation needed for each individual. The allocation is most often addressed by trials and errors.

6.4.5 Dynamics of GP

As mentioned earlier on, GP closely parallels GAs unless for the genotypic components (initialization, crossover and mutation). The evolution engine most often involves a steady-state engine based on tournament selections (section 6.3.7), accommodated along parallelized implementations.

However, the strong GP bias against exploitation has some effects on the evolution engine too. Practically, GP compensates for its limited exploitation abilities by more exploration; evolution starts with a large uniform sample of the search space, up to some hundred thousands individuals.

The barrier of introns

Strangely enough, these large populations can be efficiently evolved during a few (hundred) generations only. Afterward, the performances are found to stagnate, while evolution collapses: the size of the genetic material increases to the point that the available memory resources are exhausted, and/or the fitness evaluation becomes unbearably long. This size increase is referred to as *bloat phenomenon*.

Experimentally, the average individual size increases along evolution, not because intrinsically more complex individuals are discovered, but because deeper and wider individuals appear, involving many non-coding subtrees termed *introns*. Introns interfere with the GP mechanism in a way which is not yet fully understood.
By definition, introns have no effect whatsoever on the individual fitness; still, it is observed that removing the introns from the genetic material (all GP individuals undergoing a forced algebraic-like simplification) adversely affects the overall GP performances (Koza (1992)).

The return of exploitation

On the other hand, the effects of variation operators are significantly modified as the average individual size increases.

Precisely, crossover and mutation both modify subtrees, the root node of which is uniformly selected among the parent nodes. The probability of selecting small subtrees is exponentially increased as the size and width of the tree increases.

As crossover and mutation tend to modify small subtrees in wide and long trees, more smooth effects are obtained; the offspring fitness happens to moderately differ from that of their parents.

In other words, crossover and mutation recover some exploitation abilities as individual trees grow bigger and wider.

Not only can variation operators operate more gradual changes due to the size increase. But also better changes can be operated by crossover, due to the presence of introns. Indeed, introns constitute a pool of subtrees which once were part of relevant individuals; and crossover can tap into these relevant subtrees, making them visible again in the coding (non intron) part of the offspring. The relevance of the intron material is *a contrario* suggested by the fact that removing them impairs the final GP performances.

In this perspective, introns are viewed as an internal memory, created by GP evolution and eventually explaining its good behavior compared to pure random walk. The drawback is that the memory growth is not managed for the time being, causing evolution to collapse.

Selection and equivalence classes

The intron growth can also be analyzed with respect to the structure of the search space, and the selection effects. Consider for instance the tree representation of polynoms, and let $\{p\}$ denote the set of trees (genotypes) equivalent to a given polynom p (phenotype) up to algebraic simplification.

Let $n(p,d)$ be the number of trees in $\{p\}$ with depth d. It is clear that $n(p,d)$ exponentially increases with d; the intuition is that any tree can be transformed into many equivalent larger trees, by augmenting its intron size.

Uniformly sampling all trees in $\{p\}$ with a bounded depth will thus naturally select the longest trees associated to phenotype p.

On the other hand, crossover has the opportunity of creating new genotypes with same phenotype as their parents, for instance by swapping intron subtrees. Let p be the best phenotype represented in the current population; all offspring associated to this phenotype will likely be retained by selection (until a better phenotype is discovered). The population will be crowded by more and more genotypes associated to the fittest-so-far phenotype, making it even easier for crossover to produce other and longer genotypes in $\{p\}$.

As the equivalence class of the best phenotype is more intensively sampled due to the combined effects of crossover and mutation, the amount of introns mechanically increases in the population (Langdon & Poli (1997)).

6.4.6 *Application to Identification of Rheological Models*

The GP approach is illustrated on an example application in the field of Mechanical Modeling, concerned with non parametric identification of materials behavioral laws. The reader is referred to (Schoenauer *et al.* (1996), Sebag, Schoenauer, & Maitournam (1997)) for more details about the application.

The domain

Numerical Engineering is eagerly demanding for accurate behavioral laws of new materials (composites, polymers), as these new materials are being increasingly used in industry even though their behavior is much less documented than that of older materials (iron).

From the accuracy of the materials behavioral law depends the accuracy of numerical simulations, which in turn governs the quality of Structural Design. Structural Design is aimed at structure safety (the actual structure behavior should satisfy the same requirements as the simulated behavior, regarding e.g. the maximum displacement or stress in the structure under some typical loadings) and optimality (there should exist no lighter, hence less expensive structure, satisfying the desired requirements).

The current approaches to behavioral law modeling heavily rely on the mechanical engineer or scientist expertise. In most cases, the expert arduously guesses the shape of the law, the coefficients of which are adjusted by fitting the experimental data (Gittus & Zarka (1985)) ; clearly, the accuracy of the result critically depends on the expert guess.

An alternative is to analyze the material behavior at the microscopic scale, and then derive a macroscopic behavioral law from the microscopic model, using for instance homogenization (Sanchez-Palencia & Zaoui (1987)). However, homogenization might result in computationally

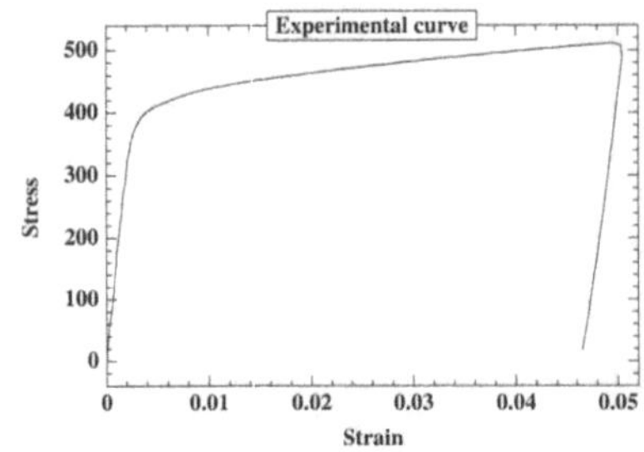

FIGURE 6.6: A typical stress-strain experimental curve

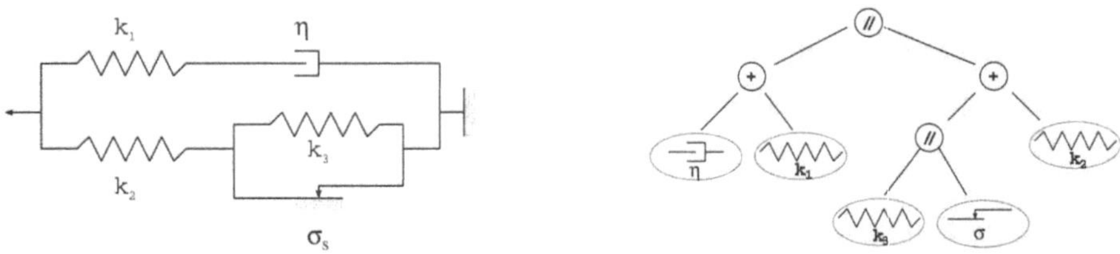

(a) Rheological model

(b) Tree representation

FIGURE 6.7: Identification of rheological models is amenable to optimization in a tree-structured space

heavy models, deterrent to structural design and the great many simulations involved.

The goal of non parametric modeling is to find both the shape and the coefficients of the materials behavioral law, from the empirical data gathered over several experiments (typically, traction-compression for a given rate of deformation, creep for a prescribed stress, and relaxation for a prescribed strain). One experiment is described as a stress-strain curve (Fig 6.6).

The search space

The target model must i) fit the data; ii) be comprehensible. The latter criterion is essential for the approach to be effectively used by mechanical engineers and scientists for the time being.

The search space considered is that of rheological models (Persoz (1960)), which are one-dimensional dynamic models. A rheological model describes a material as an assembly of elementary elastic, plastic or viscous components. Only mixed rheological assemblies composed of serial and/or parallel branches have been considered (Fig 6.7 (a)). These models are considered sufficiently expressive (modeling a wide range of one-dimensional dynamic behaviors), while they convey a *natural* interpretation about the material at hand.

The behavioral law encoded by a rheological model R (phenotype) is a partial differential equation linking the stress, the strain and their derivatives:

$$R \longrightarrow \mathcal{R}(\sigma, \varepsilon, \dot{\sigma}, \dot{\varepsilon})$$

A natural representation for rheological models is by trees (Fig 6.7(b)). The tree-structured representation of rheological models (for short, rheological trees) is defined from the node set $\mathcal{N}$ and the terminal set $\mathcal{T}$, where:

- $\mathcal{T}$ includes all elementary components in a rheological model. There exist three types of components:
  - Springs.
    To each spring is associated its stiffness k ($k \in \mathbf{R}^+$). A spring symbolizes a pure elastic behavior:
    $$\sigma(t) = k\varepsilon(t)$$
  - Dashpots.
    To each dashpot is associated its viscosity η ($\eta \in \mathbf{R}^+$). A dashpot symbolizes a pure viscous behavior:
    $$\sigma(t) = \eta\dot{\varepsilon}(t)$$
  - Sliders.
    To each slider is associated its stress threshold σ_S ($\sigma_S \in \mathbf{R}^+$). A slider symbolizes a pure plastic behavior:
    $$\dot{\varepsilon}(t) = 0 \qquad OR \qquad |\sigma(t)| = \sigma_S$$

  In summary, each terminal in $\mathcal{T}$ is a pair (t, r), where t is selected among three nominal values (spring, dashpot and slider) and r is a positive real number.

- $\mathcal{N}$ includes two binary connectors:
  - The series connector (+).
    The stress in a series node and the stress in its right and left son nodes are equal; the strain in the series node is the sum of the strains in the son nodes.
    $$\sigma_+(t) = \sigma_{left}(t) = \sigma_{right}(t)$$
    $$\varepsilon_+(t) = \varepsilon_{left}(t) + \varepsilon_{right}(t)$$
  - The parallel connector (//).
    The stress in the parallel node is the sum of the stresses in the right and left son nodes; the strain in the parallel node and in the son nodes are equal.
    $$\sigma_{//}(t) = \sigma_{left}(t) + \sigma_{right}(t)$$
    $$\varepsilon_{//}(t) = \varepsilon_{left}(t) = \varepsilon_{right}(t)$$

The fitness function

Given a rheological tree R, its fitness measures the distance between the response predicted by R and the material observed response, summed over some experiment cases.
For the sake of notational simplicity, let us assume that there exists a single, strain imposed, experiment case. This experiment case (Fig. 6.6) is described by the pairs (experimental strain $\varepsilon_e(t)$, experimental stress $\sigma_e(t)$) measured at some time steps $t_1, \dots t_N$.

The overall fitness function thus is given as:

$$\mathcal{F}(R) = \sum_{j=1}^{N} (\sigma (R, t_j, (\varepsilon_e(t_1), \dots, \varepsilon_e(t_j))) - \sigma_e(t_j))^2$$

where $\sigma(R, t_j, (\varepsilon_e(t_1), \dots, \varepsilon_e(t_j))$ stands for the response of model R at time t_j, under the loading history $(\varepsilon_e(t_1), \dots, \varepsilon_e(t_j))$.
This function computes the response of any rheological model under any loading history; in other words, the fitness function incorporates a general purpose rheological tree simulator.

The implemented simulator is a two-step process, iteratively determining the response of the model depending on its current state and on the current loading (imposed strain) applied on the structure.

At time t_j, the rheological tree is first *interpreted* as a set of partial differential equations. The main difficulty comes from the presence of sliders, which mask the presence of some other nodes and branches in the tree if the current stress in the slider is less than the slider stress threshold. The operational part of the tree (set of non masked nodes) thus is modified as the loading history goes on. As an analogy, the operational part of an electrical circuit likewise depends on the state of its interrupters; and the command applied on the electrical circuit might modify the interrupter states.
At each time step, the interpreter thus determines the operational part of the rheological tree depending on its state variables, and constructs the corresponding set of differential equations. Then, this set of equations plus the equations describing the current loading (imposed strain), is *solved numerically*, and gives the response predicted by the model at time t_j.

Genotypic operators

The initialization procedure constructs rheological trees with a bounded depth D_{Max}, by calling the recursive function *Tree_Generation*.

We used canonical GP crossover, swapping two randomly selected nodes in the parents (section 6.4.3), and canonical mutation (crossover with a random parent). A parametric mutation operator was also used on all newborn offspring, locally optimizing the real-valued coefficients attached to the tree structure in a Monte-Carlo way.

Tree_Generation(D)	D maximal depth allowed
Select uniformly n in $\{ +, //, spring, dashpot, slider\}$	
If $n \in \{+, //\}$	n is a connector
Do $n_1 = Tree\_Generation(D-1)$;	
Do $n_2 = Tree\_Generation(D-1)$;	
Return $n(n_1, n_2)$	
Else	n is a terminal
Do $v = U(0, V)$	uniform selection of the attached coefficient
Return (n, v)	

FIGURE 6.8: Initialization of rheological trees

Experimental results

The experimental setting is described in Table 6.2.

A proof of principle of the approach, illustrated on Fig 6.9, is given by the comparison between GP and Monte-Carlo optimization (retaining the best one over a uniform sample of 40,000 rheological trees). The "experimental" data are generated from a known target model, proposed for polyethylene (Kichenin, Dang-Van, & Boytard (1996)).

The best fitness obtained for a given number of fitness evaluations, averaged over 10 runs, and the best fitness obtained over all 10 runs is displayed, to be compared to the best fitness obtained for a random sample of 40,000 trees.

It is noted that the best fitness is not zero, due to rounding errors. The stop criterion is when the fitness of the current optimal individual is less than an estimate of the unavoidable rounding errors (Schoenauer *et al.* (1996)).

A model close to the target model up to a connector modification, was repeatedly found. According to the experts, the difference comes from the absence of creep in the considered experiments.

The total runtime is about 3 hours on HP-780 workstation.

Lessons

Three main lessons were gained in this application.

First of all, the development of the fitness function, particularly the rheological simulator, required intensive programming efforts. Evolution was found good at chasing and exploiting the holes in the program, and thus significantly contributed to its debugging.
As mentioned earlier on, programming the fitness function must be taken seriously, for it actually amounts to implementing a domain oracle.

TABLE 6.2: Tableau for rheological model identification

Objective:	Models fitting the available experiments
Terminal set:	Spring, Slider, Dashpot
Node set:	Parallel, Series
Fitness cases:	Loading histories
Raw Fitness:	Sum of errors over fitness cases
Evolution engine	Generational (offsprings replace parents)
Parameters	Population size: 200
	max depth of new individuals: 12
	max depth of offspring: 50
Crossover rate	.4
Mutation rate	.2
Selection method	Tournament (3)
Termination criterion	40,000 fitness evaluations

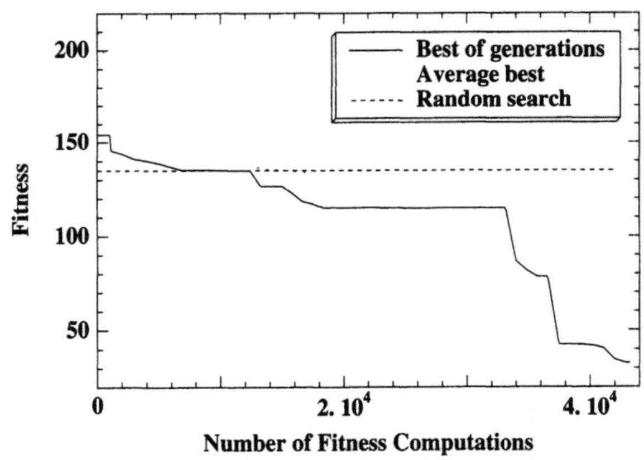

FIGURE 6.9: Dynamics of Evolution

Second, the trade-off between exploration and exploitation was significantly more difficult to adjust than for parametric evolution algorithms (genetic algorithms or evolution strategies).
In some sense, non parametric optimization addresses a noisy optimization problem; the fitness of an individual tree R noisily estimates the fitness of the best tree R^* with same structure as R. This noise can be decreased through the use of parametric mutation, which simply optimizes the coefficients in R and thus brings its fitness closer to that of R^*. However, the approach taken here spends equal refinement times on each individual; heuristics developed to deal with noisy evolutionary optimization (Beyer (1998)) could be employed for a more judicious allocation of the refinement efforts.

The third lesson concerns the interplay between model identification and the quality of the available data. Typically, the standard experiments considered in mechanical modeling are compatible with several candidate models of similar complexity. On-going work is devoted to the dual optimization problem of experiment design, aiming at experiments for deciding between a set of candidate models.

6.5 Genetic Programming with Grammars

Returning to the tree-structured search spaces considered in the previous section, an important difference appears between the polynom and rheological model spaces on one hand, and the computer program space on the other hand.

In the polynom and rheological cases, any well formed tree in the search space corresponds to a valid individual, respectively a polynom or a rheological model (regardless of the fact that the tree can be algebraically simplified).
In the computer program space, this is no longer the case; some well-formed trees do not correspond to valid programs (e.g. affectation to non l-value).

The set of criteria or priors about the validity of particular individuals can be thought of as domain background knowledge. It has been emphasized throughout the paper that the use of background knowledge is essential to *satisficing* model identification.

This section is devoted to accommodating background knowledge in GP. A most simple background knowledge available in experimental sciences, namely dimensional consistency, is chosen to illustrate the approach.

6.5.1 Background Knowledge and Constraints

Most generally, the identification goal is to find a solution complying with the domain background knowledge, which can be viewed as a set of constraints.

Many constraints on the target solutions are commonly encountered in real-world problems. Besides enforcing some properties of the final solutions, constraints can also be used to draw evolution away from *a priori* useless regions. As evolutionary algorithms and genetic programming in particular commonly face huge search spaces, any means of focusing the search is very welcome.

In the frame of model identification in experimental sciences, constraints on the individual behavior or phenotype (e.g. stability of the model under some circumstances) are referred to as functional constraints; constraints on the individual syntax or genotype are referred to as syntactic constraints.

An example of syntactic constraints is given by dimensional consistency. Assuming that problem variables are attached with some unit (time, length, mass or combination of those), dimensional consistency describes the validity of any expression of these variables (e.g. adding meters and seconds is not allowed).

6.5.2 Constraints and Evolutionary Computation

Obviously, the simplest way of dealing with constraints is to address them prior to evolution, and have them hard wired in the search space. As this is not possible in the general case, constrained optimization is achieved by driving evolution away from all undesirable regions, be they non-admissible or poorly fit.

A strong pre-requisite is to ensure that the fitness of *all* individuals in the search space can be evaluated, and that non-admissible individuals cannot cause any fatal error case. In particular, the use of protected GP operators is preconized: protected division, inactive affectation on non-terminals, loop operators with an upper time limit, and so forth (Koza (1992)).

Artificial evolution handles constraints through i) the fitness function (at the phenotypic level); ii) the initialization procedure and variation operators (at the genotypic level); iii) any hybrid heuristics combining the genotypic and phenotypic levels.

A most natural way of diverting evolution from non-admissible regions is through the fitness function, by turning a non-admissible individual into a non fit one. Typically, the constraint violations are accounted for by penalization terms in the fitness function; various schemes for adjusting the penalization factors have been proposed (Schoenauer & Xanthakis (1993), Joines & Houck (1994), Eiben & Ruttkay (1996), Eiben, Hinterding, & Michalewicz (1999), Keijzer

& Babovic (1999), BenHamida & Schoenauer (2000)).

Although adjusting the penalization schedules might prove tricky at later stages, the advantage of this option is being easy to start with; it only requires to implement constraints within the fitness program.

In counterpart, it entails a waste of computational resources as most part of the evolution time is spent in non admissible regions.

A more demanding option is to reconsider the genotypic operators. The initialization procedure is modified to produce admissible individuals to the best possible extent. The variation operators are re-designed to produce constraint-adapted individuals.

In this last option, significant development efforts are usually required to tailor the genotypic operators (crossover operator which is closed under the constraints; a repair operator repairing non admissible offspring; variation operators exploring the edges of the constraints (Schoenauer & Michalewicz (1996), Schoenauer & Michalewicz (1997))).

The option standalone might also be computationally expensive or inefficient. Typically, the initialization procedure might find it hard to sample tiny admissible regions in a huge search space; or the repair operator can be computationally heavy.

The field of constrained evolutionary computation is rapidly growing; many hybrid approaches incorporating phenotypic and genotypic heuristics have been proposed.

Still, the best way of dealing with constraints remains to redesign the search space and/or the genotypic operators, such that evolution is only faced with admissible individuals. The good news is that genetic programming can get rid of some kinds of syntactic constraints in such a way.

6.5.3 The search space as a language

An elegant approach to constrained genetic programming, first proposed by (Gruau (1996)), relies on the well studied formalism of grammars.

Grammars are concerned with the constructive characterization of a language (set of admissible expressions) satisfying syntactic constraints. A grammar is made of rules, such that any correct expression in the language can be formed from these rules.

We shall restrict ourselves to Backus Naur form grammars (BNF), described as a 4-tuple $\{S,N,T,P\}$.

- S is the start symbol in the grammar, the common origin of all expressions/sentences recognized by the grammar.

TABLE 6.3: The polynom regular grammar ≡ Trees ($\mathcal{N} = \{+, \times\}$, $\mathcal{T} = \{X, \mathcal{R}\}$)

$$N = \{<E>,\ <O>,\ <V>\}$$

$$T = \{+,\ \times,\ X,\ \mathcal{R}\}$$ //$\mathcal{R}$ stands for any real-valued constant

$$P = \left\{ \begin{array}{lcl} S & := & <E>\ ; \\ <E> & := & <O>\ <E>\ <E>\ \mid <V>\ ; \\ <O> & := & + \mid \times\ ; \\ <V> & := & X \mid \mathcal{R}\ ; \end{array} \right\}$$

- N is the set of non terminal symbols, which correspond to abstract admissible constructs in the language (e.g. noun phrase; multiplicative expression;..).
- T is the set of terminal symbols, which includes all atomic elements in the language (e.g. words; variables; operators). It must be noted that *terminal* has distinct and partly overlapping meanings in BNF grammars and in genetic programming.Typically, an operator symbol (e.g. $\times$) is a non-terminal symbol in GP, whereas it is a terminal symbol in BNF grammars (more on this later on).
- P is the set of production rules. One production rule is attached to each non-terminal symbol (among which the start symbol, by abuse of language). A production rule describes all admissible ways, termed *derivations*, of rewriting the non-terminal symbol attached to the rule. For instance, a multiplicative expression is a constant, or a variable, or the product of two expressions.

A grammar thus provides a generative model of the search space:
The expression at hand is initialized to the start symbol S. While the expression still contains non-terminal symbols, i) one non-terminal symbol is selected in the current expression; ii) the production rule associated to this non-terminal symbol is considered, and one among its derivations is selected; iii) the non-terminal symbol is replaced with the selected derivation.

As an example, the polynom search space (section 6.4.1) corresponds to the simple grammar in Table 6.3 (regular grammar in Chomsky's terminology). The non terminal symbols correspond to the notions of expression ($< E >$), operator ($< O >$) and operand ($< V >$). The production rule attached to an expression recursively describes an expression as either a variable, or the application of an operator to two expressions.

As a pedagogical example showing how easily syntactic constraints can be enforced through grammars, let us assume that the target language is restricted to polynoms trees alternating addition and multiplication symbols.

This constraint is enforced as i) the non-terminal $< E >$ symbol is replaced by two non terminal symbols $< mult\_E >$ and $< add\_E >$, respectively denoting multiplicative and additive

TABLE 6.4: Enforcing syntactic constraints on polynoms

$$
\begin{array}{lcl}
N & = & \{< mult\_E >,\ < add\_E >,\ < O >,\ < V >\} \\
T & = & \{+,\ \times,\ X,\ \mathcal{R}\} \\
P & = & \left\{ \begin{array}{ccl}
S & := & < mult\_E > \mid < add\_E > ; \\
< mult\_E > & := & \times\ < add\_E >\ < add\_E > \mid < V > ; \\
< add\_E > & := & +\ < mult\_E >\ < mult\_E > \mid < V > ; \\
< O > & := & + \mid \times ; \\
< V > & := & X \mid \mathcal{R} ;
\end{array} \right\}
\end{array}
$$

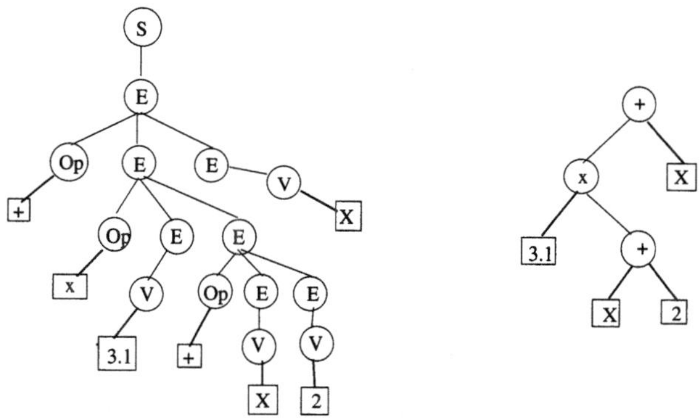

Derivation tree Parse tree

FIGURE 6.10: An example of GP tree (left) and derivation tree (right).

expressions; ii) the associated production rules respectively describe an additive expression as the addition of two multiplicative expressions, and an multiplicative expression as the product of two additive expressions (Table 6.4).

Interestingly, the construction of any admissible expression through the application of derivation rules, can itself be represented as a tree termed *derivation tree*. An example of derivation tree and the corresponding expression tree is displayed in Fig 6.10.

Note that in both the derivation and the parse trees, non-terminal symbols correspond to internal nodes while terminal symbols lie at the leaves of the tree.

The difference concerns the place of operator symbols (addition, multiplication). Operators appear as leaves of the derivation tree, as they correspond to BNF terminal symbols. This contrasts with the GP frame, where operators are non-terminal symbols and appear as internal nodes of the parse tree.

6.5.4 Grammar-Guided GP

By construction, there exists a surjective mapping between the set of derivation trees and the set of admissible expressions defined by the grammar. An elegant way for GP to satisfy syntactic constraints is by searching the derivation tree space instead of the parse tree space (Gruau (1996)).

This combination of GP and BNF grammars, or Grammar Guided GP (G3P), is achieved by designing new genotypic operators.

The G3P initialization procedure is derived from the grammar description in a straightforward way. The start symbol is rewritten using one randomly selected derivation in the associated production rule. Any non-terminal symbol in the current derivation tree is likewise rewritten using a randomly selected derivation in the associated production rule, until the expression only involves terminal symbols.

Some difficulties might arise when complex grammars are considered, as the initialization procedure also enforces a limited maximal depth of the parse trees. This point will be discussed further (section 6.5.7).

G3P crossover is restricted to ensure that admissible offspring are generated from admissible parents. Concretely, each node in the derivation tree is labeled with a grammar symbol, be it terminal symbol or not (Fig 6.10). A crossover operator closed wrt the grammar is obtained by i) selecting a node n attached to a non-terminal symbol in the first parent derivation tree; ii) selecting a node n' in the second parent, attached to the same non-terminal symbol. In case no such node could be find in the second parent, crossover simply produces as offspring the copies of the parents; iii) otherwise, crossover proceeds by swapping the subtrees rooted in nodes n and n'. Each parent is thereby perturbed as one non-terminal symbol is rewritten differently, using another expression of same type extracted from the other parent. This ensures that the offspring still belong to the language, and crossover is closed with respect to the grammar.

One important difference between G3P crossover and GP crossover regards the possible perturbations of operator symbols (e.g. addition, multiplication) in the individual phenotypes. As operator symbols are considered as non-terminal symbols in the parse tree (Fig 6.10), they cannot be modified through GP crossover without also modifying the whole subtree attached to the operator node. In contrast, G3P crossover is allowed to modify the operator symbols only, since these correspond to terminal symbols in the derivation trees.

In summary, G3P crossover allows for finer grained perturbations on the individuals.

G3P mutation is restricted along the same lines as G3P crossover. Concretely, a node n is selected with uniform probability among the nodes attached to a non terminal symbol in the parent derivation tree. The derivation subtree attached to n is replaced with a randomly generated subtree rewriting the same non-terminal symbol.

6.5.5 *Discussion*

Grammar Guided GP can be viewed as a generalization of Strongly Typed GP (Montana (1995), Haynes, Schoenefeld, & Wainwright (1996)).

Historically, strongly typed GP (STGP) appeared as a reaction against the closure hypothesis governing the early GP works (Koza (1992)). The closure hypothesis is another word for the absence of syntactic constraints; it assumes that any offspring constructed through canonical crossover and mutation, is admissible.

The main benefit of the closure hypothesis is to allow for using most simple variation operators. In counterpart, the price to pay is computationally heavy; the search space size, exponential in the maximal tree depth allowed, is huge even for problems of moderate difficulty (Whigham (1995)).

Ultimately, the general trend toward strongly typed programming languages in computer science, pushed toward using typed domain variables and operators in GP too. The rationale behind is that a better specified language filters out more irrelevant programs, thereby mechanically increasing the result quality everything else being equal.

6.5.6 *Dimensional Grammars*

The potentialities of grammars are illustrated on a common constraint in model identification, the need for dimensionally consistent models (Ratle & Sebag (2000)).

Several assumptions are done:

- The problem domain involves a finite (small) number k of elementary units (e.g.*mass, length* and *time*). Any compound unit relevant to the application can be expressed with reference to the elementary units as a integer vector (e.g. $Newton = mass \times length \times times^{-2}$ is represented by its exponent vector $(1,1,-2)$).
- All domain variables (e.g. displacement, stress) are associated a given compound unit.
- The number of relevant compound units is restricted; typically, the exponent of each compound unit with respect to any elementary unit is in a given integer range ($\{-2\ldots 2\}$ was taken in the experiments). Let D denote the set of allowed compound units (e.g. $D=\{-2,\ldots 2\}^3$).

Under these assumptions, dimensional consistency can be automatically represented as a grammar (Table 6.5). To each allowed compound unit is associated a non-terminal symbol of

Dimensional Grammar (terminal symbols T, dimension range D)

Init

$N = \{\}; P = \{\}$

For each $x \in D$

Create non terminal symbol $< N_x >$

Initialize the associate production rule

$< N_x > \; := \; exp_x$

$exp_x \; = \; + \; < N_x > \; < N_x > \; | \; - \; < N_x > \; < N_x > \; |$

$\quad exp \; < N_x >< N_e >$

For each terminal symbol V in T,

If the V unit is x,

$exp_x \; = \; exp_x \mid V;$

For each $y, z \in D$ such that $y + z = x$,

$exp_x \; = \; exp_x \mid \times < N_y > \; < N_z >;$

For each $y, z \in D$ such that $y - z = x$,

$exp_x \; = \; exp_x \mid \div < N_y > \; < N_z >;$

$N = N \cup \{N_x\}$

$P = P \cup \{N_x := exp_x\}$

Return (S, N, T, P)

TABLE 6.5: Dimensional Grammar Generation

the grammar. The associated production rule describes all ways of constructing an expression of this unit: by adding or subtracting two expressions of the same unit; by choosing a domain variable attached to this unit, if such a domain variable exists; by multiplying or dividing two expressions with appropriate units.

Note that particular constructs suggested from the domain knowledge, can easily be included in the grammar. Table 6.5 considers the multiplication of an expression with the exponential of a dimensionless expression ($< N_e >$ is the non terminal symbol associated to dimensionless expressions and $exp < N_a >< N_b >$ stands for the product of $< N_a >$ with the exponential of $< N_b >$).

It must be noted that even though the grammar size is exponential in the allowed range ($5^3 = 125$ non-terminal symbols are used in the experiments), this entails no computational overhead compared to a procedural dimensional consistency check.

6.5.7 *Initialization and Constraint Solving*

Some unexpected difficulties arose in the first experiments done with G3P and dimensional grammars. These difficulties shed some light on the early limitations of grammar guided GP (Ryan, Collins, & O'Neill (1998)).

For convenience, GP initialization commonly enforces a maximal tree depth on the generated individuals. This maximal depth can be viewed as a new constraint on the search space. On the other hand, dimensional grammars are the most complex grammars to our best knowledge, ever considered in G3P.

Uniform G3P initialization (section 6.5.4) thus faces a severely constrained satisfaction problem. Due to the complexity of the grammar, the production rule associated to most non-terminal symbols involves very few terminal derivations (derivations involving no non-terminal symbols); uniform initialization thus tends to produce *very* long and deep trees. Therefore, much time is spent in the initialization procedure, generating and then rejecting overly deep trees.

What is even worse, uniform initialization ultimately produces poorly diversified individuals; and final results are poor as evolution hardly recovers from this initial handicap.

In a first attempt to enforce the generation of short trees, selection probabilities were set on each derivation rule, borrowing to the formalism of stochastic grammars; the selection of the derivation rule for a non-terminal symbol is done on the basis of the derivation probabilities. Shorter trees are obtained by setting high probabilities on terminal derivation rules.
However, the adjustment of these probabilities proved time-consuming, without significantly improving the diversity in the initial population.

The approach finally retained is inspired from constraint satisfaction and attribute grammars (Ratle & Sebag (2000)). To each grammar symbol $< N >$ (respectively, each derivation exp in the production rules), is attached an integer value $m(N)$ (resp. $m(exp)$; this integer value is the minimal depth over all parse trees rewriting the symbol or the expression.

The minimal-depth index is recursively constructed from the following axioms:

- The minimal depth associated to a terminal symbol is 1.
- The minimal depth associated to expression $< op > \quad < N > \quad < N' >$ is defined as the max of the minimal depths of N and N', plus one.

$$m(< op >< N >< N' >) = 1 + max\ (\ m(N),\ m(N')\ )$$

- The minimal depth associated to a non terminal symbol is the minimum of the minimal depths of its derivations.

The minimum depth index can be used to uniformly sample the set of derivation trees while enforcing a maximum tree depth:

- Let D_{Max} denote the maximal tree depth allowed. If the minimum tree depth associated to the start symbol S is greater than D_{Max}, no admissible tree exists in the language within depth D_{Max}.
- Otherwise, let us consider all non-terminal symbols in the expression under construction. By induction, each occurrence of a non-terminal symbol $< N >$ situated at depth D_t in the derivation tree admits at least one derivation with depth index less than $D_{Max} - D_t$. Such derivations are *locally admissible* in the sense that they guarantee the existence of a parse subtree rewriting the current occurrence of $< N >$, and complying with the maximum tree depth D_{Max} of the final parse tree.
- Each occurrence of a non-terminal symbol in the derivation tree is rewritten by uniformly selecting a derivation among the locally admissible derivations.

Through the use of this depth index, GP constrained initialization both avoids considering any non-admissible individual, and enforces the diversity of the initial population.

6.5.8 Application to Dimensionally Consistent Identification

The G3P approach is illustrated on an example application in the field of Mechanical Modeling, concerned with dimensionally consistent identification of materials behavioral laws from indentation tests. The reader is referred to (Ratle & Sebag (2000)) for more details about the application.

The domain

Figure 6.11(a) presents a schematic view of the experimental setup. A hard indenter of a prescribed shape (usually conical or tetrahedral) is pressed against the surface of the material to be tested out. The experimenter records the reaction force along time and displacement.

The goal is to find the analytical relations between force, displacement, and the properties of the material at hand. The state of the art only provides relations for well known materials with simple behavioral laws. When the materials behavioral law is complex, the indentation model is obtained through computationally heavy simulations, applying finite element methods to simulate the material reaction under the indentation loading.

In the case of ill-known materials, only experimental data are available (Fig 6.11(b)).

The goal, as in the previous application (section 6.4.6), is to find a simple analytical model from the experimental data.

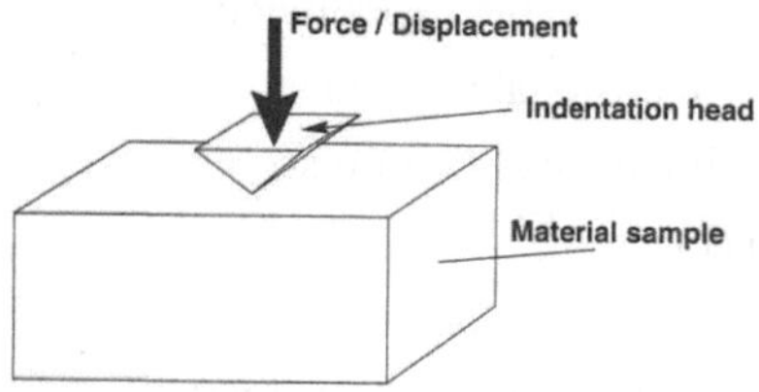

(a) Experimental setup of the indentation tests

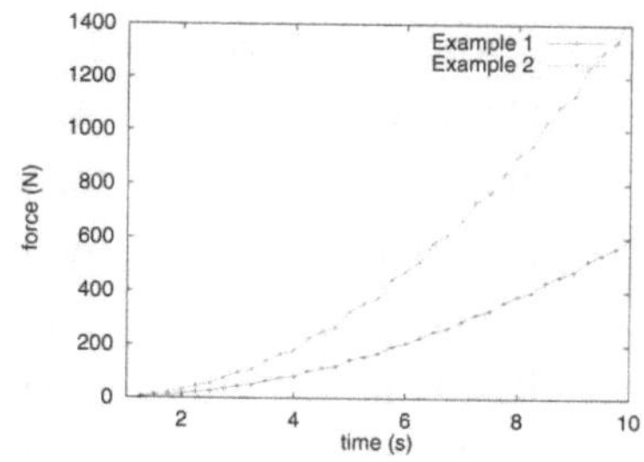

(b) Typical force vs time relations of numerical simulation

FIGURE 6.11: Mechanical Modeling from Indentation Tests

TABLE 6.6: Domain variables and physical units

Physical units			
Quantity	mass	length	time
E (Young's modulus)	+1	−1	−2
K (viscous modulus)	+1	−1	−1
n (plasticity factor)	0	0	0
S_y (yield strength)	+1	−1	−2
u (displacement)	0	+1	0
t (time)	0	0	+1
F (Indentation Force)	+1	+1	−2

The search space

What makes the difference between this goal and standard data fitting, is that the target model must be consistent with respect to the domain variable units, as given in Table 6.6.

The exponent range allowed in each elementary unit is $\{-2\ldots 2\}$. Dimension range D thus is $\{-2,\ldots 2\}^3$.

The BNF grammar describing the dimensionally admissible constructs in the prescribed range is defined from the application variables (Table 6.5).

Assuming that the target relation describes the force depending on the time and displacement, the production rule associated to the start symbol only prescribes the solution unit (Newton):

$$S := \ < N_{(1,1,-2)} >$$

The production rule associated to the start symbol can also be used to convey any prior knowledge about the shape of the relation and guide the search. For instance, knowing that the relation

TABLE 6.7: Tableau for dimensionally consistent model identification

Objective:	Models fitting the available experiments
Terminal set:	derived from Dimensional Grammar
Function set:	derived from Dimensional Grammar
Fitness cases:	20 Loading histories
Raw Fitness:	Sum of errors over fitness cases
Evolution engine	Generational (offsprings replace parents)
Parameters	Population size: 4,000
	max depth of new individuals: 12
	max depth of offspring: 50
Crossover rate	.8
Mutation rate	.2 on non-terminals; .8 on terminals
Selection method	Tournament (3)
Termination criterion	4,000,000 fitness evaluations

linking the force and the displacement is of the type $F(t) = A \times u^2 \times e^{B \times t}$, where A and B are unknown functions of the material properties, the resulting production rule is:

$$S := \; < N_{(1,-1,-2)} > \; \times \; u^2 \; \times \; exp^{<N_{(0,0,-1)}> \times t}$$

The fitness function and the setting

The fitness function associated to individual X simply is the L_2 distance between the experimental force and the force predicted from the model X in the experimental conditions. Experimental data were kindly provided by N. Tardieu and A. Constantinescu, LMS, Ecole Polytechnique.

The experimental setting is described in Table 6.7.

Experiment goal

The efficiency of G3P was investigated by comparing the results obtained with six different grammars.

- U-ND: The first grammar (universal non dimensional grammar) describes the same search space as explored by canonical GP.

$$\begin{array}{lcl} S & := & <E>; \\ <E> & := & <O><E><E> \mid <V>; \\ <O> & := & + \mid - \mid \times \mid \div \mid \exp\,; \\ <V> & := & E \mid K \mid N \mid S_Y \mid u \mid t \mid 1.0 \mid 2.0 \mid 3.0 \mid 4.0\,; \end{array}$$

- BK_1-ND: The second grammar is again a non dimensional one, but its incorporates some partial information on the shape of the target solution.

$$S := \ <E> \ \times \ exp^{<E> \times t}$$

$$\begin{array}{lcl} S & := & exp <E><F>; \\ <E> & := & <O><E><E> \mid <V>; \\ <F> & := & \times <E> t; \\ <O> & := & + \mid - \mid \times \mid \div \ ; \\ <V> & := & E \mid K \mid N \mid S_Y \mid u \mid t \mid 1.0 \mid 2.0 \mid 3.0 \mid 4.0 \ ; \end{array}$$

- BK_2-ND: The third grammar differs from the second one as it incorporates a more precise guess about the shape of the target solution, given through the production rule of the start symbol:

$$S := \ <E> \ \times \ u^2 \ \times \ exp^{<E> \times t}$$

- U-D: This grammar enforces dimensional consistency but the production rule associated to the start symbol only expresses that the target solution is expressed in Newtons.

- BK_1-D: This grammar combines dimensional consistency and the same partial information on the shape of the target solution as in BK_1-ND.

- BK_2-D: This grammar combines dimensional consistency and the same information on the shape of the target solution as in BK_2-ND.

A proof of principle of the approach is given by the comparison between the several grammars; Fig 6.12 and 6.13 show the results obtained with non-dimensional grammars and dimensional grammars respectively. For each grammar, the best fitness is averaged over 20 runs, and plotted against the corresponding number of fitness evaluations.

It is noted that the final performances are not necessarily improved by giving more insights into the target shape of the solution; on the other hand, the expert expectations are not always fulfilled by experimental data.

Overall, dimensional consistency appears to be clearly beneficial; the results obtained with dimensional grammars supersede those obtained with other grammars, by an average of 6 standard deviations (Ratle & Sebag (2000)).

These good results are partially interpreted as the search space explored along dimensional grammars is reduced by several orders of magnitude compared to non-dimensional grammars.

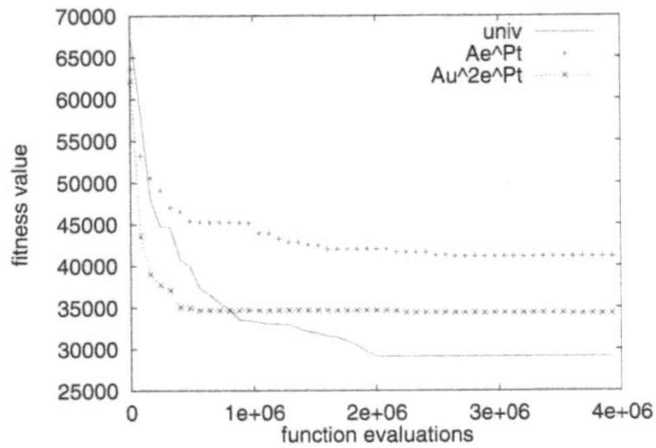

FIGURE 6.12: Average best fitness for the three non-dimensional grammars

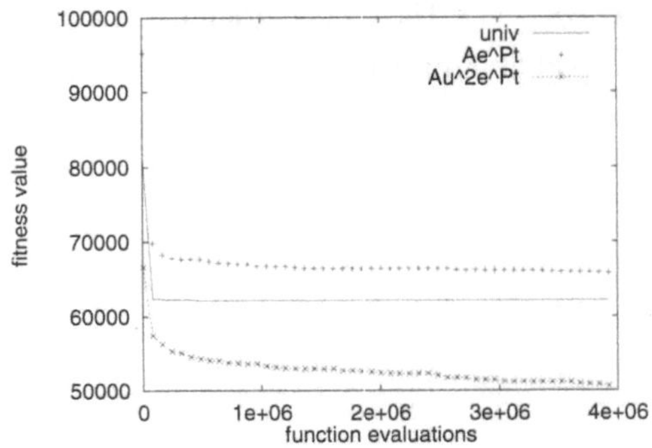

FIGURE 6.13: Average best fitness for the three dimensional grammars

6.6 Discussion and Conclusion

The goal of this paper was to present a *satisficing* approach to non parametric learning from experimental data in mechanical, physical, or chemical application domains.

It has been emphasized that real-world problems involve other quality criteria besides data fitting, which commonly leads to address model identification as an (ill-posed) optimization problem.

The practical benefits of background knowledge have been illustrated from a most simple and common rule in experimental sciences, namely dimensional consistency. Such a rule considerably trims out the search space; as expected, the search space reduction allows for improving the learning performances.

The identification approach presented, genetic programming, is a stochastic population-based optimization approach, with the ability to handle non parametric search spaces and to accommodate syntactic background knowledge.

A proof of principle of the approach in the domain of Mechanical Modeling is presented, briefly describing previous results obtained for non parametric learning of rheological models (Schoenauer *et al.* (1996)) and dimensionally consistent laws (Ratle & Sebag (2000)).

The main limitations of the approach can be understood from the two typical EC failure stories.

The first failure story reports that the canonical version of EC, which was implemented in a few days, did not solve the (difficult) problem at hand. The second failure story reports that EC could solve the problem, but required some orders of magnitude more resources than deterministic approaches.

These stories *a contrario* illustrate that i) EC should not be used when satisfactory results can be obtained otherwise, as EC is tremendously computationally heavy (up to some 10^9 fitness evaluations); ii) no evolutionary magic allows EC to solve out of the box any problem. What makes the difference is evolution to be provided with accurate domain knowledge, and efficient ways of using it. Not surprisingly, similar conclusions were drawn in the field of artificial intelligence some twenty years ago (Janikow (1993)).

In perspective, many links between artificial evolution and the rest of the material in this book are offered to investigation. Domains of particular interest to AE are density estimation and universal prediction.

Among the most interesting developments of artificial evolution are distribution-based evolutionary approaches – see among others (Baluja & Caruana (1995), Sebag & Ducoulombier (1998), Salustowicz & Schmidhuber (1997), Larranaga & Lozano (2001), Ratle & Sebag (2001)). Distribution-based evolution cuts into the genetic inspiration and evolves instead a distribution on the genotypic search space. At each step, the population is sampled after the current distribution, before being evaluated; the distribution is updated from the best and possibly the worst individuals in the sampled population. As expected, the update rule ensures that the probability associated to the best individual(s) increases.

Distribution-based evolution (DBE) can thus be set as the pursuit of a target distribution, the distribution of the fitness optima in the genotypic space. Borrowing to the domain of density estimation could provide i) better update rules than the simple additive update rules used at the moment, improving the exploitation ability of DBE; ii) further insights into better exploration heuristics (how to avoid premature convergence of the distribution).

Another perspective concerns the adjustment of the exploration/exploitation tradeoff in EC. This adjustment can be viewed as a credit allocation problem, and formalized in terms of reinforcement learning (Sutton & Barto (1998), Berny (2000)).

As noted earlier on, this credit allocation problem can be viewed as a one-arm bandit problem with non fixed reward distributions (Auer, Cesa-Bianchi, & Gentile (2001)). In this line, sequential non parametric forecasting could be used to dynamically adjust for instance the evolution operator probabilities. Each evolution operator being considered as an expert, its weight could be adjusted from its contribution to the evolution progress. Again, non parametric forecasting is directly relevant to the exploitation aspects of evolution; but some adaptations might be required, as the progress of evolution depends on the joint efforts of all operators.

Acknowledgments

I specially thank Marc Schoenauer, whose insights into evolution significantly influenced this work and who contributed many comments and suggestions about the manuscript.
I thank Jean-Pierre Nadal for his help during the final writing of the paper.
I thank Joseph Zarka and Jean Salençon, who are at the origin of this work.

I acknowledge Alain Ratle, with whom G3P was jointly developed. Helmut Hörner kindly offered the software that served as basis for G3P, and the experimental data were kindly provided by N. Tardieu and A. Constantinescu.

6.7 References

Anglano, C.; Giordana, A.; Bello, G. L.; and Saitta, L. 1998. An experimental evaluation of coevolutive concept learning. In Shavlik, J., ed., *Proceedings of the 15th International Conference on Machine Learning*, 19–27. Morgan Kaufmann.

Antonisse, J. 1989. A new interpretation of schema notation that overturns the binary encoding constraint. In Schaffer, J. D., ed., *Proceedings of the 3rd International Conference on Genetic Algorithms*, 86–91. Morgan Kaufmann.

Auer, P.; Cesa-Bianchi, N.; and Gentile, C. 2001. Adaptive and self-confident on-line learning algorithms. *Machine Learning Journal*, to appear.

Bäck, T., and Schütz, M. 1996. Intelligent mutation rate control in canonical GAs. In Ras, Z. W., and Michalewicz, M., eds., *Foundation of Intelligent Systems 9th International Symposium, ISMIS '96*, 158–167. Springer Verlag.

Bäck, T. 1995. *Evolutionary Algorithms in theory and practice*. New-York:Oxford University Press.

Baluja, S., and Caruana, R. 1995. Removing the genetics from the standard genetic algorithms. In Prieditis, A., and Russel, S., eds., *Proceedings of the 12th International Conference on Machine Learning*, 38–46. Morgan Kaufmann.

Banzhaf, W.; Nordin, P.; Keller, R.; and Francone, F. 1998. *Genetic Programming — An Introduction On the Automatic Evolution of Computer Programs and Its Applications*. Morgan Kaufmann.

BenHamida, S., and Schoenauer, M. 2000. An adaptive algorithm for constrained optimization problems. In Schoenauer, M., et al., eds., *Proceedings of the* 6th *Conference on Parallel Problems Solving from Nature*, 529–539. Springer Verlag, LNCS 1917.

Bennett-III, F.; Koza, J.; Keane, M.; and Andre, D. 1999. Genetic programming: Biologically inspired computation that exhibits creativity in solving non-trivial problems. In *AISB'99 Symposium on Scientific Creativity*, 29–38. The Society for the Study of Artificial Intelligence and Simulation of Behaviour.

Berny, A. 2000. Selection and reinforcement learning for combinatorial optimization. In Schoenauer, M., et al., eds., *Proceedings of the* 6th *Conference on Parallel Problems Solving from Nature*, 601–610. Springer Verlag, LNCS 1917.

Beyer, H.-G. 1998. Mutate large, but inherit small ! On the analysis of mutations in $(1,\lambda)$-ES with noisy fitness data. In Bäck, T.; Eiben, G.; Schoenauer, M.; and Schwefel, H.-P., eds., *Proceedings of the* 5th *Conference on Parallel Problems Solving from Nature*, 109–118. Springer Verlag, LNCS 1498.

Blickle, T., and Thiele, L. 1995. A mathematical analysis of tournament selection. In Eshelman, L. J., ed., *Proceedings of the* 6th *International Conference on Genetic Algorithms*, 9–16. Morgan Kaufmann.

Blockeel, H., and Raedt, L. D. 1998. Top-down induction of first order logical decision trees. *Artificial Intelligence* 101(1-2):285–297.

Breiman, L.; Friedman, J.; Olshen, R.; and Stone, C. 1984. *Classification and Regression by tree*. Wadsworth, Belmont California.

Breiman, L. 1998. Arcing classifiers. *Annals of Statistics* 26(3):801–845.

Caruana, R. A., and Schaffer, J. D. 1988. Representation and hidden bias : Gray vs binary coding for genetic algorithms. In *Proceedings of the* 5th *International Conference on Machine Learning*. Morgan Kaufmann.

Cerf, R. 1996. An Asymptotic Theory of Genetic Algorithms. In Alliot, J.M., et al. eds., *Artificial Evolution*, 37–53. Springer Verlag, LNCS 1063.

Cervone, G.; Kaufman, K. A.; and Michalski, R. S. 2000. Experimental validations of the learnable evolution model. In *Proceedings of International Conference on Evolutionary Computation*. MIT Press.

Cristianini, N., and Shawe-Taylor, J. 2000. *An introduction to Support Vector Machines*. Cambridge University Press.

Cun, Y. L. 1987. *Modèles connexionnistes de l'apprentissage*. Ph.D. Dissertation, Université Pierre et Marie Curie, Paris VI.

Cussens, J. 1997. Part-of-speech tagging using PROGOL. In Džeroski, S., and Lavrač, N., eds., *Proceedings of the 7th International Workshop on Inductive Logic Programming*, volume 1297, 93–108. Springer Verlag.

Daida, J. 1999. Challenges with verification, repeatability, and meaningful comparison in genetic programming: Gibson's magic. In *Proceedings of the Genetic and Evolutionary Conference 99*, 1069–1076. Morgan Kaufmann.

Das, R., and Whitley, D. 1991. The only challenging problems are deceptive: Global search by solving order-1 hyperplanes. In Belew, R. K., and Booker, L. B., eds., *Proceedings of the 4th International Conference on Genetic Algorithms*, 166–173. Morgan Kaufmann.

Dasgupta, D., and Michalewicz, Z. 1997. *Evolutionary Algorithms in Engineering Applications*. Springer Verlag.

Davis, L. 1989. Adapting operator probabilities in genetic algorithms. In Schaffer, J. D., ed., *Proceedings of the 3rd International Conference on Genetic Algorithms*, 61–69. Morgan Kaufmann.

de Garis, H. 1990. Genetic programming : building artificial nervous systems using genetically programmed neural networks modules. In Porter, R., and Mooney, B., eds., *Proceedings of the 7th International Conference on Machine Learning*, 132–139. Morgan Kaufmann.

Deb, K. 2001. *Multi-Objective Optimization Using Evolutionary Algorithms*. John Wiley.

DeJong, K., and Spears, W. 1992. A formal analysis of the role of multi-point crossover in genetic algorithms. *Artificial Intelligence* 5:1–26.

DeJong, K. A. 1992. Are genetic algorithms function optimizers ? In Manner, R., and Manderick, B., eds., *Proceedings of the 2nd Conference on Parallel Problems Solving from Nature*, 3–13. North Holland.

Dzeroski, S., and Todorovski, L. 1993. Discovering dynamics. In Utgoff, P., ed., *Proceedings of the 10th International Conference on Machine Learning*, 97–103. Morgan Kaufmann.

Eiben, A., and Ruttkay, Z. 1996. Self-adaptivity for constraint satisfaction: Learning penalty functions. In Fukuda, T., ed., *Proceedings of the Third IEEE International Conference on Evolutionary Computation*, 258–261. IEEE Service Center.

Eiben, A. E.; Hinterding, R.; and Michalewicz, Z. 1999. Parameter control in Evolutionary Algorithms. *IEEE Transactions on Evolutionary Computation* 3(2):124.

El-Beltagy, M.; Nair, P.; and Keane, A. 1999. Metamodeling techniques for evolutionary optimization of computationally expensive problems: Promises and limitations. In Goldberg, D.E., et al., eds., *Proceedings of the Genetic and Evolutionary Conference 99*, 196–203. Morgan Kaufmann.

Falkenhainer, B., and Michalski, R. 1986. Integrating quantitative and qualitative discovery : The ABACUS system. *Machine Learning* 367–401.

Fogel, D., and Stayton, L. 1994. On the effectiveness of crossover in simulated evolutionary optimization. *BioSystems* 32:171–182.

Fogel, L. J.; Owens, A. J.; and Walsh, M. J. 1966. *Artificial Intelligence through Simulated Evolution*. New York: John Wiley.

Fogel, D. B. 1995. *Evolutionary Computation. Toward a New Philosophy of Machine Intelligence*. Piscataway, NJ: IEEE Press.

Forrest, S., and Mitchell, M. 1993. What makes a problem hard for a genetic algorithms : Some anomalous results and their explanation. *Machine Learning* 285–319.

Garnier, J.; Kallel, L.; and Schoenauer, M. 1999. Rigorous hitting times for binary mutations. *IEEE Transactions on Evolutionary Computation* 7(2):167–203.

Gittins, J. C. 1989. *Multi-Armed Bandit Allocation Indices*. Chichester: John Wiley.

Gittus, J., and Zarka, J. 1985. *Modeling small deformations of polycrystals*. Elsevier Applied Sciences Publishers.

Glover, F., and Kochenberger, G. 1995. Critical event tabu search for multidimensional knapsack problems. In *Proceedings of the International Conference on Metaheuristics for Optimization*, 113–133. Kluwer Publishing.

Goldberg, D. E. 1989. *Genetic algorithms in search, optimization and machine learning*. Addison Wesley.

Gorges-Schleuter, M. 1998. A comparative study of global and local selection in evolution strategies. In Bäck, T.; Eiben, G.; Schoenauer, M.; and Schwefel, H.-P., eds., *Proceedings of the 5th Conference on Parallel Problems Solving from Nature*, 367–377. Springer Verlag, LNCS 1498.

Gruau, F. 1994. Genetic micro programming of neural networks. In Kenneth E. Kinnear, J., ed., *Advances in Genetic Programming*, 495–518. MIT Press.

Gruau, F. 1996. On using syntactic constraints with genetic programming. In Angeline, P., et al., eds., *Advances in Genetic Programming II*, 377–394. MIT Press.

Hansen, N., and Ostermeier, A. 1996. Adapting arbitrary normal mutation distributions in evolution strategies: The covariance matrix adaption. In *Proceedings of the Third IEEE International Conference on Evolutionary Computation*, 312–317. IEEE Press.

Haynes, T.; Schoenefeld, D.; and Wainwright, R. 1996. Type inheritance in strongly typed genetic programming. In Angeline, P., et al., eds., *Advances in Genetic Programming II*, 359–375. MIT Press.

Hecht-Nielsen, R. 1989. *Neuro-computing*. Addison Wesley.

Herdy, M. 1996. Evolution strategies with subjective selection. In Voigt, H.-M.; Ebeling, W.; Rechenberg, I.; and Schwefel, H.-P., eds., *Proceedings of the 4th Conference on Parallel Problems Solving from Nature*, 22–31. Springer Verlag, LNCS 1141.

Hinterding, R.; Michalewicz, Z.; and Eiben, A. E. 1997. Adaptation in evolutionary computation: A survey. In Bäck, T.; Michalewicz, Z.; and Yao, X., eds., *Proceedings of the Fourth IEEE International Conference on Evolutionary Computation*, 65–69. IEEE Press.

Rasheed, K., and Hirsh, H. 1997. Using Case Based Learning to Improve Genetic Algorithm Based Design Optimization. In Bäck, T., ed., *Proceedings of the 7th International Conference on Genetic Algorithms*, 513–520. Morgan Kaufmann.

Höhn, C., and Reeves, C. 1996. Are long path problems hard for genetic algorithms? In Voigt, H.-M.; Ebeling, W.; Rechenberg, I.; and Schwefel, H.-P., eds., *Proceedings of the 4th Conference on Parallel Problems Solving from Nature*, 134–143. Springer Verlag, LNCS 1141.

Holland, J. H. 1975. *Adaptation in natural and artificial systems*. University of Michigan Press, Ann Arbor.

Iba, H. 1996. Random tree generation for genetic programming. In Voigt, H.-M.; Ebeling, W.; Rechenberg, I.; and Schwefel, H.-P., eds., *Proceedings of the* 4th *Conference on Parallel Problems Solving from Nature*, 144–153. Springer Verlag, LNCS 1141.

Jacobi, N. 1998. *Minimal Simulations for Evolutionary Robotics*. Ph.D. Dissertation, University of Sussex, Great Britain.

Janikow, C. Z. 1993. A knowledge-intensive genetic algorithm for supervised learning. *Machine Learning* 13:189–228.

Jin, Y.; Olhofer, M.; and Sendhoff, B. 2000. On evolutionary optimisation with approximate fitness functions. In Whitley, D., et al., eds., *Proceedings of the Genetic and Evolutionary Conference*, 786–793. Morgan Kaufman

Joines, J., and Houck, C. 1994. On the use of non-stationary penalty functions to solve nonlinear constrained optimization problems with GAs. In Michalewicz, Z.; Schaffer, J. D.; Schwefel, H.-P.; Fogel, D. B.; and Kitano, H., eds., *Proceedings of the First IEEE International Conference on Evolutionary Computation*, 579–584. IEEE Press.

Jones, T., and Forrest, S. 1995. Fitness distance correlation as a measure of problem difficulty for genetic algorithms. In Eshelman, L. J., ed., *Proceedings of the* 6th *International Conference on Genetic Algorithms*, 184–192. Morgan Kaufmann.

Jones, T. 1995. Crossover, macromutation and population-based search. In Eshelman, L. J., ed., *Proceedings of the* 6th *International Conference on Genetic Algorithms*, 73–80. Morgan Kaufmann.

Kallel, L., and Schoenauer, M. 1997. Alternative random initialization in genetic algorithms. In Bäck, T., ed., *Proceedings of the* 7th *International Conference on Genetic Algorithms*, 268–275. Morgan Kaufmann.

Keijzer, M., and Babovic, V. 1999. Dimensionally aware genetic programming. In Goldberg, D.E., et al., eds, *Proceedings of the Genetic and Evolutionary Conference 99*, 1069–1076. Morgan Kaufmann.

Kichenin, J.; Dang-Van, K.; and Boytard, K. 1996. Finite element simulation of new two dissipative mechanism models for bulk medium-density polyethylene. *J. of Mat. Sc.* 31:1653–1661.

Kirkpatrick, S.; Gelatt, C. D.; and Vecchi, M. P. 1983. Optimization by simulated annealing. *Science* 220:671–680.

Koza, J. R., and al. 1999. *Genetic Programming III: Automatic Synthesis of Analog Circuits*. Massachussetts: MIT Press.

Koza, J. R. 1992. *Genetic Programming: On the Programming of Computers by means of Natural Evolution*. Massachusetts: MIT Press.

Koza, J. R. 1994. *Genetic Programming II: Automatic Discovery of Reusable Programs*. Massachussetts: MIT Press.

Koza, J. R. 1999. Human-competitive machine intelligence by means of genetic algorithms. In Booker, L.; Forrest, S.; Mitchell, M.; and Riolo, R., eds., *Festschrift in honor of John H. Holland*, 15–22. Ann Arbor, MI: Center for the Study of Complex Systems.

Lai, T., and Robbins, H. 1985. Asymptotically efficient adaptive allocation rules. *Advances in applied mathematics* 6:4–22.

Langdon, W. B., and Poli, R. 1997. Fitness causes bloat. In *Soft Computing in Engineering Design and Manufacturing*, 13–22. Springer Verlag.

Langley, P.; Simon, H.; and Bradshaw, G. 1983. Rediscovering chemistry with the BACON system. In Michalski, R.; Carbonell, J.; and Mitchell, T., eds., *Machine Learning: an artificial intelligence approach*, volume 1. Morgan Kaufmann.

Larranaga, P., and Lozano, J. A. 2001. *Estimation of Distribution Algorithms. A New Tool for Evolutionary Computation*. Kluwer Academic Publishers.

Levenick, J. R. 1991. Inserting introns improves genetic algorithm success rate : Taking a cue from biology. In Belew, R. K., and Booker, L. B., eds., *Proceedings of the* 4th *International Conference on Genetic Algorithms*, 123–127. Morgan Kauffmann.

Mahfoud, S. W. 1995. A comparison of parallel and sequential niching techniques. In Eshelman, L. J., ed., *Proceedings of the* 6th *International Conference on Genetic Algorithms*, 136–143. Morgan Kaufmann.

Mansanne, F.; Carrre, F.; Ehinger, A. .; and Schoenauer, M. 1999. Evolutionary algorithms as fitness function debuggers. In Ras, Z. W., and Skowron, A., eds., *Foundation of Intelligent Systems 12th International Symposium, ISMIS99*. Springer Verlag, LNCS 1609.

Merz, P., and Freisleben, B. 1999. Fitness landscapes and memetic algorithm design. In Corne, D.; Dorigo, M.; and Glover, F., eds., *New Ideas in Optimization*. 245–260. London: McGraw-Hill.

Michalewicz, Z., and Schoenauer, M. 1996. Evolutionary Algorithms for Constrained Parameter Optimization Problems. *IEEE Transactions on Evolutionary Computation* 4(1):1–32.

Michalewicz, Z. 1996. *Genetic Algorithms+Data Structures=Evolution Programs*. New-York: Springer Verlag. 3rd edition.

Mitchell, M. 1996. *An Introduction to Genetic Algorithms*. MIT Press.

Mitchell, T. 1997. *Machine Learning*. McGraw-Hill.

Montana, D. J. 1995. Strongly typed genetic programming. *Evolutionary Computation* 3(2):199–230.

Muggleton, S., and De Raedt, L. 1994. Inductive logic programming: Theory and methods. *Journal of Logic Programming* 19:629–679.

Muggleton, S. 1995. Inverse entailment and PROGOL. *New Gen. Comput.* 13:245–286.

Naudts, B., and Kallel, L. 2000. Comparison of summary statistics of fitness landscapes. *IEEE Transactions on Evolutionary Computation* 4(1):1–15.

Persoz, B. 1960. *Introduction à la rhéologie*. Dunod, Paris.

Poloni, C., and Pediroda, V. 1997. GA coupled with computationally expensive simulations: tools to improve efficiency. In *Genetic Algorithms and Evolution Strategies in Engineering and Computer Sciences*, 267–288. John Wiley.

Quadraglia, D.; Périaux, J.; Poloni, C.; and Winter, G., eds. 1997. *Genetic Algorithms and Evolution Strategies in Engineering and Computer Sciences*. John Wiley.

Quarles, A.; Pederson, D.; and Sangiovanni-Vincentelli, A. 1994. SPICE version 3f5 user manual. Technical report, Univ. Berkeley, Cal, USA.

Quinlan, J. 1990. Learning logical definition from relations. *Machine Learning* 5:239–266.

Rabinovich, Y., and Wigderson, A. 1999. Techniques for bounding the convergence rate of genetic algorithms. *Random Structures Algorithms* 2:111–138.

Radcliffe, N. J., and Surry, P. D. 1994. Formal memetic algorithms. In Fogarthy, T., ed., *Evolutionary Computing: AISB Workshop*, 1–16. Springer Verlag, LNCS 865.

Radcliffe, N. J. 1991. Equivalence class analysis of genetic algorithms. *Complex Systems* 5:183–20.

Ratle, A., and Sebag, M. 2000. Genetic programming and domain knowledge: Beyond the limitations of grammar-guided machine discovery. In Schoenauer, M., et al., eds, *Proceedings of the 6th Conference on Parallel Problems Solving from Nature*, 211–220. Springer Verlag, LNCS 1917.

Ratle, A., and Sebag, M. 2001. Avoiding the bloat with stochastic grammar-based genetic programming. In Collet, P., et al., eds., *Artificial Evolution 2001*. Springer Verlag. to appear.

Ratle, A. 1998. Accelerating the convergence of evolutionary algorithms by fitness landscape approximation. In Bäck, T.; Eiben, G.; Schoenauer, M.; and Schwefel, H.-P., eds., *Proceedings of the 5th Conference on Parallel Problems Solving from Nature*, 87–96. Springer Verlag, LNCS 1498.

Rechenberg, I. 1973. *Evolutionstrategie: Optimierung Technisher Systeme nach Prinzipien des Biologischen Evolution.* Stuttgart: Fromman-Hozlboog Verlag.

Richards, N.; Moriarty, D.; and Miikkulainen, R. 1998. Evolving neural networks to play go. *Applied Intelligence* 8:85–96.

Rosca, D., and Ballard, D. 1994. Hierarchical self-organization in genetic programming. In Cohen, W., and Hirsh, H., eds., *Proceedings of the 11th International Conference on Machine Learning*. Morgan Kaufmann.

Rudolph, G. 1994. Convergence analysis of canonical genetic algorithm. *IEEE Transactions on Neural Networks* 5(1):96–101.

Rudolph, G. 1997. Asymptotical convergence rates of simple evolutionary algorithms under factorizing mutation distributions. In Hao, J.-K.; et al., eds., *Artificial Evolution'97*. Springer Verlag, LNCS 1363.

Rumelhart, D., and McClelland, J. 1986. *Parallel Distributed Processing*. MIT Press.

Ryan, C.; Collins, J.; and O'Neill, M. 1998. Grammatical evolution: Evolving programs for an arbitrary language. In Banzhaf, W.; Poli, R.; Schoenauer, M.; and Fogarty, T., eds., *Genetic Programming, First European Workshop, EuroGP98*, 83–96. Springer Verlag, LNCS 1391.

Salustowicz, R., and Schmidhuber, J. 1997. Probabilistic incremental program evolution. *Evolutionary Computation* 5(2):123–141.

Sanchez-Palencia, E., and Zaoui, A. 1987. *Homogenization techniques for composite media.* Lecture Notes in Physics, Springer Verlag.

Sareni, B., and Krähenbühl, L. 1998. Fitness sharing and niching methods revisited. *IEEE Transactions on Evolutionary Computation* 2(3):97–106.

Schapire, R. 1999. Theoretical views of boosting. In *Proceedings of EuroCOLT-99, European Conference on Computational Learning Theory*, 1–10.

Schnier, T., and Gero, J. 1998. From Mondrian to Frank Lloyd Wright: Transforming evolving representations. In Parmee, I., ed., *Adaptive Computing in Design and Manufacture*, 207–220. Springer Verlag.

Schoenauer, M., and Michalewicz, Z. 1996. Evolutionary computation at the edge of feasibility. In Voigt, H.-M.; Ebeling, W.; Rechenberg, I.; and Schwefel, H.-P., eds., *Proceedings of the 4th Conference on Parallel Problems Solving from Nature*, 245–254. Springer Verlag, LNCS 1141.

Schoenauer, M., and Michalewicz, Z. 1997. Boundary operators for constrained parameter optimization problems. In Bäck, T., ed., *Proceedings of the 7th International Conference on Genetic Algorithms*, 322–329. Morgan Kaufmann.

Schoenauer, M., and Xanthakis, S. 1993. Constrained GA optimization. In Forrest, S., ed., *Proceedings of the 5th International Conference on Genetic Algorithms*, 573–580. Morgan Kaufmann.

Schoenauer, M.; Sebag, M.; Jouve, F.; Lamy, B.; and Maitournam, H. 1996. Evolutionary identification of macro-mechanical models. In Angeline, P. J., and K. E. Kinnear, J., eds., *Advances in Genetic Programming II*, 467–488. Cambridge, MA: MIT Press.

Schwefel, H.-P. 1981. *Numerical Optimization of Computer Models*. New-York: John Wiley & Sons. 1995 – 2nd edition.

Sebag, M., and Ducoulombier, A. 1998. Extending population-based incremental learning to continuous search spaces. In Bäck, T., et al., eds., *Proceedings of the 5th Conference on Parallel Problems Solving from Nature*, 418–427. Springer Verlag, LNCS 1498.

Sebag, M., and Rouveirol, C. 2000. Resource-bounded relational reasoning: Induction and deduction through stochastic matching. *Machine Learning* 38:41–62.

Sebag, M.; Schoenauer, M.; and Maitournam, H. 1997. Parametric and non-parametric identification of macro-mechanical models. In Quadraglia, D.; Périaux, J.; Poloni, C.; and Winter, G., eds., *Genetic Algorithms and Evolution Strategies in Engineering and Computer Sciences*, 327–340. John Wiley.

Simon, H.A. 1982. *Models of Bounded Rationality*. Cambridge, MIT Press.

Soule, T., and Foster, J. A. 1997. Code size and depth flows in genetic programming. In Koza, J., et al., eds., *Genetic Programming 1997: Proceedings of the Second Annual Conference*, 313–320. Morgan Kaufmann.

Srinivasan, A.; Muggleton, S.; Sternberg, M.; and King, R. 1996. Theories for mutagenicity: a study in first order and feature-based induction. *Artificial Intelligence* 85:277–299.

Starkweather, T.; Whitley, D.; and Mathias, K. 1991. Optimization Using Distributed Genetic Algorithms. In Schwefel, H., and Manner, R., eds., *Parallel Problem Solving from Nature*, 176–185. Springer Verlag.

Sutton, R., and Barto, A. G. 1998. *Reinforcement learning*. MIT Press.

Syswerda, G. 1989. Uniform crossover in genetic algorithms. In Schaffer, J. D., ed., *Proceedings of the 3rd International Conference on Genetic Algorithms*, 2–9. Morgan Kaufmann.

Todorovski, L., and Dzeroski, S. 2001. Using domain knowledge on population dynamics modeling for equation discovery. In de Raedt, L., and Flach, P., eds., *Proceedings of ECML-01, European Conference on Machine Learning*, 478–490. Springer Verlag, LNAI 2167.

Vapnik, V.; Levin, E.; and Cun, Y. L. 1994. Measuring the VC-dimension of a learning machine. *Neural Computation* 6(5):851–876.

Vapnik, V. N. 1995. *The Nature of Statistical Learning*. Springer Verlag.

Vapnik, V. N. 1998. *Statistical Learning Theory*. Wiley.

Venturini, G.; Slimane, M.; Morin, F.; and de Beauville, J. A. 1997. On using interactive genetic algorithms for knowledge discovery in databases. In Bäck, T., ed., *Proceedings of the 7th International Conference on Genetic Algorithms*, 696–703. Morgan Kaufmann.

Washio, T., and Motoda, H. 1997. Discovering admissible models of complex systems based on scale-types and identity constraints. In *Proc. of Int. Joint Conf. on Artificial Intelligence*, 810–817.

Washio, T.; Motoda, H.; and Niwa, Y. 2001. Discovering admissible simultaneous equation models from observed data. In de Raedt, L., and Flach, P., eds., *Proceedings of ECML-01, European Conference on Machine Learning*, 539–551. Springer Verlag, LNAI 2167.

Whigham, P. 1995. Inductive bias and genetic programming. In *IEEE Conference on Evolutionary Computation*, 461–466. IEEE Press.

Yao, X.; Liu, Y.; and Lin, G. 1999. Evolutionary programming made faster. *IEEE Transactions on Evolutionary Computation* 3(2):82–102.

MIX
Papier aus verantwortungsvollen Quellen
Paper from responsible sources
FSC® C105338

If you have any concerns about our products,
you can contact us on
ProductSafety@springernature.com

In case Publisher is established outside the EU,
the EU authorized representative is:
Springer Nature Customer Service Center GmbH
Europaplatz 3, 69115 Heidelberg, Germany

Printed by Libri Plureos GmbH
in Hamburg, Germany